IET TELECOMMUNICATIONS SERIES 69

5G Wireless Technologies

Other volumes in this series:

Volume 9	**Phase noise in signal sources** W.P. Robins
Volume 12	**Spread spectrum in communications** R. Skaug and J.F. Hjelmstad
Volume 13	**Advanced signal processing** D.J. Creasey (Editor)
Volume 19	**Telecommunications traffic, tariffs and costs** R.E. Farr
Volume 20	**An introduction to satellite communications** D.I. Dalgleish
Volume 26	**Common-channel signalling** R.J. Manterfield
Volume 28	**Very small aperture terminals (VSATs)** J.L. Everett (Editor)
Volume 29	**ATM: The broadband telecommunications solution** L.G. Cuthbert and J.C. Sapanel
Volume 31	**Data communications and networks, 3rd edition** R.L. Brewster (Editor)
Volume 32	**Analogue optical fibre communications** B. Wilson, Z. Ghassemlooy and I.Z. Darwazeh (Editors)
Volume 33	**Modern personal radio systems** R.C.V. Macario (Editor)
Volume 34	**Digital broadcasting** P. Dambacher
Volume 35	**Principles of performance engineering for telecommunication and information systems** M. Ghanbari, C.J. Hughes, M.C. Sinclair and J.P. Eade
Volume 36	**Telecommunication networks, 2nd edition** J.E. Flood (Editor)
Volume 37	**Optical communication receiver design** S.B. Alexander
Volume 38	**Satellite communication systems, 3rd edition** B.G. Evans (Editor)
Volume 40	**Spread spectrum in mobile communication** O. Berg, T. Berg, J.F. Hjelmstad, S. Haavik and R. Skaug
Volume 41	**World telecommunications economics** J.J. Wheatley
Volume 43	**Telecommunications signalling** R.J. Manterfield
Volume 44	**Digital signal filtering, analysis and restoration** J. Jan
Volume 45	**Radio spectrum management, 2nd edition** D.J. Withers
Volume 46	**Intelligent networks: Principles and applications** J.R. Anderson
Volume 47	**Local access network technologies** P. France
Volume 48	**Telecommunications quality of service management** A.P. Oodan (Editor)
Volume 49	**Standard codecs: Image compression to advanced video coding** M. Ghanbari
Volume 50	**Telecommunications regulation** J. Buckley
Volume 51	**Security for mobility** C. Mitchell (Editor)
Volume 52	**Understanding telecommunications networks** A. Valdar
Volume 53	**Video compression systems: From first principles to concatenated codecs** A. Bock
Volume 54	**Standard Codecs: Image compression to advanced video coding, 3rd edition** M. Ghanbari
Volume 59	**Dynamic Ad Hoc Networks** H. Rashvand and H. Chao (Editors)
Volume 60	**Understanding Telecommunications Business** A Valdar and I Morfett
Volume 65	**Advances in Body-Centric Wireless Communication: Applications and state-of-the-art** Q.H. Abbasi, M.U. Rehman, K. Qaraqe and A. Alomainy (Editors)
Volume 67	**Managing the Internet of Things: Architectures, theories and applications** J. Huang and K. Hua (Editors)
Volume 68	**Advanced Relay Technologies in Next Generation Wireless Communications** I. Krikidis and G. Zheng
Volume 70	**Cloud and Fog Computing in 5G Mobile Networks** Dr. Evangelos Markakis, Dr. George Mastorakis, Dr. Constandinos X. Mavromoustakis and Dr. Evangelos Pallis (Editors)
Volume 73	**Network as a Service for Next Generation Internet** Qiang Duan and Shangguang Wang (Editors)
Volume 905	**ISDN applications in education and training** R. Mason and P.D. Bacsich

5G Wireless Technologies

Edited by
Angeliki Alexiou

The Institution of Engineering and Technology

Published by The Institution of Engineering and Technology, London, United Kingdom

The Institution of Engineering and Technology is registered as a Charity in England & Wales (no. 211014) and Scotland (no. SC038698).

First published 2017

The Institution of Engineering and Technology
Michael Faraday House
Six Hills Way, Stevenage
Herts, SG1 2AY, United Kingdom

www.theiet.org

British Library Cataloguing in Publication Data
A catalogue record for this product is available from the British Library

ISBN 978-1-78561-061-5 (hardback)
ISBN 978-1-78561-062-2 (PDF)

Typeset in India by MPS Limited
Printed in the UK by CPI Group (UK) Ltd, Croydon

Contents

Chapter 1

The road to 5G – visions and challenges

Angeliki Alexiou[1]

1.1 A vision for 5G (in a nutshell)

"From the smallest personal items to the largest continents, everything, everywhere will be digitally connected, and responsive to our wants and likes" is a typical vision for the future of communication networks as stated in "The World in 2025: 10 Predictions of Innovation" (Thomson Reuters). In particular, wireless communications are expected to dominate everything, everywhere, and transform everyday life, mainly by means of revolutionary fifth-generation (5G) technologies characterized by concepts such as: cell-less architectures, massive spatial processing, tactile response times, big data processing and virtualization, to name a few. Toward this end, research, performance evaluation, experimentation, and standardization activities have started [1,2]: international industry-academia consortia (METIS2020, IWPC, 5G Korea), public–private partnership associations (5G-PPP), regulatory authorities (ITU-R WP5D), and operators fora (NGMN) have recently focused their attention on setting the requirements, exploring the system concept characteristics, and investigating the enabling technologies for 5G [3].

Wireless evolution has come a long way over the last 40 years, starting from a technology-centric all the way to a technology-transparent future, shifting the focus of innovation from components and radio transceiver technologies (e.g., turbo coding, MIMO) to resource allocation and interference management (e.g., CDMA, OFDMA, Coordinated Multipoint transmission/reception-CoMP) and, finally, to user-centric and device-driven communications, in the "internet of everything." Accelerated by the dramatic impact of applications of the kinds of social networking, the center of gravity for 5G will be shifted further to application-driven connectivity (transparently deployed over technologies, infrastructures, users, and devices), which will be mainly enabled by means of dense access/serving node deployments and exploitation of proximal communications, concepts that are typically referred to as *network densification* [4]. In contrast to previous cellular evolution steps, which were mainly driven by increasing system bandwidth and improving (multi-user) spectral efficiency, network densification is expected to be

[1]Department of Digital Systems, University of Piraeus, Piraeus 18534, Greece

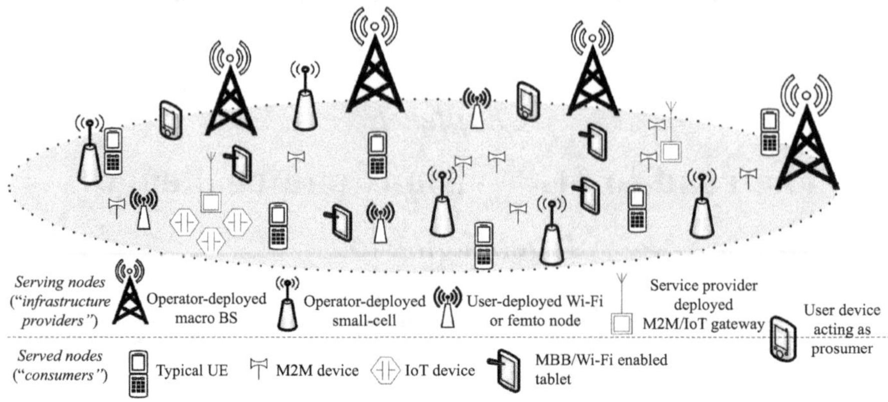

Figure 1.1 AUDN infrastructure composed by operator- and user-deployed heterogeneous serving ANs, multiple types of user and machine served nodes, and disruptive devices acting as prosumers

the cornerstone of future wireless networks. Even though network densification was also considered in past cellular generations, it was used only as an "add-on" measure for locally coping with heavily loaded parts of the system and not as a fundamental, large-scale design/system concept.

Toward realizing the ambitious goals set for 5G, the density of access/serving nodes is expected to increase up to the point where it is comparable to or even much higher than the (also increasing) density of user equipments (UEs) [5], thus introducing the ultra-dense network (UDN) paradigm. Realization of such a disruptive network topology, with respect to the evolution path of previous cellular generations, will be achieved by utilizing, in addition to densified traditional access nodes (ANs) such as small cells, user deployed ANs (e.g., Wi-Fi, femtocells), as well as "infrastructure prosumer" UEs, i.e., (mobile) devices with computational and storage capabilities allowing them to act as infrastructure ANs (Figure 1.1).

Clearly, a successful UDN deployment is not an easy task due to the challenging interference landscape introduced by the dense and, to a large extent, irregular AN deployment with transmissions under a universal frequency reuse scheme. In this sense, UDNs can be viewed as the new wireless frontier that can be made a reality only through re-thinking of current cellular practices in terms of: (i) *underlying technology enablers*, namely, air-interface, including optimized frame structures and associated overhead signaling, multiple access schemes and resources management strategies, connectivity modes, access topology, and backhaul design [5], and (ii) *implementation implications*, such as business modeling, cost/complexity, backward compatibility, and market uptake issues. In order to successfully and efficiently address these issues, a clear understanding of UDN fundamental performance limits, obtained using appropriate mathematical models, is of utmost importance. Toward this end, two relevant research challenges can be identified:

1. How should densification scale with increasing traffic load, rate, and reliability requirements taking into account various interference, multiple access, resources management, and cost/complexity issues?
2. Is densification *alone* sufficient to accommodate future generation network requirements? In that respect, should one consider cooperation, coordination, and cognition (between nodes) as integral part of the UDN design that could offer significant performance gains?

Motivated by the above questions, this book aims at revealing, by means of addressing 10 significant wireless research ideas/topics, some of the most crucial fundamental limitations, promising technology enablers, system concept paradigm shifts, and critical deployment design factors for the realization of 5G and present a qualitative and quantitative assessment of their potential.

1.2 Key disruptive system concept trends

Many groundbreaking technological advances are currently considered as promising enablers for realizing the 5G expectations, such as massive MIMO, full duplexing, device-to-device communications, caching, and mm-Wave communications. They are often investigated within the context of traditional value-chain principles, mainly because new appropriate system models, reflecting the critical parameters of 5G usage scenarios, with respect to network architectures, infrastructure elements functionalities, and resources ownership, have not been specified. Much of the mainstream understanding of 5G sees communication services provision as an enhanced version of today's service provision model as offered by mobile operators. Extrapolating on the same "next generation" evolution path (from 2G to 3G, from 3G to 4G, and so on) may not offer the optimal route toward meeting 5G expectations.

Usage, deployment, and technology trends currently emerging give the first signs of evidence that *the most decisive factor for 5G success is the system concept on how to integrate the various advanced technology building blocks*. Some of these key system concept trends include:

1. **The edge of the network is being increasingly densified.** The extensive deployment of small cells, not only by means of operator but also of user-deployed infrastructure, in an unplanned manner, changes the ownership and control of the network edge.
2. **The core network is being transformed through the distribution of network functionalities**, and their migration to the network edge and the emergence of "plug and play" core network systems.
3. **The notion of "network resource" is being re-defined.** Commoditization of network resources signifies a new era in optimizing, architecturing, and deploying networks: processing power, storage, backhaul, spectrum, bandwidth,

etc. can all be, due to technological advances, produced/offered/deployed/utilized in a manner that allows units of these resources to be rented, traded, and sold. In that way, the technology and value chains are transformed along with the underlying business models.

4. **The virtualization of the network functionalities, and the resources sharing involved, change the meaning/importance of proximity, OPEX/CAPEX split and ownership models.** There are opportunities for designing infrastructure to minimize cost/complexity, maximize openness, etc. that introduce several degrees of freedom in the network optimization and deployment.

1.3 Performance limitations, new design principles, and three paradigm shifts

It is commonly agreed among stakeholders that the future wireless network should offer **order(s) of magnitude greater performance levels** in terms of area capacity, density of connected devices, peak and per user rate, latency, and reliability. Nevertheless, the most critical characteristic is that the future network will have to accommodate for **an extremely diversified user, device, and service environment.** Examples include low-rate and bursty Internet-of-Things (IoT) communications, mission-critical control applications with strong quality of service requirements, ubiquitous mobile Internet access, and bandwidth-hungry real time-critical immersive experience applications (such as ultra-high-definition video, remote 3D display, i.e., "virtual presence"). Toward achieving these ambitious targets, wireless research, as an initial step, attempts to "linearly extrapolate" on the approach of previous cellular network generations (2G, 3G, 4G) and commonly considers the following three (obvious) technical directions as a means to address the new 5G requirements [5,6]:

- **Incorporating more spectrum** would most probably take into consideration the cm-Wave licensed bands (such as the 3.4–3.8 GHz band, which is emerging as a new small-cell band), unlicensed bands (such as the 5 GHz Wi-Fi band, which is also considered for the unlicensed band LTE, also known as LTE-U/LAA), and even the unallocated, higher frequency bands in the mm-Wave regime (28, 38, 70 GHz).
- **Densifying infrastructure** would benefit from higher spatial reuse, by deploying hyper-dense small-cell clusters and/or exploiting Wi-Fi technology for offloading significant amounts of traffic in hotspot and indoor scenarios.
- **Inventing new radio technologies** would attempt to maximize spectral efficiency by means of relaxing orthogonality in the air-interface design (e.g., GFDM waveforms), enriching current systems with asynchronous and non-orthogonal multiple-access (SCMA, NOMA schemes), advancing inter-cell cooperation and coordination for interference management (e.g., CoMP), exploiting massive antenna elements to create spatially isolated streams over the same spectral/temporal resource unit (Massive/FD-MIMO), and incorporating Multi-User MIMO to emerging Wi-Fi standards (802.11ad, ax).

While these solutions can be proved to offer substantial performance gains when deployed alone and under idealized network conditions (in terms of signaling, backhaul requirements, and processing/cost complexity), **it is highly questionable whether they can indeed be included as an integral part of the network before certain inherent limitations are accounted for first:**

- **Cellular System Concept Limitations:** The cellular paradigm targets the provision of certain Quality of Service levels for a variety of applications over large geographical regions through a well-planned deployment of base station nodes. **Limited flexibility** is assumed in allocated power, spectral, and processing resources (RF, baseband), and modest inter-node cooperation capabilities are supported, in order to ensure feasibility and scalability in terms of cost/complexity. Nevertheless, given the extreme network densification trend [7], the network of the future is expected to achieve *capacity scaling* requirements by adopting *highly irregular topologies*, also supporting a number of *hierarchical cellular layers* (i.e., smaller cells clusters within larger cells), rendering the "traditional" cellular network planning practically irrelevant. The new complex and *unpredictable interference conditions* would eventually question the feasibility of the expected-to-be-achieved spatial reuse levels. In addition, current cellular systems are **generic-purpose designed networks.** Their poor parameterization and limited degrees of freedom set substantial constraints in their ability to adapt to the *huge dynamic range of applications, services, and devices* expected in 5G.
- **Closed Radio Network Architecture Limitations:** Existing and upcoming releases of radio networks are tightly bound to the underlying hardware infrastructure, imposing **heavy constraints on network upgradeability** (cost, downtimes) and **limiting flexibility in supporting multiple radio technologies**. In the past this was not considered a major hindrance since initiating research on a new technology, standardizing it, and finally deploying it were expected to take more than a decade. The wireless evolution time cycles are now accelerating: while we are still at the beginning of 4G deployments, we expect to have the first 5G pilot trials by the Tokyo Olympics of 2020. Moreover, the vast dynamic range of 5G requirements [8] will require multiple design solutions, in terms of coverage, base station specifications, power consumption, etc., most probably, offered by several vendors. Closed hardware architecture models would therefore require huge capital and operational expenditures to be able to support **interoperability** and **network expandability**.
- **Mobile Networks Value Chain Limitations:** From a business modeling perspective, the complete mobile network chain (including spectrum, physical radio access infrastructure, backhaul, core network, content, applications, billing, etc.) has been traditionally owned and controlled by a single entity, the Mobile Network Operator (MNO).

  Alternative concepts have recently emerged [9], namely (i) the Virtual Network Operator (MVNO) who owns and controls only a part of the chain (e.g., the core network) and relies on service-level-agreements (SLAs) with

MNOs for making use of the radio access infrastructure and (ii) the over-the-top (OTT) service providers, who extract value from the current MNO chain ("for free"). Nevertheless, **these business models, heavily relying on hard ownership models, are not flexible enough to offer new niche opportunities**. This is mainly because the cost of deploying, operating, and managing the massive number of infrastructure elements required by the future network, while at the same time offering low-cost services to end users, is not affordable by current operators (let alone promising in terms of revenue making). A common characteristic of all traditional business models is the limited *resource/cost sharing among network stakeholders*.

To overcome these limitations, 5G will have to take the challenge to "rethink" wireless network design, in terms of the following *three paradigm shifts*:

1. **Rethink wireless system concept**, by introducing a novel **heterogeneous UDN environment** [10–17], whereby massive numbers of ANs per area unit are deployed over heterogeneous topologies and technologies, including:
   - Operator-driven hyper-dense small-cell deployments, offering multiple orders of magnitude, increase in the number of available infrastructure elements per user device.
   - Complementary radio access technology networks (such as Wi-Fi, Directional Antenna Systems-DAS) operated by alternative providers (e.g., in airports, shopping malls, conference centers).
   - User-deployed home infrastructure, such as wireless routers for Internet access, femtocells, and machine-to-machine gateways.
   - "Crowdsourced" high-end user devices (such as smart-phones or tablets) equipped with various wireless interfaces, and acting as "prosumers" of network access in highly dense device scenarios (e.g., auditorium, stadium, cinema).

   In Figure 1.2, the heterogeneous UDN system concept is illustrated for various radio access technologies, topologies and multiple access schemes, including D2D, M2M, etc. The underlying core idea is that of *Proximal Communications*, i.e., high wireless connectivity availability by means of a large number of nodes in the close proximity of every node. The basic access topology is realized here by a "2-hop" (or "2-layer") link consisting of an "access" hop connecting every node to a gateway and a second "backhaul" hop, connecting the gateway to the main wireless infrastructure access point.
2. **Rethink radio network infrastructure design and (hardware) implementation, by cloudifying the wireless network functions.** *Wireless Network Cloudification (WCN)* can provide on-demand, cost-efficient, and service-oriented networks designed on-the-fly, by means of decoupling the hardware infrastructure and the supported functionalities. This is mainly achieved by utilizing general-purpose (commodity) hardware (e.g., general purpose processing units or GPPs) and relevant facilities (e.g., IT data-centers) and by introducing radio access softwarization/virtualization, in order to implement in software all system functionalities, including baseband processing, radio

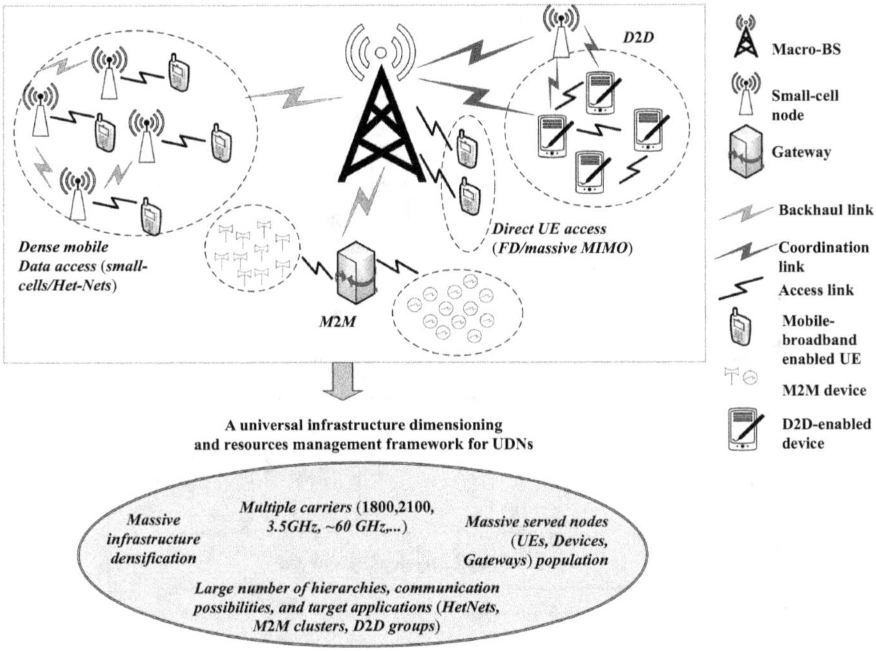

Figure 1.2 A heterogeneous ultra-dense network system concept based on proximal access links of diverse radio technologies

resources scheduling, network routing, etc. WCN allows for extremely fast technology upgrades, efficient maintenance, minimum service interruption, and dynamic on-demand network configuration and management during run-time, creating the perception of "infinite" and "elastic" network scalability.

In Figure 1.3, the Wireless Network Cloudification system concept is illustrated. The underlying idea is the specification of two substrates: the physical substrate, where all physical resources reside, and the virtual substrate, where all softwarized network functionalities are implemented. The convergence "virtualization" layer in between realizes the mapping between physical resources and virtual entities (softwarized network functionalities).

3. **Rethink the notion of "resource element"** by introducing a *generalized and extended resource concept*, potentially incorporating every possible network component/element, such as antennas, base-stations, backhaul links/resources, spectrum bands/chunks, computing processing units, etc., breaking the network into various sub-components and transforming resources into "commodities," which introduce novel ownership and control models, thus *revolutionizing the current Mobile Network Value Chain*. Catalyzed by virtualization, redefining the notion of resource and resources ownership and access, creates a whole new ecosystem, where it is possible to build a network from commodities, tailored to meet the particularities of each application to be provided.

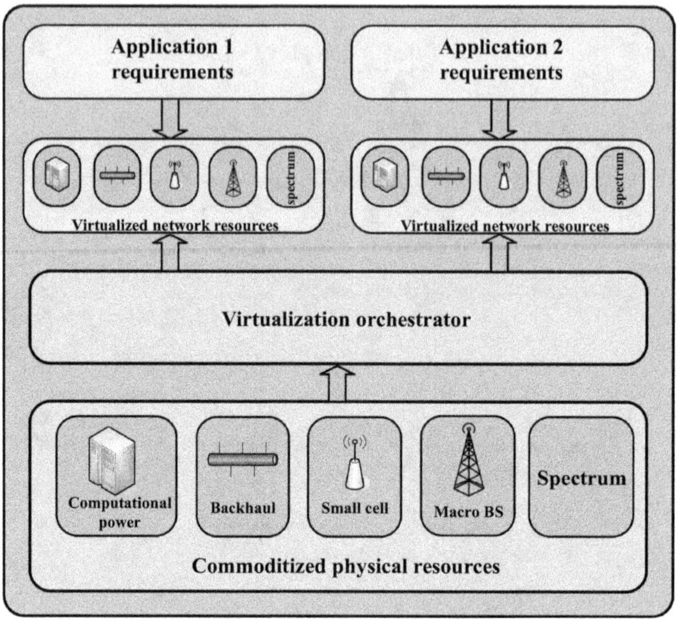

Figure 1.3 A wireless network cloudification system concept, based on the commoditization of physical resources, and an orchestrator sublayer responsible to optimally implement the mapping between the physical and the virtual substrates

1.4 Critical usage scenarios in 5G

In order to lay the foundations of a radical and revolutionary change in current cellular thinking brought about by 5G and at the same time, enhance its impact and facilitate its adoption, the selection, identification, and specification of appropriate usage scenarios is essential. One should also consider evolutionary aspects and follow the study items (in progress) carried out by standardization bodies, regulation, and technology fora, such as 3GPP, ITU, NGMN, etc. A set of well-defined scenarios, reflecting both the revolutionary trends critical parameters and the evolutionary paths aspects, should include:

- A potential **deployment topology** (e.g., planned, unplanned, on-demand/event-driven, deterministic/canonical geometry, random);
- Certain **geographical characteristics** (e.g., large-scale wide area, indoor, hotspot);
- Relevant/critical **user-profile scenarios** (e.g., rate-demand, real-time sensitive, control critical, high level of reliability, mobility).

Although a considerable amount of research effort has been recently dedicated to the identification and specification of the most representative 5G scenarios, most of the proposed approaches agree, more or less, that the most critical technology challenges and 5G differentiators are expected to appear in one of the following three usage cases (also described in a similar fashion in the ITU IMT 2020 and beyond Vision recommendation document [18]).

1.4.1 Crowded local access

The proliferation of mobile broadband access (first delivered by 3G-HSPA and now taken to a next level by 4G LTE-A networks), supported by the advent of high-end smart-phone devices, along with the rapidly evolving social networking and cloud services in the Internet world, has further catalyzed the convergence between communications and computer networking. A new paradigm shift has emerged: the collapse of the borders between content and communication medium. We are now moving from the scenario where a user consumes static information posted on the Internet (e.g., web browsing, file downloading), to a new era where the Internet information is created dynamically and at any time by users who are both creating and consuming the Internet content (e.g., through video and image uploading, posting/interacting with other people in social networks). This leads to ever-increasing levels of mobile data volume that today's networks seem inadequate to support. What is even more challenging is supporting the large temporal and spatial dynamic range and volatility of mobile data. In particular, extreme peaks of mobile traffic could occur in a specific place (e.g., a bus-stop, a stadium hosting a football match, special entertainment event/happening) and at a specific time (while waiting for the bus, when a goal is scored, New Year count down at Times Square in New York). *Massive data volume local access for dynamic crowd demand is a key scenario for the network of the future.*

1.4.2 Bursty IoT

A transformative vision for our society consists in connecting practically everyone and everything to the Internet ("Internet-of-Everything" or IoE). New applications involving machine-initiated communications are emerging, including smart metering (sensors/actuators), e-health (monitoring of patients health condition), and intelligent transportation (fleet tracking and traffic management). Differently from the commonly adopted design rules of "conventional" mobile broadband networks that focus on increasing the achieved (peak) data rate per device, the machine communications paradigm primarily targets the support of massively increasing densities of devices generating low-rate sporadic and/or bursty data. This traffic activity can create geographically and temporarily concentrated data bursts, which drive current networks to overloading and outage. *Accommodating for bursty IoT communications entails the challenge of handling the vastly diversified quality-of-service requirements and volatile traffic and usage parameters.*

1.4.3 Ultra-reliable and low latency communications

Wireless control of industrial manufacturing or production processes (Cyber Physical Systems in Industry 4.0), remote medical surgery, distribution automation in smart grids, transportation safety monitoring, autonomous driving, real-time gaming, virtual presence and augmented reality are only just a few of new challenging applications that would bring future networks far beyond the perception of "infinite" capacity and "infinite" connectivity in the IoE. "Tactile Internet" [19] as it is now commonly referred to, is expected to *bring communications to the speed of human senses and introduce the co-design of communication, control and computation systems*. Different from the previous two scenarios, the challenge in this one is to fundamentally rethink the way wireless infrastructure has been designed so far (commonly focusing on boosting throughput and capacity figures by orders of magnitude) and place the emphasis on achieving *"zero" latency, very short response/reaction time (in the order of a millisecond), high availability/reliability (a second of outage per year) and high level of security*.

1.5 This book: motivation and structure

The objective of this book is to present, elaborate and assess, with respect to the impact of critical parameters, the most promising system concepts, architectures, and technology enablers that are expected to play a differentiator's role in the realization and adoption of 5G. To this end, 10 promising advanced technology topics are investigated in the following corresponding book chapters (Chapters 2–11), each describing a 5G innovation that is expected to transform wireless design understanding. A new era is commencing, following "technology-driven," "user-centric," and "Everything-as-a-Service" eras: the era of "Connectivity-as-a-Utility."

More specifically, the book chapters are organized into three thematic entities.

The first four chapters (Chapters 2–5) present the most important *technology enablers* that are expected to boost spectral efficiency by several orders of magnitude. Namely:

- In Chapter 2, "New 5G air interface: challenges for efficient multi-service coexistence for 5G below 6 GHz," a view of the current service and use case landscape for 5G networks is presented. A selection methodology of appropriate use cases for a flexible air interface design is proposed. Challenges for this design on link and system levels are then discussed along with associated open research questions. It is emphasized that the design of a new air interface is heavily relying on standardization-related issues, such as waveform and frame design, PHY layer procedures, transceiver processing (interference management, multiuser reception, etc.), channel coding, adaptive coding and modulation, etc. These are all topics that imply suitable interfaces, signaling, and protocols (to be standardized) rather than proprietary algorithmic solutions. Timely interaction and exchange between research and standardization is of paramount importance in that respect.
- In Chapter 3, "New quasi-deterministic approach to channel modeling in millimeter-wave bands," a novel quasi-deterministic modeling methodology for millimeter-wave channels is proposed. The methodology makes use of a number of previous experimental works and a few special measurement campaigns

at 60 GHz performed during the MiWEBA project in different scenarios and environments. This methodology is based on the representation of the millimeter-wave channel impulse response function by a few strong scenario-specific deterministic rays and a certain number of relatively weak random rays that represent random or undefined components of the environment. This flexible combination of the deterministic and statistical approaches has reasonable complexity and accuracy comparable with—site dependent—deterministic models. This approach is accepted as a basic approach for channel modeling in the upcoming IEEE 802.11ay standard and is being currently further investigated and extended to support various MIMO modes.

- In Chapter 4, "Large-scale antenna systems," an overview of one of the most prominent 5G candidate technologies is provided. Although, multi-antenna systems constitute the basis of current mobile communication standards (such as Long-Term Evolution), as a major enabler for achieving high spectral efficiency, power efficiency is considered key in the realization of 5G. By means of improvements in both the hardware and signal processing components, LSAS, also known as "massive MIMO systems," aim at simultaneously improving the spectral and energy efficiency by incorporating hundreds of antennas at the Base Stations to serve dozens of users simultaneously. The main LSAS benefits along with the practical implementation and operation issues are analyzed in this chapter. Massive communications system modeling, optimization and design approaches are further elaborated.
- In Chapter 5, "Effects of densification and randomness of infrastructure deployment in cellular networks," another promising 5G technology based on the concept of *extreme spatial (infrastructure) reuse*, by means of densification and randomness of access point deployment, is analytically assessed. In such a new operational environment, where 5G user nodes are envisioned to enjoy "exclusive" access provided by one (or more) serving access points in their vicinity, interference levels are no longer controllable by careful access point placement and can potentially become extremely large under certain (random) deployment configurations. Quantitative and qualitative insights are obtained on the interplay between fundamental performance metrics, such as user rate and signal-to-interference ratio and densification, topology randomness, and large-scale effects.

In the next four chapters of the book, four *system concept paradigm shifts* are presented that—if successfully adopted—are expected to fundamentally change wireless networks, mainly because they are based on novel disruptive wireless access schemes and alternative (often non-conventional) design and optimization principles. More specifically:

- In Chapter 6, "Wireless device-to-device (D2D) links for machine-to-machine (M2M) communication," motivated by the synergy between D2D and M2M, a number of technologies are presented that enable M2M-via-D2D communication to operate as an underlay to the ordinary cellular transmissions. D2D communications are expected to play an important role in 5G cellular networks, by increasing the spatial reuse of spectrum resources and enabling communication links with low latency. D2D consists of two main building blocks: proximity discovery and direct communication between nearby users. On the other hand, M2M communications are often characterized by fixed, low transmission rates that could take advantage of the D2D topology, and access schemes to overcome latency, aggregation load, and other challenges. In this chapter, this novel system concept is further elaborated and explored.

- In Chapter 7, "Caching in large wireless networks," the notion of caching, originating from web caching and Content Delivery Networks (CDN), is explored and elaborated as a means to improve the sustainability of wireless networks. Moving from CDN to wireless CDN introduces new paradigms of content replication. These new schemes promise to improve efficiency by combining caching with properties of PHY layer, while new challenges are imposed by mobility, geometrical properties of wireless signal propagation, and interference.
- In Chapter 8, "Full duplexing," full-duplex transmission is considered as a promising air interface technique for future wireless systems as it tackles key issues such as spectral efficiency, link level capacity and latency. Until recently, the use of full-duplex transmission in wireless communications systems was believed to be infeasible due to the self-interference phenomenon which occurs as part of the transmitted signal leaks or couples to the co-located receiver. In recent years, full-duplex communications have attracted significant attention and various designs have been proposed for full-duplex transceivers. This chapter focuses on full-duplex transceiver solutions for compact form factor wireless devices, such as handheld ones and small access points/base stations (e.g., femto cell). It further proposes full-duplex transmission solutions in small area radio communication systems.
- In Chapter 9, "Decoupled uplink and downlink access in heterogeneous networks," Downlink and Uplink Decoupling (DUDe) is proposed as a means to allow the user to independently transmit and receive to and from different base stations. DUDe breaks the hard and classical constraint of cell selection based on downlink received power, and provides the network with the flexibility to associate users to different eNBs in the DL and UL. In this chapter, radio planning and interference management challenges in HetNets are discussed and decoupling between UL and DL is recognized as a new technique that can effectively improve the HetNet performance. It is also shown that DUDe can positively impact the performance of other radio access technologies, such as Carrier Aggregation (CA), Coordinated Multipoint transmission and reception (CoMP), and millimeter wave.

In the last part of the book, two novel wireless network architectures and management approaches are discussed. They address the issues of virtualization of network functions and flexible/dynamic spectrum management and ownership models. They describe potential avenues toward the materialization of the network resources commoditization and liberalization of spectrum (and other network assets). More specifically:

- In Chapter 10, "Wireless networks virtualization," virtualization is viewed as one of the critical technologies to maximize network flexibility, scalability, and efficiency and, in that way, accommodate diverse applications and scenarios. Despite the simplicity of the idea of adopting virtualization—in principle an IT concept—to telecom industry, the implementation is more complicated in practice. Wireless communication is distinct from IT data centers in that wireless communications have extremely stringent real-time processing requirements. Given that RAN functions, such as baseband processing, require intensive computation and real-time processing, the RAN virtualization requires joint optimization of various system modules, including the operating systems, the hypervisor, etc. It also requires enhancement and optimization on existing traditional IT technologies including the network acceleration, live migration, VM communication, support of OpenStack, to name just a few.
- In Chapter 11, "Licensed shared access and three-tier spectrum sharing models: regulation, business and technology perspectives," spectrum-sharing concepts are addressed from regulatory, business, and technology perspectives. The European Licensed Shared Access (LSA) spectrum-sharing concept is presented, covering its regulatory, business, and technical aspects, with focus on the LSA case example of sharing between LTE and incumbents program making and special events (PMSE) in the 2.3–2.4 GHz band. The LSA evolution including more dynamic sharing toward the US three-tier spectrum sharing model for Citizen Broadband Radio Service (CBRS) based on Spectrum Access System (SAS) is depicted in terms of its ecosystem, business benefits, and technical approaches. Future outlook is given to envisage the development of sharing models for the mobile broadband in 5G and beyond.

At the dawn of a new wireless generation, there is hope and hype around meeting ambitious targets and measuring success against performance bounds ranging from orders of magnitude of improvements to "infinite" capacity to a new era of "wireless connectivity as a utility." These targets are expected to be reached by means of inventing novel technologies that—one way or another—are able to overcome the current network limitations and significantly optimize the trade-off between performance and complexity/cost by introducing new and (potentially) inexpensive degrees of freedom.

Before 5G shows its "true colors," it is attempted in this book to present promising technology differentiators, which could offer the "delta" factor capable of elevating 5G up to the high expectations set at this point. It is also hoped that, through the chapters of this book, the transformational power of research innovation can be assessed against wireless engineering and market relevance impact, as wireless networks are moving away from a technology-driven past to a utility-perceived future.

References

[1] Andrews, J.G.; Buzzi, S.; Choi, W. *et al.*, "What will 5G be?," *IEEE Journal on Selected Areas in Communications*, vol. 32, no. 6, pp. 1065–1082, Jun. 2014.

[2] Li, Q.C.; Niu, H.; Papathanassiou, A.T.; and Wu, G., "5G network capacity: key elements and technologies," *IEEE Vehicular Technology Magazine*, vol. 9, no. 1, pp. 71–78, Mar. 2014.

[3] Osseiran, A.; Boccardi, F.; Braun, V. *et al.*, "Scenarios for the 5G mobile and wireless communications: the vision of the METIS project," *IEEE Communications Magazine*, vol. 52, no. 5, pp. 26–35, May 2014

[4] Bhushan, N.; Li, J.; Malladi, D. *et al.*, "Network densification: the dominant theme for wireless evolution into 5G," *IEEE Communications Magazine*, vol. 52, no. 2, pp. 82–89, Feb. 2014.

[5] Mogensen, P.; Pajukoski, K.; Tiirola, E. *et al.*, "Centimeter-wave concept for 5G ultra-dense small cells," *2014 IEEE 79th Vehicular Technology Conference (VTC Spring)*, 18–21 May 2014.

[6] Hwang, I.; Song, B.; and Soliman, S.S., "A holistic view on hyper-dense heterogeneous and small cell networks," *IEEE Communications Magazine*, vol. 51, no. 6, pp. 20–27, Jun. 2013.

[7] Yunas, S.; Valkama, M.; and Niemelä, J., "Spectral and energy efficiency of ultra-dense networks under different deployment strategies," *IEEE Communications Magazine*, vol. 53, no. 1, pp. 90–100, Jan. 2015.

[8] Baldemair, R.; Irnich, T.; Balachandran, K. *et al.*, "Ultra-dense networks in millimeter-wave frequencies," *IEEE Communications Magazine*, vol. 53, no. 1, pp. 202–208, Jan. 2015.

[9] Third Generation Partnership Project, Technical Specification Group Radio Access Network, "Scenarios and requirements for small cell enhancements for E-UTRA and E-UTRAN (Release 12)," http://www.3gpp.org

[10] Andrews, J.G.; Baccelli, F.; and Ganti, R.K., "A tractable approach to coverage and rate in cellular networks," *IEEE Transactions on Communications*, vol. 59, no. 11, pp. 3122–3134, Nov. 2011.

[11] ElSawy, H.; Hossain, E.; and Haenggi, M., "Stochastic geometry for modeling, analysis, and design of multi-tier and cognitive cellular wireless networks: a survey," *IEEE Communications Surveys & Tutorials*, vol. 15, no. 3, pp. 996–1019, Third Quarter 2013.

[12] Hu, L.; Sanchez, L.L.; Maternia, M. *et al.*, "Heterogeneous LTE-advanced network expansion for 1000× capacity," *2013 IEEE 77th Vehicular Technology Conference (VTC Spring)*, pp. 1–5, 2–5 June 2013.

[13] Gelabert, X.; Legg, P.; and Qvarfordt, C., "Small cell densification requirements in high capacity future cellular networks," *2013 IEEE International Conference on Communications Workshops (ICC)*, pp. 1112–1116, 9–13 June 2013.

[14] Hu, R.Q.; and Qian, Y., "An energy efficient and spectrum efficient wireless heterogeneous network framework for 5G systems," *IEEE Communications Magazine*, vol. 52, no. 5, pp. 94–101, May 2014.

[15] Stefanatos, S.; and Alexiou, A., "Access point density and bandwidth partitioning in ultra dense wireless networks," *IEEE Transactions on Communications*, vol. 62, no. 9, pp. 3376–3384, Sep. 2014.

[16] Gotsis, A.G.; Stefanatos, S.; and Alexiou, A., "Spatial coordination strategies in future ultra-dense wireless networks," *2014 11th International Symposium on Wireless Communications Systems (ISWCS)*, pp. 801–807, Aug. 2014.

[17] Gotsis, A.G.; and Alexiou, A., "On coordinating ultra-dense wireless access networks: optimization modeling, algorithms and insights," Technical Report, Dec. 2013, online: http://arxiv.org/abs/1312.1577.

[18] Recommendation ITU-R M.2083-0, "IMT vision – framework and overall objectives of the future development of IMT for 2020 and beyond," Sep. 2015.

[19] Simsek, M.; Aijaz, A.; Dohler, M.; Sachs, J.; and Fettweis, G., "5G-enabled tactile Internet," *IEEE Journal on Selected Areas in Communications*, vol. 34, no. 3, pp. 460–473, Mar. 2016.

Chapter 2

New 5G air interface: challenges for efficient multi-service coexistence for 5G below 6 GHz*

Frank Schaich[1], Salah El Ayoubi[2], Berna Sayrac[2], Gerhard Wunder[3], Martin Schubert[4] and Andreas Georgakopoulos[5]

5G will have to cope with a high degree of heterogeneity in terms of: (a) services (mobile broadband, massive machine and mission critical communications, broad-/multicast services and vehicular communications); (b) device classes (low-end sensors to high-end tablets); (c) deployment types (macro and small cells); (d) environments (low-density to ultra-dense urban); and (e) mobility levels (static to high-speed transport).

Consequently, diverse and often contradicting key performance indicators need to be supported, such as high capacity/user-rates, low latency, high reliability, ubiquitous coverage, high mobility, massive number of devices, low cost/energy consumption. 4G is not designed to meet such a high degree of heterogeneity efficiently. Moreover, having multiple radio access technologies for multi-service support below 6 GHz will be too costly.

A new multi-service air interface (AI) for below 6 GHz through a modular design needs to be developed. To allow the system to adapt to the anticipated heterogeneity, the pursued properties are: flexibility, scalability, versatility, efficiency, future-proofness. To this end, the technical AI components (e.g., flexible waveform and frame design, scalable multiple access procedures, adaptive retransmission schemes, enhanced multi-antenna schemes with/without cooperation, advanced multi-user detection, interference coordination, support for ultra-dense cell layouts, multi-cell radio resource management, device-to-device) need

[1]Nokia Bell Labs, Lorenzstr. 10, 70435 Stuttgart, Germany
[2]Orange Labs, 44 avenue de la République, 92326 Chatillon, France
[3]Fraunhofer Heinrich Hertz Institut, Einsteinufer 37, D-10587 Berlin, Germany & FU Berlin, Takustr. 9, 14195 Berlin, Germany
[4]Huawei German Research Center, Riesstraße 25, 80992, Munich, Germany
[5]WINGS ICT Solutions, 336 Syggrou Avenue, 17673 Athens, Greece

*The views expressed are those of the authors and do not necessarily represent the project. The consortium is not liable for any use that may be made of any of the information contained therein.

to be developed and integrated into an overall AI framework where adaptation to the above described sources of heterogeneity will be accomplished and validated through system-level simulations.

2.1 Introduction

Cellular technology has been renewed about every decade since the introduction of GSM, which can be considered as the main representative of the Second Generation (2G) cellular technology. In the 1990s, GSM became widespread and offered voice services all around the globe. With the rising success of the World Wide Web, data services became the key driver for the next generation of cellular technology. Around 2000, the introduction of the Third Generation (3G) cellular technology enabled the rapid growth of mobile data services. Around 2010, the introduction of 4G (LTE – Long-Term Evolution) has brought major improvements to wireless broadband access to Internet services.

The next generation is expected to arrive around 2020. There are five *trends*, which are pushing the limits of 4G, pointing towards the need for a Fifth generation cellular technology:

- **Trend 1** – *Increase in capacity*: The demand for wireless data is predicted to increase significantly, resulting in 1,000 times higher mobile data volumes and 10–100 times higher end user data rates.
- **Trend 2** – *Increase in the number of connected devices*: The number of connected devices is predicted to increase by a factor of 10–100, which means that up to 300,000 devices need to be served per access point.
- **Trend 3** – *Increase in reliability*: Wireless connectivity will be applied to new use cases that require extremely reliable connections (typically 99.999% availability and reliability) and mission-critical communications, such as vehicle-to-vehicle coordination, critical control of the power grid, etc.
- **Trend 4** – *Decrease in latency*: Remote presence and tactile Internet impose stringent latency constraints on the end-to-end connection, including the wireless part. Forecasts indicate that the latency should be decreased by a factor of 5 in order to enable such services.
- **Trend 5** – *Increase in efficiency*: Efficiency in terms of resource utilization (e.g., energy, spectrum) is becoming more and more pronounced. It is seen as an indispensable ingredient for a healthy/sustainable ICT market/business and environment.

The challenge for 5G is therefore not only to increase the user rates or the capacity (as it has been so far the case for former generations), but also to master the wide variety of requirements/conditions as observed in the above trends. Consequently there is a need to work towards a new 5G air interface that will offer much more than just a faster variant of 4G.

The 5G air interface ***below 6 GHz*** is envisioned to provide the following *key characteristics*:

- ***flexibility*** to support the *broad class of services* with their associated (broad class) of Key Performance Indicators (KPIs),
- ***scalability*** to support the high number of devices,
- ***versatility*** to support the diverse device types and traffic/transmission characteristics,
- ***efficiency*** to support the requirements on energy consumption and resource utilization, and
- ***future-proofness*** to support easy integration of new features.

In order to bring this vision into reality the following objectives need to be pursued:

- In-band coexistence of highly differing services, device types and traffic/transmission characteristics by means of a highly flexible, versatile and scalable air interface.
- Ubiquitous coverage and high capacity where and when required.
- High efficiency in terms of energy and resource consumption.
- Future-proof compared to former generations by facilitating the simple introduction of new features.
- Evaluation and validation of the developed concepts by means of system-level simulations and hardware proof of concepts for selected components.

2.1.1 Chapter contributions

This chapter aims at shedding some light on the 5G landscape from service and use case perspectives and to identify key research challenges for designing an air interface that satisfy the service requirements in various use cases. This chapter is structured as follows. In Section 2.2, we present the core services and their associated KPIs and requirements. Section 2.3 shows how these core services have been captured in the literature in use cases that facilitate integration of different services. Section 2.4 then presents the main selected 5G use cases. Section 2.5 presents the challenges when designing the air interface for the different services, on both link level and system level. Section 2.6 shows how the technical solutions can be evaluated using system-level simulators. Section 2.7 describes the prospective impact and timelines in 3GPP, and, finally Section 2.8 concludes the chapter.

2.2 Core services and their associated KPIs

2.2.1 Core services

The key characteristics of the 5G air interface (flexibility, scalability, versatility, efficiency and future-proofness) will enable new services with diverse application areas e.g., connected devices, vehicle-to-vehicle coordination, critical control of the power grid, remote presence, tactile Internet and more. These services can be grouped into the following five categories called the *5G core services*:

- **Mobile BroadBand (MBB[2]):** 5G will have to support mobile broadband connections as of today, but with increased capacity, efficiency and data transmission rates. Typical use cases corresponding to MBB are multimedia streaming, combinational services including Voice over IP (VoIP) as one of the services, Internet browsing, videoconferencing, file downloads and uploads to the cloud, some gaming services, etc.
- **Massive Machine Communications (MMC[3]):** MMC can be considered as one part of Machine Type Communications (MTC) [1–3]. In MMC there is a massive amount of sensors/metres/actuators that are deployed anywhere in the landscape (e.g., fire sensors in areas with high danger of forest fires, sensors for water quality inspection/management, smoke alarm detectors, power failure notifications, tamper notifications, actuators to control access, temperature, lighting in buildings or traffic lights in the context of smart cities, etc.) and that need to access the wireless network. In most of the cases, MMC traffic involves relatively small packets per connection and therefore requires low throughput per device. In addition, a typical characteristic of MMC is that the sensors must be robust, simple (i.e., low-cost), able to operate on substantially long battery lives and reachable in challenging coverage conditions, e.g., indoor and basements. Today, this is also known as "low-end Internet of Things (IoT)". Typical use cases corresponding to MMC are smart metering, natural ecosystem monitoring, remote maintenance/control, fleet/parcel tracking/tracing, etc.
- **Mission Critical Communications (MCC[4]):** MCC includes another type of MTC which implies a wireless connectivity, where sensor messages need to be transmitted between the respective communication partners while satisfying one or several of the following: very low response times, very high reliability and very high availability. Typical use cases of MCC are private safety and security applications (e.g., video surveillance and intrusion detection), vital sign monitoring, factory automation, etc. It has to be noted that MCC also addresses person-to-person (P2P) communications in the framework of e.g., natural disasters and public safety.
- **Broadcast/Multicast Services (BMS):** BMS involve simultaneous content delivery in a "one-to-many" or "many-to-many" mode. Efficient integration of BMS into 5G will render valuable spectrum resources available for other (unicast) services. Typical examples of BMS are not only the well-known mobile TV and digital radio cases, but also software/firmware updates/downloads; targeted use cases could be "Stadium 2.0" dedicated coverage or "Smart Cities" (e.g., digital signature). In the specific case of multimedia content delivery, BMS is known as Multimedia Broadcast/Multicast Service (MBMS).
- **Vehicle-to-vehicle and vehicle-to-infrastructure communications (V2X):** V2X implies direct wireless connectivity either between vehicles (Vehicle-to-Vehicle, V2V) or between vehicles and the roadside (Vehicle-to-Infrastructure,

[2]Also called eMBB for enhanced MBB [5] or xMBB for extreme MBB [6].
[3]Also called mMTC for massive Machine Type Communications [6].
[4]Also called uMTC for ultra-reliable Machine Type Communications [6].

V2I) [4]. Today, also known as Intelligent Transportation Systems (ITS), V2X consists of use cases where the other four core services (MBB, MMC, MCC and BMS) involve nodes with high speeds. Typical examples are road safety (e.g., collision avoidance) which is the combination of V2X with MCC, pay-as-you-drive which is the combination of V2X with MMC and infotainment which is the combination of V2X with BMS and/or MBB.

2.2.2 Key performance indicators

In relation to the above five core services, the following *main KPIs* for 5G are identified:

- **KPI 0:** User experienced data rate
- **KPI 1:** Traffic density (to achieve high system capacity)
- **KPI 2:** Latency
- **KPI 3:** Coverage (to provide ubiquitous access)
- **KPI 4:** Mobility
- **KPI 5:** Connection density
- **KPI 6:** Reliability/availability
- **KPI 7:** Complexity reduction
- **KPI 8:** Energy efficiency

Even though many KPIs are self-explanatory, further clarifications are provided in the next paragraph:

- **KPI 0:** the data rate experienced by the user directly relates to the bandwidth respectively spectral efficiency allocated to the service. In most cases, the allocated service bandwidth is shared with other users, depending on the traffic load, and the fraction of the actually allocated bandwidth is evaluated on average over the entire service duration rather than instantaneously. It is worth noting that it is meaningful to consider KPI 0 in conjunction with other KPIs, e.g., KPI 1, KPI 4 and KPI 5, due to the trade-offs between them. The actual (range of) values of the corresponding requirement are expected to strongly depend on the specific use case at hand.
- **KPI 2:** Latency is defined as in the NGMN white paper [5]. Two distinct cases should be considered:
  - E2E Latency (more related to the user experience): Measures the duration between the transmission of a small data packet from the application layer at the source node and the successful reception at the application layer at the destination node plus the equivalent time needed to carry the response back (if foreseen).
  - User Plane Latency (more related to the air interface design): Measures the time it takes to transfer a small data packet from user terminal to the Layer 2/Layer 3 interface of the 5G system destination node, plus the equivalent time needed to carry the response back.

- **KPI 3:** coverage may be defined in terms of the reachable area/distance around/from the transmitter(s) where an adequate sensitivity level is still available, in order to guarantee that the wanted signal is received above or at least at the reference sensitivity level of the receiver (it applies to both Uplink (UL) and Downlink (DL)). Depending on the considered core service, like e.g., MMC, coverage may also be defined as the Maximum Coupling Loss (MCL) allowed both on the UL and on the DL (assuming the minimum), in terms of the difference between the transmit power of the transmitter and the reference sensitivity level of the receiver (taking into account thermal noise density, the receiver noise figure, the interference margin, the occupied channel bandwidth, the required SINR and the receiver processing gain).
- **KPI 6:** availability may be assumed to be equal to (1 – service blocking probability), where service blocking probability is due to lack of enough resources to access, grant and provide the service, even in case of adequate coverage. Additionally, reliability may be assumed equal to (1 – service dropping rate[5]), where service dropping may occur e.g., in case of quite severe multiple interferers received together with the wanted signal, even in case of adequate coverage. Depending on the considered core service, reliability may also be defined as the probability of transferring a given payload and receiving the corresponding acknowledgement with some predefined reliability and latency requirement.
- **KPI 7 and 8:** These two KPIs will be assessed in comparison with current standards. The estimated complexity and energy consumption should be comparable to today's metropolitan deployments. Keeping the energy consumption constant while improving the system performance implies the need for better energy efficiency.

The above KPIs can be classified into two categories:

- **User Experience KPIs**: 0, 2, 3[6], 4, 6
- **System Performance KPIs**: 1, 3, 5, 7, 8

The User Experience KPIs directly impact the Quality of Service (QoS) and the overall Quality of Experience (QoE) for the user of a given service, whilst the System Performance KPIs mostly relate to the service delivery efficiency from the mobile network operator's (MNO's) perspective. Both categories are equally important: failing to meet the User Experience KPIs leads to lack of customers for delivering a service (i.e., meeting the System Performance KPIs), even though the service is delivered efficiently. Failing to meet the System Performance KPIs leads to lack of MNOs providing a service despite meeting users' expectations (i.e., meeting the User Experience KPIs).

[5]Service dropping rate is linked to the degradation of service quality due to the erroneous reception (of packets/frames/symbols/bits).

[6]Even if KPI3 is considered a System Performance KPI, it could be considered a User Experience KPI as well since it is a crucial element which impacts both aspects.

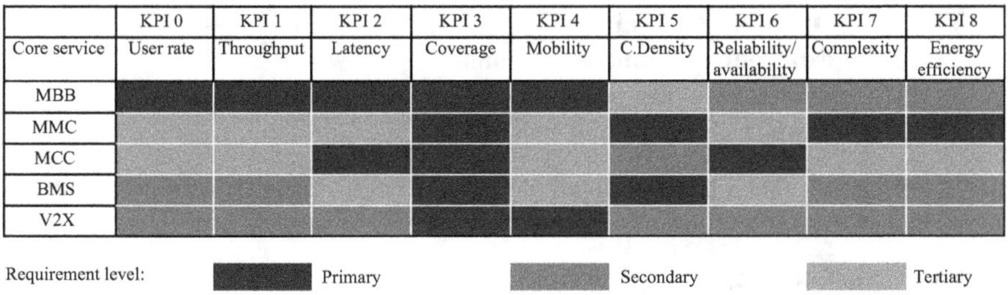

Figure 2.1 Core services and related KPIs with different requirement levels

2.2.3 KPI relevance to core services

A closer look on the core services introduced above can help identifying the KPIs that are mostly relevant for them. The primary, secondary and tertiary requirement levels are shown in Figure 2.1 for each core service. In this figure, the higher the relative requirement level of the corresponding KPI on a per core service basis, the more challenging the corresponding (range of) values and consequently the requirement itself for that KPI is.

2.3 Use cases from the 5G literature

It is widely recognized that considering each service separately and building a 5G network accordingly will neither be economically nor environmentally sustainable [7]. For this reason, the 5G network design has to be performed towards a set of use cases that typically combine multiple service types. In this section, we briefly introduce the 5G use cases from the literature (mainly from METIS project [6], NGMN [5] and 3GPP [8,9]) and show how representative use cases can be identified.

2.3.1 Use cases from METIS/METIS-II projects

METIS project first introduced 12 Test Cases (TC) in [6], where a test case is a practical scenario formulated from end-users' perspective and contains a set of assumptions, constraints and requirements. These test cases were selected such that they essentially sample the space of future applications, implying that some of them have applications that traditionally have not been considered within studies of telecommunication systems (as of April 2013). The 12 test cases are the following [6]:

- **TC1: Virtual reality office**
- **TC2: Dense urban information society**
- **TC3: Shopping mall**
- **TC4: Stadium**
- **TC5: Teleprotection in smart grid network**
- **TC6: Traffic jam**
- **TC7: Blind spots**
- **TC8: Real-time remote computing for mobile terminals**

- **TC9: Open air festival**
- **TC10: Emergency communications**
- **TC11: Massive deployment of sensors and actuators**
- **TC12: Traffic efficiency and safety**

Subsequently, METIS introduced nine use cases inspired by recent technology trends and future projections. The nine use cases are the following [10]:

- **UC1: Gaming**
- **UC2: Marathon**
- **UC3: Media on demand**
- **UC4: Unmanned aerial vehicles**
- **UC5: Remote tactile interaction**
- **UC6: eHealth**
- **UC7: Ultra-low cost 5G network**
- **UC8: Remote car sensing and control**
- **UC9: Forest industry on remote control**

METIS-II project has selected a subset of the above 21 use cases in total (renaming the test cases to use cases) for TC1 (same requirements as defined in METIS), TC2 (updated METIS requirements to ensure that indoor is included with NGMN requirements of more than 300 Mbps for 95% of users), TC11 (same requirements as defined in METIS) and TC12 (updated METIS requirements to ensure that low latency extends to V2I, so that solutions can help for other uMTC applications, and added MBB capabilities for on-board entertainment). METIS-II has introduced one more use case [7]

- **UC22**: Broadband access everywhere

focusing on coverage in rural and suburban areas with NGMN requirements of 50 Mbps+ in DL and 25 Mbps+ in UL.

2.3.2 NGMN use cases

NGMN has developed 24 Use Cases for 5G [5] as representative examples, which are grouped into 8 Use Case Families. The use cases and the use case families serve as inputs for setting the requirements and defining the building blocks of the 5G architecture. The use cases are not meant to be exhaustive, but rather to be treated as a tool to ensure that the level of flexibility required in 5G is well captured. Figure 2.2 shows the eight use case families with one example use case given for each family.

The 5G use cases demand for very diverse and sometimes extreme requirements. It is anticipated that a single solution to satisfy all the extreme requirements at the same time may lead to over-specification and high cost, unless a versatile framework is offered to the network operator. Nevertheless, several use cases are anticipated to be active concurrently in the same operator network, thus requiring a high degree of flexibility and scalability of the 5G network.

Table 2.1 shows the mapping between METIS/METIS II use cases and NGMN use cases.

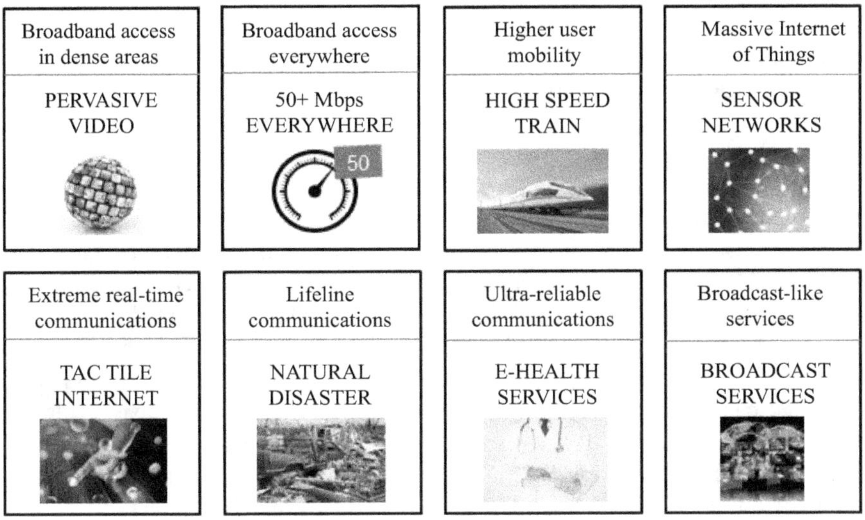

Figure 2.2 *5G use case families and related examples – NGMN Source (Photos: © Shutterstock.com/Leigh Prather, Alexey Stiop, SOMMAI, Syda Productions, flydragon, BackgroundStore, Ortodox, tovovan.)*

Table 2.1 *Mapping between METIS/METIS-II (highlighted) use cases and NGMN use cases*

No.	METIS/METIS II test/use cases	NGMN
1	Virtual reality office	4
2	Dense urban information society	3
3	Shopping mall	3, 5
4	Stadium	5
5	Teleprotection in smart grid network	13
6	Traffic jam	2
7	Blind spots	7
8	Real-time remote computing for mobile terminals	8, 10
9	Open air festival	5
10	Emergency communications	16
11	Massive deployment of sensors and actuators	13
12	Traffic efficiency and safety	17
13	Gaming	2, 15
14	Marathon	2, 5, 12
15	Media on demand	1, 4
16	Unmanned aerial vehicles	18, 22
17	Remote tactile interaction	15, 19, 20
18	eHealth	20
19	Ultra-low cost 5G network	7
20	Remote car sensing and control	10, 13, 17, 19
21	Forest industry on remote control	19
22	Broadband access everywhere	6

2.3.3 3GPP SMARTER SI use cases

The Feasibility Study on New Services and Markets Technology Enablers (FS_SMARTER) carried out by 3GPP SA1 aims to identify the market segments and verticals whose needs should be met by 3GPP, and to develop the use cases and requirements that the 3GPP eco-system would need to support in the future [8,9].

The focus of this work is on the use cases and corresponding requirements that cannot be met with the Evolved Packet System (EPS). Nevertheless, SMARTER also includes analysis on which legacy services and requirements from the existing 3GPP systems need to be included and whether fallback mechanisms need to be developed.

Twenty SMARTER use cases have been identified as highlighted in Table 2.2.

From the above table it can be seen that 15 SMARTER use cases may be mapped to the corresponding NGMN use cases, whose numbering refers to the NGMN use cases' list reported in [5], starting from use case 1 (Pervasive video) up to use case 24 (Broadcast-like services: Local, Regional, National).

The remaining five SMARTER use cases (highlighted in Table 2.2) are not covered by any NGMN use case. Those five use cases may be characterized in two ways. On the one hand, some of them are related to the legacy services provided by earlier generations/legacy systems (such as SMARTER use cases 4 and 16). On the other hand, some others are addressing functional characteristics of the overall air interface, such as flexibility and scalability, network architecture (network slicing) and flexible application traffic routing (such as SMARTER use cases 2, 8 and 9). In this way, they do not directly map to 5G core services.

Table 2.2 Mapping between SMARTER use cases and NGMN use cases

No.	SMARTER use cases	NGMN u.c.
1	Ultra reliable communications	17–22
2	Network slicing	–
3	Lifeline communications/natural disaster	16
4	Migration of services from earlier generations	–
5	Mobile broadband for indoor scenario	4
6	Mobile broadband for hotspots scenario: dense urban area scenario	3
7	On-demand networking	5
8	Flexible application traffic routing	–
9	Flexibility and scalability	–
10	Mobile broadband services with seamless wide-area coverage	2, 5–8
11	Virtual presence	4
12	Connectivity for drones	22
13	Industrial control	15, 18
14	Tactile Internet	15
15	Localized real-time control	18
16	Coexistence with legacy systems	–
17	Extreme real-time communications and the tactile Internet	15
18	Remote control	19, 22
19	Light weight device configuration	13
20	Wide area sensor monitoring and event driven alarms	13–14

2.4 The 5G use cases from FANTASTIC-5G

FANTASTIC-5G project has selected a subset of 5G use cases that are mostly representative and that can be satisfied with spectrum below 6 GHz (the primary focus of the project).

2.4.1 Detailed 5G use cases for RAN design

Based on the analysis performed above, a potential regrouping of the overall 5G use cases is presented in Table 2.3, as well as their mapping to 5G core services.

2.4.2 Selected FANTASTIC-5G use cases

To design an air interface that satisfies the most stringent requirements of different core services in realistic scenarios, a subset of use cases that are mostly representative needs to be selected.

Step 1: select use cases not covered by LTE and LTE-Advanced or even earlier RATs

UC23 is already covered by LTE and LTE-Advanced, based on a single frequency network (SFN) mode of operation, whilst UC24 requires high flexibility and scalability implying new components to be included, compared to the available ones in LTE and LTE-Advanced. In fact, as described in [5] while personalization of communication will lead to a reducing demand for legacy broadcast as deployed today, e.g., TV, the fully mobile and connected society will nonetheless need efficient distribution of information from one source to many destinations. These services may distribute content as done today (typically only DL), but also provide a feedback channel (in the UL) for interactive services or acknowledgement information. Both, real-time or non-real-time services should be possible and can have a wide distribution or focused in smaller regions.

Following this consideration, UC23 is ruled out.

Step 2: select use cases not covered by higher frequencies (e.g., above 6 GHz) RATs

In case of higher frequencies RATs (e.g., above 6 GHz), multiple RATs with potentially different air interfaces could complement each other for providing broadband access in dense areas. This approach allows for multiple RATs with complementary use cases, which could be used to provide capacity and high data rates in dense urban areas, as well as indoors, to support the mobile broadband use case categories. To provide broadband access in dense areas with lower frequencies RATs (below 6 GHz) the use case "Dense urban society below 6 GHz" is defined. As already stated above, this use case is deemed important to boost capacity in 5G and is relevant for 5G broadband services, particularly for the frequency range between 3 and 6 GHz.

Following these considerations, UC1, UC2, UC4, UC5 are ruled out.

Table 2.3 List of the overall 5G use cases and mapping to the core services

NGMN categories	FANTASTIC-5G use cases	Core service
Broadband access in dense area	**1. Pervasive video**	**MBB**
	2. Operator cloud services	**MBB**
	3. Dense urban society	**MBB**
Indoor ultra-high broadband access	**4. Smart Office**	**MBB**
Broadband access in a crowd	**5. HD video/photo sharing in stadium/open-air gathering**	**MBB**
50+ Mbps everywhere	**6. 50 Mbps everywhere**	**MBB**
Ultra low-cost broadband access for low ARPU areas	**7. Ultra-low cost networks**	**MBB**
Mobile broadband in vehicles	**8. High-speed train**	**MBB, V2X**
	9. Moving hot spots	**MBB, V2X**
	10. Remote computing	**MBB, V2X**
Airplanes connectivity	**11. 3D connectivity: aircrafts**	**V2X**
Massive low-cost/long-range/low-power MTC	**12. Smart wearables (clothes)**	**MMC**
	13. Sensor networks	**MMC**
Broadband MTC	**14. Mobile video surveillance**	**MCC**
Ultra low latency	**15. Tactile Internet**	**MCC**
Resilience and traffic surge	**16. Natural disaster**	**MCC**
Ultra-high reliability and ultra-low latency	**17. Automatic traffic control/driving**	**MCC, V2X**
	18. Collaborative robots	**MCC**
	19. Remote object manipulation – remote surgery	**MCC**
Ultra-high availability and reliability	**20. eHealth: extreme life critical**	**MCC**
	21. Public safety	**MCC**
	22. 3D connectivity: drones	**MCC**
Broadcast like services	**23. News and information**	**BMS**
	24. Broadcast like services: local, regional, national	**BMS**

Step 3: select use cases that encompass and are more demanding than other use cases (in terms of KPIs and corresponding requirements) in the framework of the same core services

In the framework of the MBB core service, UC6 is more demanding than UC7, in terms of almost all the KPIs (ranging from KPI 0 to KPI 6, with the only exceptions of KPI 7 and KPI 8) and corresponding requirements. Nevertheless, since KPI7 and KPI8 are important metrics for MNOs, we will pay special attention to those KPIs while working on UC6.

In the framework of the MBB + V2X core services, UC8 may be considered more demanding than UC9 and UC10, in terms of at least KPI 3 and KPI 4 which are crucial KPIs in case of vehicular applications.

From the V2X core service perspective, UC8 is more demanding than UC11 since besides KPI 3 and KPI 4 (crucial in case of vehicular applications), constraints coming from mobile broadband applications need to be considered at the same time and may be challenged and jeopardized by coverage and mobility.

In the framework of the MMC core service, UC13 is more demanding than UC12, in terms of KPI 3, KPI 5, KPI 7 and KPI 8, crucial in case of massive low-cost/long-range/low-power MTC.

In the framework of the MCC core service, UC15 is more demanding than UC14, in terms of KPI 2, KPI 3 and KPI 6 which are crucial KPIs in case of mission critical communications.

From the MCC core service perspective, UC17 is more demanding than UC16, UC18, UC19, UC20, UC21 and UC22 since, besides KPI 2, KPI 3 and KPI 6 (crucial in case of mission critical communications), KPI 4 has to be taken into account at the same time and mobility usually poses remarkable challenges on meeting requirements for coverage and reliability/availability.

Following these considerations, UC7, UC9, UC10, UC11, UC12, UC14, UC16, UC18, UC19, UC20, UC21, UC22 are ruled out.

Based on the overall criteria, the resulting selected use cases for FANTASTIC-5G are UC3, UC6, UC8, UC13, UC15, UC17, UC24 (now referred to as Use Cases 1–7):

- **Use Case 1:50 Mbps everywhere** **MBB**
- **Use Case 2:High speed train** **MBB + V2X**
- **Use Case 3:Sensor networks** **MMC**
- **Use Case 4:Tactile Internet** **MCC**
- **Use Case 5:Automatic traffic control/driving** **MCC + V2X**
- **Use Case 6:Broadcast like services: local, regional, national** **BMS**

  Use Case 7:Dense urban society below 6 GHz **MBB**

The detailed description with KPI definitions for each of the use cases can be found in Tables 2.4–2.10. It is worth noting that the above seven selected use cases represent all the core services either as a single core service (MBB, MMC, MCC, BMS) or as a combination of two core services (V2X in conjunction with either MBB or MCC). We note that, the NGMN use case "Dense urban society" has been modified to "Dense urban society below 6 GHz" focusing on the frequency range between 3 and 6 GHz. As a consequence, the NGMN requirement of 300 Mbps has been reduced to 50 Mbps. It is also important to highlight that the selected use cases by METIS II are mostly taken into account by the above list of selected use cases, in terms of UC6, UC13 and UC17. Note that UC4 (smart office) cannot be taken into account, since it is a use case which is typically covered (also) by higher frequencies RATs.

Table 2.4 Use case 1: 50 Mbps everywhere – requirements definition

	Use case 1: 50 Mbps everywhere					
	KPI no.	**KPI definition**	**Req. Lev.**	**Requirement definition**	**UoM**	**Notes**
User Exp.	**KPI 0**	User experienced data rate		50 Mbps (DL), 25 Mbps (UL)	Mbps	
Syst. Perf.	**KPI 1**	Traffic Density		Suburban: DL: 20 Gbps/km^2 Suburban: UL: 10 Gbps/km^2 Rural: DL: 5 Gbps/km^2 Rural: UL: 2.5 Gbps/km^2	Gbps/km^2	Traffic density is the total expected traffic volume per area which takes into account both the user experienced data rate (KPI0) and also the connection density (KPI5)
User Exp.	**KPI 2**	Latency		10 ms	ms	User plane latency is considered
Syst. Perf.	**KPI 3**	Coverage		99.999% of the area where the service is provided	%	The wanted signal needs to be received everywhere, above or at least at the reference sensitivity level of the receiver (it applies both on the UL and on the DL), in order to guarantee the required minimum user data rate
User Exp.	**KPI 4**	Mobility		0–120 km/h	km/h	
Syst. Perf.	**KPI 5**	Connection density		Suburban: 400 users/km^2 Rural: 100 users/km^2	Users/km^2	

User Exp.	**KPI 6**	Reliability/availability		Reliability: 95% Availability: 99%	%	Reliability may be assumed equal to (1 – service dropping rate), where service dropping may occur e.g., in case of quite severe multiple interferers received together with the wanted signal, even in case of adequate coverage. Depending on the considered service, reliability may be defined as the probability of correctly transferring a given payload and correctly receiving ACK within a 10 ms latency Non-availability may occur due to lack of coverage. But this is reflected by KPI 3. Therefore we consider that availability may be assumed equal to (1 – service blocking probability), where service blocking probability is due to lack of enough resources to access, grant and provide the service, even in case of adequate coverage
Syst. Perf.	**KPI 7**	Complexity reduction		50%	%	Reduction percentage compared to legacy networks complexity
Syst. Perf.	**KPI 8**	Energy efficiency		50%	%	Reduction percentage compared to legacy networks energy consumption

Req. Lev. **Primary** **Secondary** **Tertiary**

Table 2.5 Use case 2: high speed train – requirements definition

	Use case 2: High speed train					
	KPI no.	**KPI definition**	**Req. Lev.**	**Requirement definition**	**UoM**	**Notes**
User Exp.	**KPI 0**	User experienced data rate		50 Mbps (DL), 25 Mbps (UL)	Mbps	
Syst. Perf.	**KPI 1**	Traffic density		DL: 100 $Gbps/km^2$ UL: 50 $Gbps/km^2$	$Gbps/km^2$	Traffic density is the total expected traffic volume per area which takes into account both the user experienced data rate (KPI0) and also the connection density (KPI5).
User Exp.	**KPI 2**	Latency		10 ms	ms	User plane latency is considered.
Syst. Perf.	**KPI 3**	Coverage		99.5% of the area where the service is provided	%	The transmitted signal needs to be received above or at least at the reference sensitivity level of the receiver (it applies both on the UL and on the DL), in order to guarantee the required minimum user data rate.
User Exp.	**KPI 4**	Mobility		0–500 km/h	km/h	
Syst. Perf.	**KPI 5**	Connection density		2,000 $users/km^2$ (500 active users per train × 4 trains)	$Users/km^2$	Source: NGMN white paper [NGMN-5G]

User Exp.	**KPI 6**	Reliability/availability		Reliability: 95% Availability: 99%	%	Reliability may be assumed equal to (1 – service dropping rate), where service dropping may occur e.g., in case of quite severe multiple interferers received together with the wanted signal, even in case of adequate coverage. Depending on the considered service, reliability may be defined as the probability of correctly transferring a given payload and correctly receiving ACK within a 10 ms latency Availability may be assumed equal to (1 – service blocking probability), where service blocking probability is due to lack of enough resources to access, grant and provide the service, even in case of adequate coverage
Syst. Perf.	**KPI 7**	Complexity reduction		-	-	Same complexity as of today on both the infrastructure side and the device side
Syst. Perf.	**KPI 8**	Energy efficiency		25%	%	Reduction percentage compared to legacy networks and devices energy consumption Low-energy operation of all radio nodes is expected due to energy cost and EMF reasons Reduced battery consumption on the device side

Req. Lev. **Primary** **Secondary** **Tertiary**

Table 2.6 Use case 3: sensor networks – requirements definition

	Use case 3: Sensor networks					
	KPI no.	**KPI definition**	**Req. Lev.**	**Requirement definition**	**UoM**	**Notes**
User Exp.	**KPI 0**	User experienced data rate		Not critical 160 bps–1 Mbps (mostly UL biased)	Mbps	
Syst. Perf.	**KPI 1**	Traffic density		Not critical UL: 100 Mbps/km^2	Mbps/km^2	(Ultra-)low data rate per device, small net payload (in the order of 20 to 125 bytes per message or even more in some special cases) and asynchronous devices lead to non-critical traffic density
User Exp.	**KPI 2**	Latency		Not critical 10 s	S	User plane latency is considered
Syst. Perf.	**KPI 3**	Coverage		20 dB more than legacy systems coverage (both UL and DL) with 100% coverage as a baseline	dB	Coverage may be defined as the Maximum Coupling Loss (MCL) allowed both on the UL and on the DL (assuming the minimum), in terms of the difference between the transmit power of the transmitter and the reference sensitivity level of the receiver (taking into account the thermal noise density, the receiver noise figure, the interference margin, the occupied channel bandwidth, the required SINR and the receiver processing gain)

User Exp.	**KPI 4**	Mobility		Not critical (almost stationary) 95% of the devices: 0 km/h 5% of the devices: 0–120 km/h	km/h	The majority of the devices is expected to be stationary. In case of tracking purposes the mobile speed is expected to be in the range of 0–120 km/h.
Syst. Perf.	**KPI 5**	Connection density		Up to 600,000 devices/km^2	Devices/km^2	Around 60,000 devices/km^2 are expected to be supported by Cellular IoT solutions currently being standardized within 3GPP. FANTASTIC-5G will provide 10 times to 100 times higher connection density. In case of just 10 times, this leads to 600,000 devices/km^2
User Exp.	**KPI 6**	Reliability/availability		Not critical to be met due to sporadic traffic, small payload per message, (ultra-)low data rate per device, asynchronous devices Reliability: 95% Availability: 99%	%	Reliability may be defined as the probability of correctly transferring a given payload and correctly receiving ACK within the required latency (if foreseen) Availability may be assumed equal to (1–service blocking probability), where service blocking probability is due to lack of enough resources to access, grant and provide the service, even in case of adequate coverage

(Continues)

Table 2.6 (Continued)

	Use case 3: Sensor networks					
	KPI no.	**KPI definition**	**Req. Lev.**	**Requirement definition**	**UoM**	**Notes**
Syst. Perf.	**KPI 7**	Complexity reduction	Primary	Devices: 90%	%	Reduction percentage compared to legacy devices complexity A sensor device must have a significantly lower cost than a regular handset device, i.e., not more than a few euros for the radio part of the sensor The infrastructure cost is expected to be kept on the same level as today
Syst. Perf.	**KPI 8**	Energy efficiency	Primary	Ten-year device's battery life-time	years	The device's battery lifetime is requested to be at least 10 years with battery capacity of 5 Wh, since it may equal the device lifetime itself, in case of disposable devices placed in deeply indoor locations with e.g., metal boxes packaging, making the battery replacement a quite uncomfortable and expensive procedure No specific constraints for the infrastructure

Req. Lev. **Primary** **Secondary** **Tertiary**

Table 2.7 Use case 4: tactile Internet – requirements definition

	Use case 4: Tactile Internet					
	KPI no.	**KPI definition**	**Req. Lev.**	**Requirement definition**	**UoM**	**Notes**
User Exp.	**KPI 0**	User experienced data rate		50 Mbps (DL), 25 Mbps (UL)	Mbps	
Syst. Perf.	**KPI 1**	Traffic density		Potentially high	Tbps/km^2	Traffic density is the total expected traffic volume per area which takes into account both the user experienced data rate (KPI0) and also the connection density (KPI5)
User Exp.	**KPI 2**	Latency		$\leq$ 1 ms	ms	User Plane latency is considered
Syst. Perf.	**KPI 3**	Coverage		99.999% of the area where the service is provided	%	The transmitted signal needs to be received everywhere, above or at least at the reference sensitivity level of the receiver (it applies both on the UL and on the DL), in order to guarantee the required minimum user data rate.
User Exp.	**KPI 4**	Mobility		0–120 km/h	km/h	
Syst. Perf.	**KPI 5**	Connection density		Not critical	Users/km^2	

(Continues)

Table 2.7 (Continued)

	Use case 4: Tactile Internet					
	KPI no.	**KPI definition**	**Req. Lev.**	**Requirement definition**	**UoM**	**Notes**
User Exp.	**KPI 6**	Reliability/availability		Reliability: 99.999% Availability: 99.999%	%	Reliability may be assumed equal to (1–service dropping rate), where service dropping may occur e.g., in case of quite severe multiple interferers received together with the wanted signal, even in case of adequate coverage. Depending on the considered service, reliability may be defined as the probability of correctly transferring a given payload and correctly receiving ACK within at most 1 ms latency Availability may be assumed equal to (1–service blocking probability), where service blocking probability is due to lack of enough resources to access, grant and provide the service, even in case of adequate coverage
Syst. Perf.	**KPI 7**	Complexity reduction		Not critical	-	Complexity reduction not required
Syst. Perf.	**KPI 8**	Energy efficiency		Not critical	-	Energy efficiency is not required

Req. Lev. **Primary** **Secondary** **Tertiary**

Table 2.8 Use case 5: automatic traffic control/driving – requirements definition

	Use case 5: Automatic traffic control/driving					
	KPI no.	**KPI definition**	**Req. Lev.**	**Requirement definition**	**UoM**	**Notes**
User Exp.	**KPI 0**	User experienced data rate		DL: 50 kbps–10 Mbps UL: ~50 bps–10 Mbps	Mbps	
Syst. Perf.	**KPI 1**	Traffic density		DL: 30 $Gbps/km^2$ UL: 30 $Gbps/km^2$	$Gbps/km^2$	Considered the worst case of Connection density for Traffic density computation
User Exp.	**KPI 2**	Latency		1–10 ms	ms	User plane latency is considered 1 ms latency is required for cooperative manoeuvre planning, which cannot be covered by 802.11p but there are scenarios where 50 ms is sufficient, for example intersection collision warning In D2D communications the support of low latency communications, while leveraging the varying degrees of network assistance, is introduced in localized scenarios with focus on public security, emergency and vehicular communications
Syst. Perf.	**KPI 3**	Coverage		99.999% of the area where the service is provided	%	The wanted signal needs to be received everywhere, above or at least at the reference sensitivity level of the receiver (it applies both on the UL and on the DL), in order to guarantee the required minimum user data rate. Out of coverage scenarios can be considered as well for some applications.

(Continues)

Table 2.8 (Continued)

	Use case 5: Automatic traffic control/driving					
	KPI no.	**KPI definition**	**Req. Lev.**	**Requirement definition**	**UoM**	**Notes**
User Exp.	**KPI 4**	Mobility		Highway: 250 (abs.)–(500) (rel.) km/h Rural: 120 (abs.)–240 (rel.) km/h 1. Urban: 60 (abs.)–120 (rel.) km/h VRUs: 3–30 km/h	km/h	Abs.: absolute speed Rel.: relative speed between the vehicles VRUs: Vulnerable Road Users 3 km/h: pedestrian 30 km/h: bicycle
Syst. Perf.	**KPI 5**	Connection density		Urban: 3,000 devices/km^2 Suburban: 2,000 devices/km^2 Rural: 1,000 devices/km^2	Devices/km^2	Three different values are considered for different scenarios, based on EAB members input
User Exp.	**KPI 6**	Reliability/availability		Reliability: 99.999% Availability: 99.999%	%	Reliability may be assumed equal to (1 – service dropping rate), where service dropping may occur e.g., in case of quite severe multiple interferers received together with the wanted signal, even in case of adequate coverage. Depending on the considered service, reliability may be defined as the probability of correctly transferring a given payload and correctly receiving ACK within the required latency In D2D communications the support of ultra-reliable communications, while leveraging the varying degrees of network assistance, is introduced in localized scenarios with focus on public security, emergency and vehicular communications

						Availability may be assumed equal to (1 – service blocking probability), where service blocking probability is due to lack of enough resources to access, grant and provide the service, even in case of adequate coverage
Syst. Perf.	**KPI 7**	Complexity reduction		-	-	The cost of V2X infrastructure should not exceed the cost of traditional cellular modems The cost of V2X devices/chipsets should not exceed the cost of traditional terminals/chipsets
Syst. Perf.	**KPI 8**	Energy efficiency		25%	%	Reduction percentage compared to legacy networks and devices energy consumption Low-energy operation of all radio nodes is expected due to energy cost and EMF reasons Energy consumption needs to be minimized and to not exceed that of conventional non-V2X devices, since V2X devices mainly depend on battery power supply

Req. Lev. | **Primary** | **Secondary** | **Tertiary**

Table 2.9 Use case 6: broadcast like services: local, regional, national – requirements definition

	Use case 6: Broadcast like services: local, regional, national					
	KPI no.	**KPI definition**	**Req. Lev.**	**Requirement definition**	**UoM**	**Notes**
User Exp.	**KPI 0**	User experienced data rate		200 Mbps (DL), 0.5 Mbps (UL)	Mbps	
Syst. Perf.	**KPI 1**	Traffic density		DL: 200 Mbps UL: 0.5 Mbps	Mbps	The data throughput corresponds to the user experienced data rate, since it is independent from the connection density in a given area, due to the usage of shared resources
User Exp.	**KPI 2**	Latency		< 100 ms	ms	User Plane latency is considered
Syst. Perf.	**KPI 3**	Coverage		99.999% of the area where the service is provided	%	Everywhere the wanted signal needs to be received above or at least at the reference sensitivity level of the receiver (it applies above all on the DL and, whenever a feedback/ACK is foreseen, also on the UL), in order to guarantee the required minimum user data rate
User Exp.	**KPI 4**	Mobility		Not critical (majority is stationary) 65% of the devices: 0 km/h 30% of the devices: 0–120 km/h 5% of the devices: 0–500 km/h	km/h	The majority of the devices (65%) is expected to be stationary. The remaining 35% is expected to experience a mobile speed in the range of 0–120 km/h (aboard cars – 30%) or in the range of 0–500 km/h (on trains – 5%)

Syst. Perf.	**KPI 5**	Connection density		Broadcast service: Not critical Multicast service: It depends on the technology, in order to select the connection density to be used as a threshold above which to establish an MBMS session and under which to establish p2p connections	devices	The number of users potentially involved in a multicast MBMS session is expected to be important and a devices' counting procedure is therefore useful before starting the MBMS session, in order to decide whether and which type of radio resources to allocate for that session in case of multicast mode (i.e., either p2p or p2m connection)
User Exp.	**KPI 6**	Reliability/availability		Reliability: Not critical (for a broadcast "one-to-many" service) Availability: 99.999%	%	In the framework of a broadcast "one-to-many" service, reliability is not requested and therefore is not critical, since there is not any possibility to guarantee reliability of a specific connection between the transmitter (e.g., BMSC) and a single receiver (e.g., user's device) Availability may be assumed equal to (1 – service blocking probability), where service blocking probability is expected to be equal to 0, due to the usage of shared resources, expected to be always available
Syst. Perf.	**KPI 7**	Complexity reduction		-	-	Same complexity as of today on both the infrastructure side and the device side
Syst. Perf.	**KPI 8**	Energy efficiency		25%	%	Reduction percentage compared to legacy networks and devices energy consumption Low-energy operation of all radio nodes is expected due to energy cost and EMF reasons Reduced battery consumption on the device side

Req. Lev. **Primary** **Secondary** **Tertiary**

Table 2.10 Use case 7: dense urban society below 6 GHz

	Use case 7: Dense urban society below 6 GHz					
	KPI no.	**KPI definition**	**Req. lev.**	**Requirement definition**	**UoM**	**Notes**
User Exp.	**KPI 0**	User experienced data rate		50 Mbps (DL), 25 Mbps (UL)	Mbps	Following METIS D1.1 [MET1.1] (defined for outdoor)
Syst. Perf.	**KPI 1**	Traffic density		125 [Gbps/km^2] (DL) 62.5 Gbps/km^2	Mbps/km^2	Source: NGMN white paper [NGMN-5G], use case ("Broadband access in dense areas")
User Exp.	**KPI 2**	Latency		10 ms	ms	User Plane latency is considered
Syst. Perf.	**KPI 3**	Coverage		99.999% of the area where the service is provided	%	The wanted signal needs to be received above or at least at the reference sensitivity level of the receiver (it applies both on the UL and on the DL), in order to guarantee the required minimum user data rate
User Exp.	**KPI 4**	Mobility		On demand, 0–100 km/h	km/h	Source: NGMN white paper [NGMN-5G], use case ("Broadband access in dense areas")
Syst. Perf.	**KPI 5**	Connection density		200–2,500/km^2	users/km^2	Total device density is 2,000–25,000/km^2, a 10% activity factor is assumed [NGMN-5G]
User Exp.	**KPI 6**	Reliability/availability		95%	%	Following METIS D1.1 [MET1.1] (defined for outdoor) Reliability may be assumed equal to (1–service dropping rate), where service dropping may occur e.g., in case of quite severe multiple interferers received

						together with the wanted signal, even in case of adequate coverage. Depending on the considered service, reliability may be defined as the probability of correctly transferring a given payload and correctly receiving ACK within a 10 ms latency Availability may be assumed equal to (1–service blocking probability), where service blocking probability is due to lack of enough resources to access, grant and provide the service, even in case of adequate coverage
Syst. Perf.	**KPI 7**	Complexity reduction		Comparable to today	-	
Syst. Perf.	**KPI 8**	Energy efficiency		The network energy consumption should be comparable to the energy consumption of today's metropolitan deployments, despite the drastically increased amount of traffic.	%	Following METIS D1.1 [MET1.1] (defined for outdoor)

Req. Lev. **Primary** **Secondary** **Tertiary**

2.5 Challenges for 5G design below 6 GHz

2.5.1 Design methodology

Based on the selected use cases and requirements, an overall air interface framework is to de designed. To this end, the most relevant enabling technologies (e.g., new waveforms, channel coding, frame structure/numerology, device-to-device (D2D), enhanced MIMO, multi-cell cooperation, control channels, layer 1 procedures, etc.) are to be identified and their relevance for achieving the requirements of each use case is to be assessed. This design has to be performed on both link and system levels, as detailed below.

To achieve this single solution, the following three-stage approach can be adopted:

- **Stage 1:** Dedicated research activities on service-specific (air interface) components for each one of the five core services outlined in Section 2.2 need to be performed. For example, channel coding for MMC, waveform and frame design for BMS, MIMO solutions for V2X, channel access for MMC, interference management solutions for MBB, etc. At the same time, an air interface framework needs to be developed for each service, which defines the functional descriptions of the components and the interfaces in-between.
- **Stage 2:** To connect and harmonize the work being done for each service, the concept of ***Service Integration Drivers (SIDs)*** can be introduced. There is one SID for each core service (that carries the same name with the service), plus one SID for composite services (called the "Overall SID"). The main role of the core service SIDs is to collect different components optimized for each service and to integrate them into a service-specific air interface solution (including waveform, frame, control signalling, coding, modulation, procedures, retransmission schemes, MIMO, etc.). This integration will be done using the air interface framework developed in Stage 1. Note that all air interface components do not have to be re-designed from scratch. All research items need to be checked against the respective solution in LTE as a benchmark. LTE components will be re-used where appropriate, but since LTE is a (non-flexible) custom-made design for MBB, it is expected that this will not hold for most of the components of MCC, MMC, BMS and V2X.
- **Stage 3:** Once the air interface solutions for each service are designed by the core service SIDs, a *sensitivity analysis* needs to be carried out by all the SIDs, to see how the different designs need to be adapted so that they conform with more than one service. In this way, one can explore the limits of the degrees of flexibility of the service-specific solutions. For example, the waveform and frame design for MMC will be adapted to MCC through a redesign/reconfiguration which is often a choice between few configurations. In that way, it is possible to converge to a single air interface solution which is able to adapt itself to the requirements of more than one service. Additionally, the impact of multiplexing different services each being optimized for a specific service in frequency and/or time will be assessed.

In the following, a more detailed description of the SIDs is provided.

2.5.2 Service integration drivers

As can be seen in Figure 2.3, five of the six SIDs are directly linked to the core services mentioned above: MBB, MMC, MCC, BMS and V2X. Each one of these five SIDs shall develop a service-specific air interface with the respective service characteristics and needs in mind. By doing so, a draft of an air interface framework becomes available (stage 1). This framework includes (1) functional descriptions of the air interface components required for implementing the air interface of the respective service (e.g., some services need means for UL sounding while others do not, some services allow for Hybrid Automatic Repeat Request (HARQ) retransmissions while others do not), (2) the interfaces between the different building blocks (e.g., different services will require different cross-layer interactions) and (3) the applicability to different deployment settings. Within each core service SID, some components will be (re-) designed and optimized for that service (shown by different highlighted balls in Figure 2.3). Some others will be readily used from the existing technologies (shown by dark grey balls in Figure 2.3). Then, they will be compiled into the service-specific air interface with the help of the framework for that service (Stage 2).

As for the sixth SID, it is aimed at analysing the feasibility of *merging* and harmonizing the service-specific solutions into a single air interface providing the *multi-service* air interface which simultaneously achieves the requirements of the five core services. For doing so, the sixth SID have to (1) investigate the dependencies, (2) find synergies, (3) carry out a sensitivity analysis in cooperation

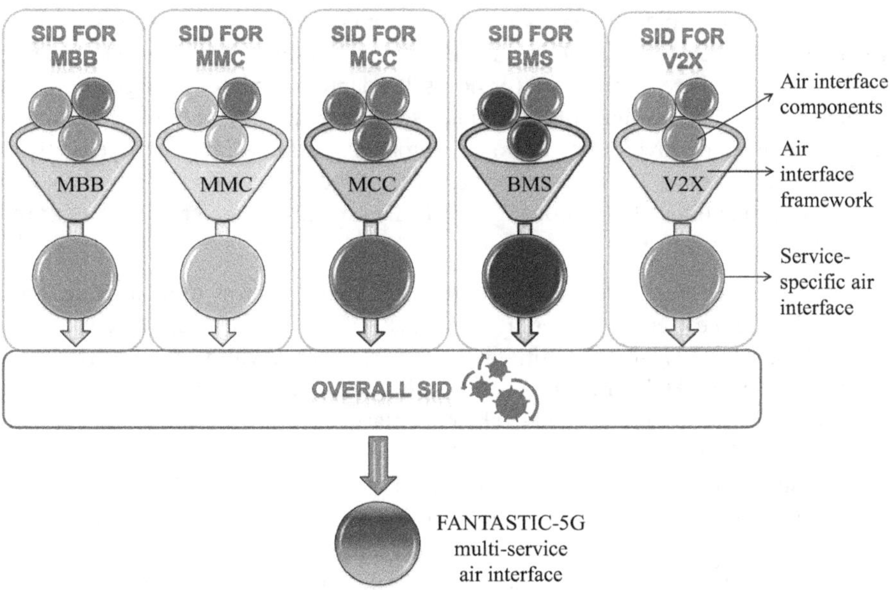

Figure 2.3 The role of service integration drivers in air interface design

with the other SIDs and (4) provide an overall air interface design taking the requirements of all services into account. The sixth SID is one of the key means to achieve a single multi-service air interface. Figure 2.3 depicts the role of the SIDs in the air interface design.

Optimally, the outcome of this activity is a single highly flexible, versatile and scalable air interface design enabling the support for all foreseen (and even not yet foreseen) services of the 5G era with highest efficiency (in terms of spectrum and energy). Even if a single multi-service air interface proves to be inefficient, a set of optimized air interface designs, each covering a subset of the aforementioned services can be delivered, thanks to the SIDs.

2.5.3 Link level challenges

The link design covers several technical components such as modulation and coding, waveform, frame and control channel structure, physical layer procedure, etc. Given the diverse and often mutually conflicting requirements of 5G, the link design is based on two general questions: (i) How can new features/services be added in later releases without imposing heavy constraints and without bothering legacy devices? and (ii) What is the best balance between flexibility and fixed pre-configuration? Due to the wide scope of 5G services, as well as the important objective of enabling future proofness, the 5G design shall possess high degree of flexibility from the very beginning. To relax the burden of backwards compatibility, the system has to be designed with a minimum amount of fixed and predefined elements. For highest flexibility the only static/predefined element is the Initial Access Channel (IAC). Every subsequent item should be open for configuration between a device and the network according to the actual needs, thereby relying on a given set of options. This way new features and services may be implemented at later releases without affecting legacy devices. In addition, the second question challenges the current mobile systems where rather fixed (or non-self-adaptive/flexible) configurations are applied. Such a non-flexible concept cannot support multi-service 5G in an efficient way, since it can provides merely a compromise in addressing the needs of all envisaged services. So, the technology components being developed require flexible adaptation capabilities. The following technical components are critical from a link level perspective:

- Waveform design: Multi-carrier waveforms with filtering functionalities are considered key enablers for a flexible PHY design of a multi-service air interface, as they allow partitioning the system bandwidth into separate subbands, whose PHY parameters can be individually configured according to the requirements of a service [11]. The candidate waveforms can be grouped into two categories, namely
    - subcarrier-wise filtered multicarrier, comprising FBMC with QAM/OQAM signalling [12], pulse shaped OFDM (P-OFDM) [13] and Flexibly Configured OFDM (FC-OFDM) [14] and
    - subband-wise filtered multicarrier, comprising UF-OFDM [15] and F-OFDM [16]

Subcarrier-wise filtered multicarrier systems use prototype filters with steep power roll-off in frequency domain for pulse shaping the subcarrier signals. Subband-wise filtered multicarrier systems (UF-OFDM and f-OFDM), on the other hand, employ FIR filtering per group of subcarriers after the classical OFDM modulation which conceptually ensures a close similarity (in fact actual implementations can apply the filtering functionality before the transformation to the time domain for reduced complexity).

- Service-specific PHY procedures: In order to ensure a reasonable signalling overhead for 5G supporting multiple services concurrently, notably the MBB service and MMC service with massive access, flexibility is also required for random access procedures. There are multiple proposals for redesigning the access procedure for MMC. One example is contention-based access applying either one-stage or two-stage transmission formats. The former follows the principle of (coded) slotted ALOHA (i.e., the device directly transmits its data in conjunction with a respective preamble specifying the used resources) and thus needs very lean control effort, while the second allows the base station to somewhat control the transmissions (i.e., the device having data to transmit initiates a service request in the form of an UL preamble) and introduces longer delays but is more resource-efficient. One-shot procedure has the following characteristics:
  - Faster, if successful, significantly less DL feedback
  - High collision probability reduces throughput
  - Coded random access for scalability
  - Envisaged solution for very small packets and low traffic load
  - Sparse signal processing detectors

  Two stage protocols, on the other hand, have
  - Additional delay
  - Depending on configuration certain amount of DL feedback is required
  - Reduced collision probability through service request over-provisioning increases throughput
  - Envisaged solution for bigger packets and higher traffic load

- Flexible frame design: The frame is the key element for aggregating user-specific transmission modes into one holistic solution and its dynamic reconfiguration is key for 5G. These configurations may vary in terms of e.g., waveform characteristics (applied filter, subcarrier spacing), transmission time interval (TTI) length and further frame design elements such as pilot density and resource partitioning:
  - MMC: moderate bandwidth with longer TTIs optimized for low cost and better coverage
  - MCC: higher bandwidth with shorter TTIs optimized for low latencies with reasonable overheads
  - MBB: short TTIs to quickly start TCP session followed by medium size TTI for minimized overhead
  - BMS: long TTIs to maximize FEC gains from time-diversity

  The unified frame concept is depicted in Figure 2.4.

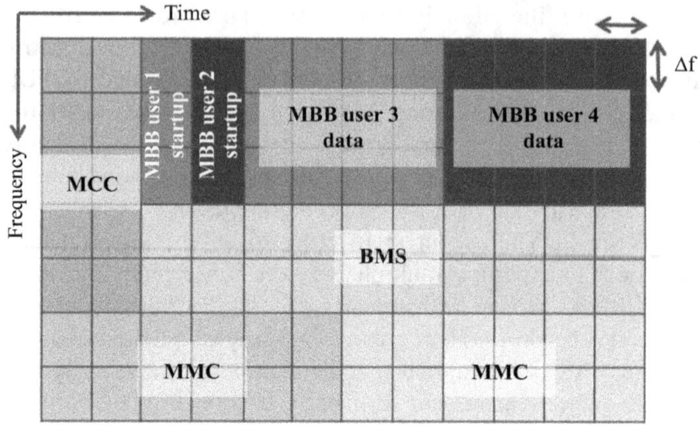

Figure 2.4 Unified frame concept: One tile corresponds to the smallest user allocation

2.5.4 System level challenges

From system and network design perspective, satisfying the requirements of the various use cases described above is challenging from the following aspects:

- Radio Resource Management (RRM): Efficient scheduling of the considered services requires the support of scheduling formats with different TTIs. For instance, users with tight latency constraints (e.g., MCC) require short TTIs, while MMC users scheduled on a narrow bandwidth are most efficiently served with longer TTIs.
- Advanced connectivity options: For satisfying the stringent requirements of 5G core services, advanced connectivity options like direct D2D [17], massive access and broadcast/multicast communications are to be considered.
- Spectral efficiency enablers: When dealing with MBB service, the high data rate expectations call for developing technology enablers that increase the spectral efficiency like advanced receivers for, e.g., Non-Orthogonal Multiple Access (NOMA) and advanced MIMO techniques with or without cooperation between different base stations [18]. This opens up a whole range of research aspects to solve problems faced, when trying to exploit the potential of a large number of streams and associated pilots and receivers. Other aspects that need to be addressed include the channel state feedback and the signalling overhead for a proper use of MIMO.

It is worth noting that other challenges exist for a system design integrating spectrum below 6 GHz with mmWave bands.

2.6 System-level evaluation of technical components

To achieve the ambitious targets, many key technology components need to be devised. To rate the designs, link- and system-level simulations need to be carried out.

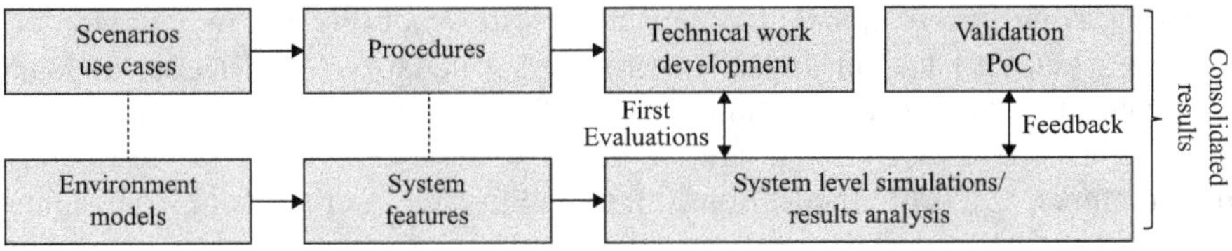

Figure 2.5 Overall role of system-level simulation

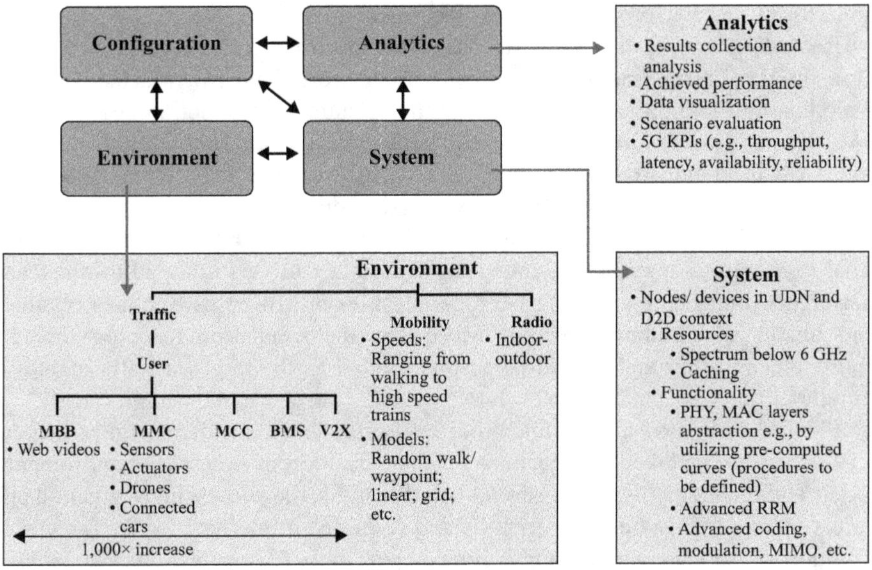

Figure 2.6 High-level architecture

Additionally, selected technology components must be integrated and demonstrated. The overall role of system-level simulations is depicted in Figure 2.5.

The selected scenarios and use cases are dictating the environment models and KPI targets. Technology components (procedures) being developed and indicating promising performance gains are to be implemented via certain system features. In a first phase during the development of the technology components first evaluations via system-level simulations are executed. These results will be used to refine the concepts. Finally, in a second phase comprehensive system level simulations will be used to analyse the feasibility of the designs in achieving its goals and KPI targets.

System-level simulators must have the capability to model and simulate the five core services of the project. In this respect a high-level architecture (see Figure 2.6) takes into account:

- Environment models
- System features
- Analysis

Regarding *environment models* the impact of traffic, mobility and radio conditions are to be taken into account in the system-level simulators. Specifically, it should be possible to generate in the simulators 1,000× more traffic (e.g., compared to 3GPP Release 12 baseline), usage of massive number of devices, ultra-dense infrastructures, D2D communications, high mobility, etc. Certain components must be developed that control the configuration of network topology, traffic distribution to cells, distribution of cells and users in space, traffic distribution in space and time, etc. Compliance with benchmarks and 3GPP standards (Release 12 plus selected relevant features from Release 13) is critical in order to ensure valid comparisons and performance merits assessments.

Regarding *system features*, the usage of spectrum e.g., in frequency bands below 6 GHz is taken into account along with network deployment density, impact of modulation, coding, MIMO, etc. Moreover, suitable abstractions of PHY and MAC layers in order to define system behaviour models and limit complexity aspects of simulations need to be considered. For instance, PHY layer aspects can be modelled with the simulator tools by utilizing pre-computed mapping curves e. g., of spectral efficiency versus signal quality indicators, block error rate versus signal to noise and interference ratio, physical data rate versus signal-to-noise and interference ratio, and so on. These lookup tables are filled with values obtained from link level simulation results. Moreover, the simulation tools provide the means for analysing and providing results related to the targeted KPIs including throughput, latency, packet losses, bit error rate, low energy consumption, etc.

Finally, it should be mentioned that well-established standards and references [5,19–22] are to be taken into account for the definition of necessary requirements, models and technologies which should be utilized for the successful realization and evaluation of a highly flexible, versatile and scalable air interface for the forecasted multitude of service classes and device types with reasonable complexity and highest efficiency for frequency bands below 6 GHz.

2.7 Conclusions

This chapter has provided a view of the current service and use case landscape for 5G networks. It also performs a selection among these use cases so that a flexible air interface design for spectrum below 6 GHz can be performed. Challenges for this design on link and system levels are then briefly discussed along with their research topics and the evaluation methodology for the corresponding technical components. We finally note that the design of a new air interface is heavily composed of topics requiring standardization, e.g., waveform and frame design, PHY layer procedures, transceiver processing (interference management, multiuser reception, etc.), channel coding, adaptive coding and modulation, etc. These are all topics that imply interfaces, messages, signalling, protocols (all of which are standardized in 3GPP) rather than proprietary algorithmic solutions and an important challenge for researchers is to be able to interact with standardization bodies, especially 3GPP, in a timely manner.

Acknowledgement

This work has been performed in the framework of the Horizon 2020 project FANTASTIC-5G (ICT-671660) receiving funds from the European Union. FANTASTIC-5G project (http://fantastic5g.eu/) undertakes the work above described and sets out a vision for a 5G air interface ***below 6 GHz***. The project also shall build up consensus on reasonable options for 5G standardization among the major industrial partners of the project that are also voting members in 3GPP and to push the innovations of the project for standardization (through study items).

References

[1] 3GPP TR 45.820, "Cellular system support for ultra low complexity and low throughput Internet of Things (Release 13)", June 2015.

[2] 3GPP TSG GERAN: Study Item – GP-150354, "Cellular system support for ultra low complexity and low throughput Internet of Things", 2015, FS_IoT_LC.

[3] 3GPP TSG RAN: Work Item – RP-141660, "Further LTE physical layer enhancements for MTC", 2014.

[4] 3GPP TSG RAN: Study Item – RP-151109, "Feasibility study on LTE-based V2X services", 2015.

[5] NGMN Alliance, "NGMN 5G White Paper" – v1.0, 17 February 2015.

[6] METIS, Deliverable D1.1, "Scenarios, requirements and KPIs for 5G mobile and wireless system", April 2013.

[7] ICT-671680 METIS-II, Deliverable 1.1, "Refined scenarios and requirements, consolidated test cases, and qualitative techno-economic assessment", January 2016.

[8] 3GPP TSG SA: Study Item – SP-150142, "Study on new services and markets technology enablers (SMARTER)", 2015.

[9] 3GPP TSG SA1: S1-151623, 3GPP TR 22.891 v0.1.1 – "Feasibility study on new services and markets technology enablers; stage 1 (Release 14)", 2015.

[10] METIS, Deliverable D1.5 "Updated scenarios, requirements and KPIs for 5G mobile and wireless system with recommendations for future investigations", April 2015.

[11] Wunder, G., Jung, P., Kasparick, M., *et al.*, "5GNOW: non-orthogonal, asynchronous waveforms for future mobile applications", *IEEE Communications Magazine*, 5G Special Issue, vol. 52, no. 2, pp. 97–105, February 2014.

[12] Farhang-Boroujeny, B., "OFDM versus filter bank multicarrier", *IEEE Signal Processing Magazine*, vol. 28, no. 3, pp. 92–112, May 2011.

[13] Zhao, Z., Schellmann, M., Wang, Q., Gong, X., Boehnke, R., and Xu, W., "Pulse shaped OFDM for asynchronous uplink access", *Asilomar Conference on Signals*, Systems and Computers, Monterey, USA, November 2015.

[14] Lin, H., "Flexible configured OFDM for 5G air interface", *IEEE Access*, vol. 3, pp. 1861–1870, 2015.

[15] Wild, T., and Schaich, F., "A reduced complexity transmitter for UF-OFDM", in *2015 IEEE 81st Vehicular Technology Conference (VTC Spring)*, Glasgow, Scotland, pp. 1–6, 11–14 May 2015.

[16] Abdoli, J., Jia, M., and Ma, J., "Filtered OFDM: a new waveform for future wireless systems", in *2015 IEEE 16th International Workshop on Signal Processing Advances in Wireless Communications (SPAWC)*, Stockholm, Sweden, pp. 66–70, 28 June 2015–1 July 2015.

[17] Fodor, G., Dahlman, E., Mildh, G., *et al.*, "Design aspects of network assisted device-to-device communications", *IEEE Communications Magazine*, vol. 50, no. 3, pp. 170–177, 2012.

[18] Kasparick, M., and Wunder, G., "Autonomous algorithms for centralized and distributed interference coordination: a virtual layer based approach", *EURASIP Journal on Wireless Communications and Networking*, April 2014:120, http://arxiv.org/abs/1304.3016.

[19] 3GPP TS 23.179, "Functional architecture and information flows to support mission critical communication", March 2017.

[20] 3GPP TR 23.887, "Study on machine-type communications (MTC) and other mobile data applications communications enhancements", December 2013.

[21] 3GPP TR 22.885, "Study on LTE support for Vehicle to Everything (V2X) services", December 2015.

[22] 3GPP TS 23.246, "Multimedia broadcast/multicast service (MBMS): architecture and functional description", December 2016.

Chapter 3

New quasi-deterministic approach to channel modeling in millimeter-wave bands

Alexander Maltsev[1,2], Andrey Pudeyev[1] and Ilya Bolotin[1]

3.1 Introduction

Increasing interest in using millimeter-wave bands for next-generation mobile wireless networks [1–5] leads to the increasing importance of the propagation channel properties investigation and proper channel model development. The new communication systems and standards require millimeter-wave channel models applicable to multiple usage cases and environments as well as a wide frequency range from 30 GHz up to 90 GHz. However, despite the large number of experimental measurement campaigns and results nowadays [6–11], there are a few practical millimeter-wave channel models available. One of the released channel models [12,13] was developed for the IEEE 802.11ad standard in the 60 GHz band (57–64 GHz). It is based on experimental measurements and ray tracing studies [11,14–17] and focused on a limited number of indoor scenarios with site-specific parameterization. The most recent METIS 2020 deliverables [18,19] suggest to exploit different channel modeling approaches including both stochastic (generic) and map-based (site-specific) models with parameterization from measurement results.

Signal propagation in the bands lower than 6 GHz is rather adequately investigated; a number of accurate and realistic modeling approaches exist as basis for both link and system level evaluations. The millimeter-wave band needs to be thoroughly investigated for wireless communication, since the 10-fold increase of carrier frequency and 100× bandwidth increase lead to qualitative changes in the propagation properties.

Firstly, the short wavelength results in a significantly higher propagation loss according to Friis equation. To overcome that, high-gain antennas are required. To support them, the channel model shall take into account spatial properties of the channel rays at TX and RX and also include wide range of adaptive antenna technologies, from RF beamsteering to baseband MIMO processing.

[1]Intel Corporation, 2200 Mission College Blvd., Santa Clara, CA 95054-1549, USA
[2]Lobachevsky State University of Nizhny Novgorod, Nizhny Novgorod 603950, Russia

Secondly, as confirmed by a number of works [7–9], the 60 GHz propagation channel has a quasi-optical nature. The fraction of power arriving at the receiver due to diffraction and transmission through objects is practically not usable. Most of the transmission power is propagated between the transmitter and the receiver through the LOS and the first- and second-order reflected rays. To establish a communication link, steerable highly directional antennas have to be used, pointing toward the direction of the LOS path (if available) and/or one or more of the strongest reflected paths. An additional consequence of the quasi-optical propagation nature is that the channel model should be fully three-dimensional, taking into account signal propagation in a real environment. For example, map-based ray tracing can be an effective means for prediction of spatial and temporal properties of the channel and may be used to assist the channel modeling.

Thirdly, it should be noted that with ideal reflections, each propagation path would include only a single ray. However, as demonstrated by experimental investigations [14–16] each reflected path actually may consist of a number of rays closely spaced to each other in the time and angular domains due to roughness and complex structure of the reflecting surfaces. Hence, the clustering approach is directly applicable to channel models for millimeter-wave indoor and outdoor systems with each cluster of the model corresponding to the LOS or NLOS reflected path.

Fourthly, another important aspect of millimeter-wave propagation is polarization. As demonstrated by experimental studies [20] the power degradation is mainly caused by the polarization mismatch between the antennas. The cross-polarization discrimination (XPD) coefficient caused by the reflections on the propagation path can be about −15 to −20 dB.

Finally, the direct experimental measurements have shown that even small movements of the TX or RX just for several centimeters may significantly change the channel transfer function in the frequency domain.

In summary, the 3D channel model for millimeter-wave bands shall in particular:

- Provide accurate space-time characteristics of the propagation channel (basic requirement) for the main usage models of interest;
- Support beamforming with steerable directional antennas on both TX and RX sides with no limitation on the antenna technology;
- Account for polarization characteristics of antennas and signals;
- Support non-stationary characteristics of the propagation channel.

Similar tasks were already addressed during the IEEE 802.11ad (WiGig) standard development and resulted in the set of indoor channel models with strict model-specific parameterization [12,13]. However, further development both of millimeter-wave technologies and mobile networks infrastructure opens a large amount of new scenarios and use cases, both indoor and outdoor. In the MiWEBA (Millimeter-Wave Backhaul and Access) [21] project, the millimeter-wave heterogeneous small cell network is considered as a basis for future 5G communication systems. Corresponding scenarios for access, backhaul and device-to-device (D2D) links were developed for the outdoor open area, street canyon and indoor hotel lobby scenarios. The upcoming new IEEE 802.11ay (NG60) standard, considered as an extension of the existing 802.11ad, has introduced new applications and use cases, including mobile offloading, wireless backhauling, mass video distribution systems, virtual- and augmented reality, etc. [22]. Such abundance of scenarios and use cases makes traditional approaches (stochastic, geometry-based stochastic, site-specific deterministic, map-based) to channel modeling impractical due to the requirements of very large amount of experimental measurements or extensive ray-tracing simulations for each scenario. A new approach to channel modeling in millimeter-wave bands for new indoor and outdoor environments, based on experiments and theoretical approximations, is required.

3.2 Q-D channel modeling approach

The state of the art for mobile communications channel characterization includes separate descriptions of the path loss models and spatial-temporal channel characteristics, typically comprised of the clustered channel impulse responses and angular spread statistics [23]. Latest works for millimeter-wave channel models also follow such an approach [24,25] and use different cluster analysis techniques to the experimental data processing. However, such approaches work well for NLOS conditions only, which is not the main usage case for the millimeter-wave communication system. At the same time, the propagation loss features of millimeter-wave signals lead to the weakness of distant reflections and the domination of rays that are close to the direct path. This requires new approaches to characterize non-stationary channel and Doppler effects.

To provide adequate modeling of the channel propagation aspects mentioned above, the Quasi-Deterministic (Q-D) approach for channel modeling is proposed and developed in this work. It should be noted that in this chapter, the authors make focus more on the methodology itself, rather than on the specific channel models and their parameters. However, some examples of the Q-D channel models are presented here for common and most important scenarios.

From the very early experimental and theoretical works in the millimeter-wave area by its pioneers Lebedew [26] and Bose [27], the quasi-optic nature of the millimeter-wave signal propagation was revealed. For practical analysis of the communication systems in that band, the diffraction effects can be neglected (of course, they still exist and can be found by means of special measurements [28]).

Under this approach for each of the channel propagation scenario, several strongest propagation paths (rays which produce the substantial part of the received useful signal power) are determined first, and signal propagation over them is calculated based on the geometry of the deployment, locations of Base Stations (BS) and User Equipment (UE) in a deterministic manner. Signal power delivered over each of the rays is calculated in accordance to theoretical formulas taking into account free space losses, reflections, polarization properties and UE mobility effects: Doppler shift and user displacement.

At the same time, some of the parameters in these calculations may be considered as random values (e.g., reflection coefficients) or even as random processes (e.g., UEs motion). It should be noted, that the number of such quasi-deterministic rays (D-rays), which should be taken into account may be different since in different scenarios, the number of the strongest rays varies. For example, for outdoor open space scenarios it has been shown, that signal propagation properties are mostly determined by two D-rays—the LOS ray and the one reflected from the ground. For outdoor street canyon scenario the propagation is determined mostly by the LOS, the ground and wall reflected rays. For the hotel lobby (indoor access large public area) scenario more D-rays need to be modeled. For this scenario it is proposed to consider all rays with up to second-order reflections as D-rays (in a similar way as was adopted in the IEEE 802.11ad evaluation methodology for indoor scenarios).

In a real environment, in addition to the strong D-rays a lot of other reflected waves are coming to the receiver from different directions. For example, there are cars, trees, lampposts, benches, far big reflectors (buildings), etc. All these rays are considered in the Q-D channel models as secondary or random rays (R-rays) and are described as random clusters with specified statistical parameters extracted from available experimental data and detail ray tracing modeling.

This chapter is focused on the rational description of the new channel modeling methodology and the way it was developed. More details about the Q-D channel modeling methodology aspects and models descriptions can be found in the works [29–31]. Examples of practical application of the developed models to the millimeter-wave 5G system evaluation are presented in [32–34].

The methodology may be used for both system- and link-level analysis of the perspective millimeter wave communication system presented. Moreover, a new model for different scenarios can be developed using the Q-D approach, especially based on the new experimental measurements. The Q-D methodology may be used to coordinate measurement campaigns directions, specify experiments conditions and equipment to measure specific parameters that may be used to build proper models within the Q-D approach.

3.3 Millimeter-wave channel experimental measurements and results interpretation

This section presents the experimental measurement results for 60 GHz band in "real world" outdoor environment. The measurement campaign was carried out by Fraunhofer Heinrich Hertz Institute (HHI) in several locations in Berlin, Germany, and by Intel/University of Nizhny Novgorod in Russia. In previous works, the results of similar measurements were traditionally used to evaluate the propagation path loss exponent and channel power delay profile. In the present work, the experimental measurements were complemented by the ray tracing simulation results in the reconstructed 3D measurement environment. As we will demonstrate in the following sections, this hybrid approach to channel investigation, based on both measurements and simulations, allows in-depth analysis of the millimeter-wave propagation mechanisms in real-world non-stationary environments.

3.3.1 Experimental measurement description and results

Urban street canyon measurements

The urban street canyon measurements were performed during working hours in an urban street canyon, located at Potsdamer Straße in downtown Berlin, Germany. The novelty of these experiments consists in the long-term investigation of non-stationary environment, which is characterized by the presence of relatively large static objects as well as multiple dynamic (moving) obstacles and reflectors typical for dense outdoor urban area. During the measurements, omni-directional antennas were used at both the TX and RX sides. Measurements were performed primarily for mmWave channel sounding in a number of stationary TX and RX positions at a fixed distance. Further experiments, but with moving RX and stationary TX, were also carried out.

The TX antenna had a height of 3.5 m above ground to represent a typical small cell base station that is being added to existing lamp posts. The RX was mounted on a mobile cart with a typical user antenna height of 1.5 m. At the first stage, multiple long-term measurements were performed for stationary TX and RX positions separated by 25 m from each other as shown in Figure 3.1. At the second stage, mobile measurements were conducted as well, where the RX was moving at a constant speed at the sidewalk along a distance of 50 m to each side of the stationary TX.

The channel sounder used in this measurement campaign was based on a purpose-built FPGA platform [35] with key parameters as listed in Table 3.1. The primary output of the sounder is a channel impulse response (CIR) for each measurement snapshot taken every 800 μs.

During the first stage, 62,500 CIR snapshots were obtained for each static position of the TX and RX pair which results in a single observation period of 50 s.

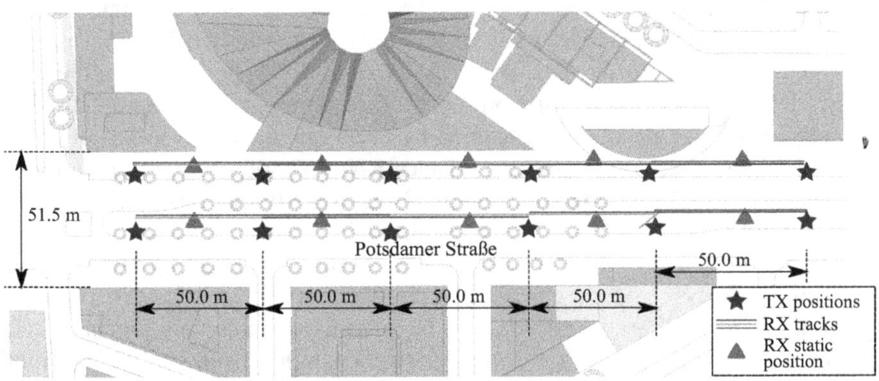

Figure 3.1 Outdoor access measurement scenario

Table 3.1 Measurements with omni-directional antenna parameters

Parameter	Value
Carrier frequency	60 GHz
Bandwidth	250 MHz
Output power	15 dBm
Antenna gain	2 dBi
Antenna pattern	Omni-directional (in horizontal plane) $>\pm 30°$ half power beamwidth (in elevation)

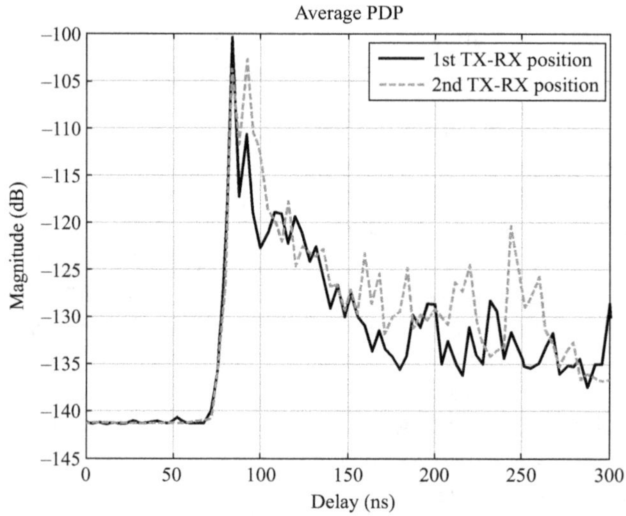

Figure 3.2 APDP from the measurements at the first stage

Figure 3.2 shows a typical averaged channel impulse response obtained from these static measurements. The delay of the highest peak in the figure is approximately at 83 ns which correspond to the line-of-sight (LOS) path between the static TX and RX separated by 25 m from each other. From the measured CIRs, the channel power delay profile was obtained by averaging the received power values over observation time of 50 s (averaging factor $N = 62{,}500$). This Average Power Delay Profile (APDP) is shown in Figure 3.2 for two different positions of the TX and RX pair.

The variation in power level between the two receiver positions can be caused by one or more rays (or paths) arriving at the RX from different angles of arrival, e.g., coming from different reflecting surfaces with approximately the same total path length between the TX and RX. Due to the limited bandwidth of the measurements (250 MHz), those multipath components cannot be resolved explicitly and produce fast fading effects.

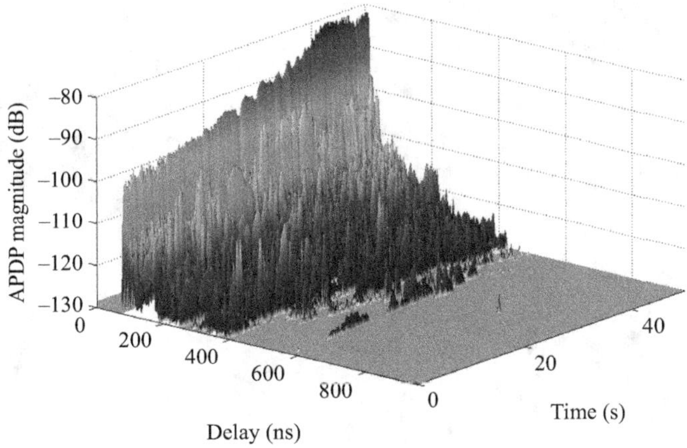

Figure 3.3 APDP of a measurement with moving receiver

During the second stage with moving receiver, a number of 40 measurement series with 62,500 snapshots each were performed. Based on the obtained experimental results, the APDP can be calculated over a certain number of snapshots to remove the fading effects. An example result of such a measurement run is shown in Figure 3.3. In this measurement the receiver was moved toward the transmitter. The fast fading effect is clearly visible on the LOS component on the left side of the plot.

University campus measurements

A study in the open area scenario environment was performed in the University of Nizhny Novgorod campus. The general experimental measurement setup is shown in Figure 3.4. The TX antenna was mounted on the vestibule of the university campus building at 6.2 m height above ground to represent the BS. The RX antenna was mounted on a moving platform with an antenna height of 1.5 m to represent the mobile UE.

A schematic illustration of this scenario is shown in Figure 3.5, where the following main parameters are used: D_0 is the direct LOS distance between the TX and RX antennas; L_0 is the horizontal distance between the transmitter and receiver, H_1 and H_2 is the height of transmitter and receiver, respectively.

For the experimental data acquisition and processing, a specially designed measurement platform was used with the key technical parameters listed in Table 3.2. Depending on the distance L_0, the transmitter was equipped with antennas of different gain. For distances less than 35 m the rectangular horn antenna 14 × 18 mm with 19.8 dBi gain was exploited, while for greater distances the highly directive lens antenna with large aperture (100 mm) and 34.5 dBi gain was used (see Figure 3.4). The receiver was equipped with the round horn antenna of diameter $d = 14$ mm and 12.3 dBi gain.

It should be noted that the sounding signal bandwidth used in these experiments was equal to 800 MHz which provided maximum time resolution of 1.25 ns.

Figure 3.4 Snapshot of the university campus experimental scenario

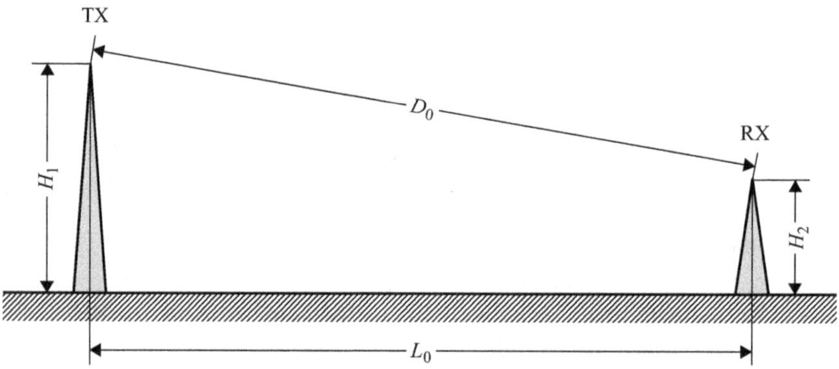

Figure 3.5 Schematic illustration of the experimental setup

The use of directional antennas at both TX and RX assisted to discover that the considered experimental scenario was characterized by only two strongest propagation paths (or rays). The first path corresponds to the LOS component and the second path to the reflection from the ground (asphalt) surface. All other propagation paths caused by reflections from surrounding objects were more than 15–20 dB lower than these two strongest components.

An example of the measured channel impulse response with two clearly distinguishable peaks is presented in Figure 3.6. The peaks correspond to the LOS and ground-reflected components are found 2.5 ns apart from each other. The power

Table 3.2 Key technical parameters of the measurement platform developed to study the fast channel fading

Parameter	Value
Frequency	60 GHz
Bandwidth	800 MHz
Output power	2.4 dBm
Platform sensitivity	−75 dBm
TX antennas	Lens: d = 100 mm/ Horn: 14 × 18 mm
TX antennas gain	34.5 dBi/19.8 dBi
TX antennas HPBW	3°/14°–18°
RX antenna	Round horn d = 14 mm
RX antenna gain	12.3 dBi
RX antenna HPBW	30°

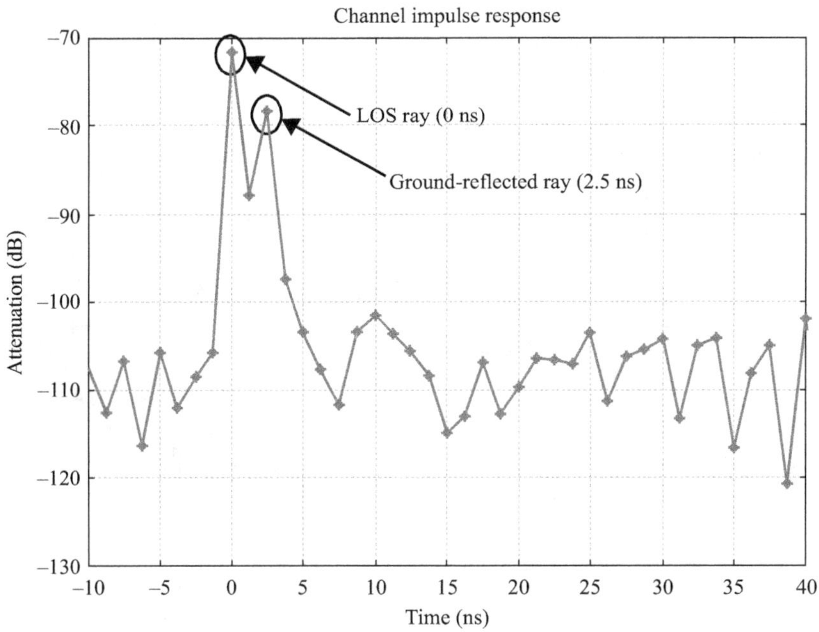

Figure 3.6 Channel impulse response, L_0 = 30.6 m

difference between those two peaks is approximately equal to the ground reflection coefficient (−6 dB).

Initially, the dependency of channel transfer function (CTF) on the RX vertical motion was investigated for distance L_0 = 30.6 m. In Figure 3.7, the CTF of 800 MHz bandwidth is presented when the RX height changes from 1.34 m to 1.5 m. A few separate CTF samples are given in Figure 3.7. In these experiments the

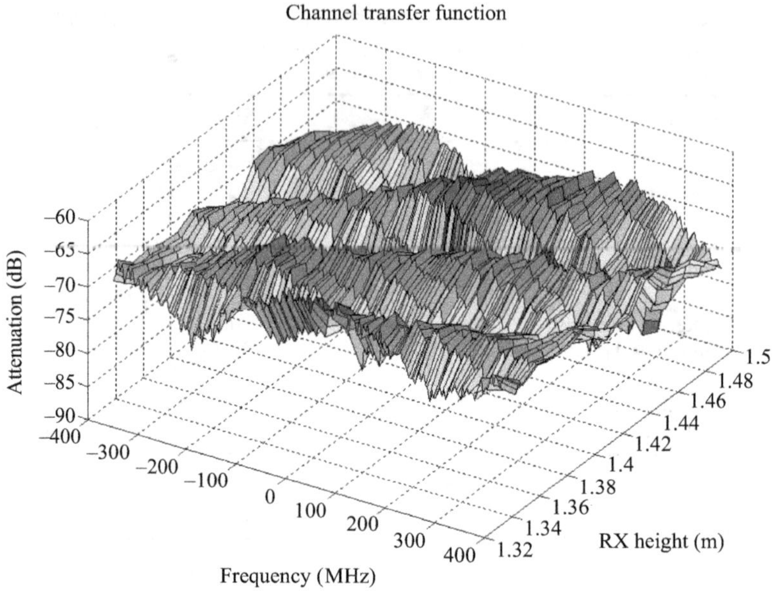

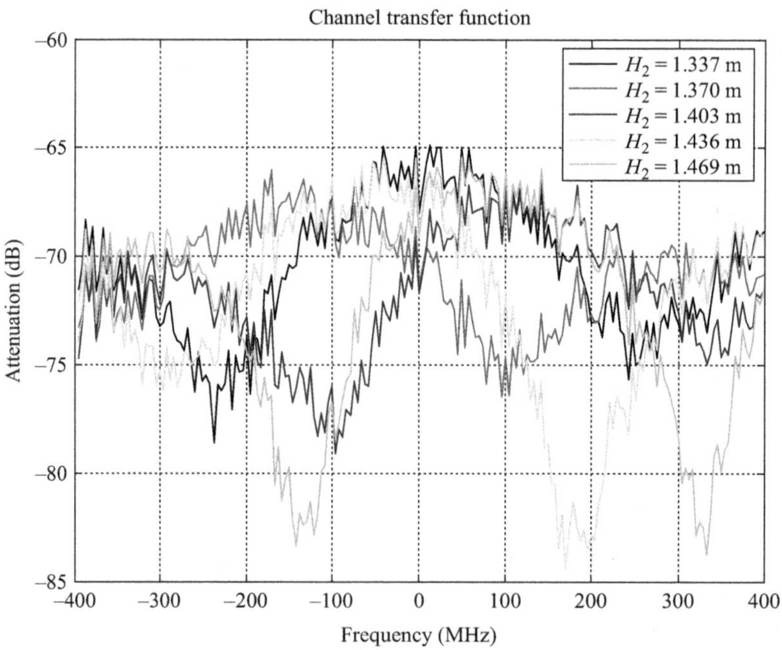

Figure 3.7 Scenario H–H, L_0 = 30.6 m, H_1 = 6.2 m. CTF dependency on the RX height H_2

Horizontal-to-Horizontal (H-H) antenna polarization configuration was used at both TX and RX antennas. The measurements for other polarization configurations were also carried out, but the use of H-H configuration allowed us to exclude additional effects associated with the Brewster angle impact. As illustrated in the figures, the channel has high frequency selectivity in 800 MHz bandwidth. But the most important outcome from these results is the very fast variations of the millimeter-wave channel when the RX height is changing. Just RX moving in the vertical direction by 2–3 cm produces significant variations of the channel.

The cross-polarization ratio (XPR) was also measured for the ground-reflected ray in the considered experimental scenario. For that purpose, orthogonal antenna polarization configurations were used at the transmitter and receiver. According to the obtained experimental results, the value of XPR was less than −25 dB.

Broadband experimental measurements with directional antennas have revealed the fine structure of the millimeter-wave channel in outdoor open-area environment. It was observed that typical mobile user antenna will receive at least two rays (the direct LOS ray and the ground-reflected ray) with rather a small time delay (2.5 ns for 30 m). It should be noted, that as the distance between the TX and RX sites increases, the time delay and angle of arrival between these rays decrease. The interference of the direct and the ground-reflected rays justifies the fast fading effects observed in the experiments with omnidirectional antennas in the beginning of current section and should be addressed in the channel modeling methodology.

Ground reflection investigation

As was demonstrated in the previous section, the ground reflection has a major effect on the signal propagation in the outdoor environment. For detailed investigation of this effect at larger distances, a special measurement trial was performed [36]. A former airport field in Berlin Gatow provided a flat and uniform asphalt surface for transmitter to receiver distances up to 1,000 m as shown in Figure 3.8. The channel sounder was the same as previously described. Different parameters of the street canyon campaign are given in Table 3.3.

The measurement channel snapshots are taken at equidistant positions of the RX. For each snapshot (channel impulse response) the instantaneous path loss is

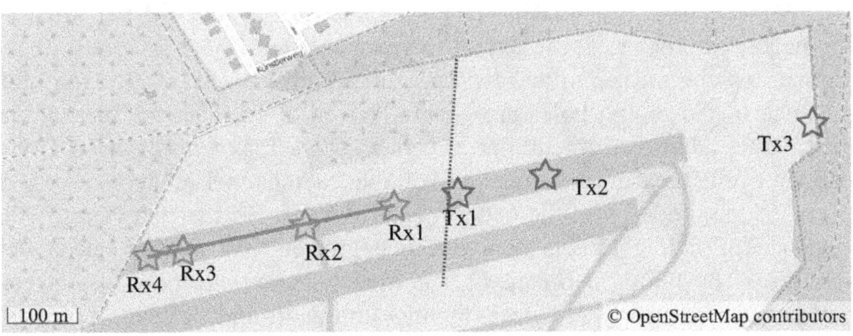

Figure 3.8 Map of airport measurement location

Table 3.3 Measurement system and environmental parameters

Parameter	Value
Tx height	4 m
Distance	40–1,000 m
Rx and Tx antenna	20 dBi horn
Half-power beamwidth	19.5° E-plane horizontal 18.1° H-plane, vertical
Rx and Tx antenna polarization	Vertical, horizontal
Atmospheric pressure	1,015 hPa
Relative humidity	50%–60%
Air temperature	28 °C

calculated. Figure 3.9 shows the measured path loss over a range from 40 to 1,000 m on the tarmac (transmitter at Tx1, Tx2, Tx3; receiver moved between Rx1 and Rx4, horizontal polarization). A moving average filter with a width of 20 samples has been applied to reduce noise effects. A distance of 60 m is covered in one measurement run. The figure therefore shows the combination of 16 subsequent measurement runs. Some artifacts can be observed at the transitions between the individual measurement runs. This can be caused by slight variations in antenna height above ground and positioning errors when the transmitter was relocated from Tx1 to Tx2 and Tx3. The measurement results for vertical polarization are not shown here but exhibit the same behavior.

A simple two-ray approach is used to model the propagation. The line of sight path (LOS) and a single ground reflection path are calculated in a polarimetric way. The ground reflection properties are modeled according to Fresnel reflection laws and the oxygen absorption is taken into account. The discrete impulse response is binned and computed into a band limited channel impulse response and the path loss is calculated, as was done with the measurement data.

Figure 3.9 shows the path loss over a range from 40 to 1,000 m with a bandwidth of 250 MHz for oxygen absorption rates of 0 dB and 14 dB per km. The latter value gives a good match between measurement and simulation. It is well in line with the MPM model [37] that gives an attenuation of 13.9 dB per km for the environmental parameters at the day of the measurement.

Comparing the measured and the simulated path loss shows that the number and distance of the fading holes are generally well aligned. In the measurements however, some additional disturbing effects exist. A major observation from the comparison of the measurement results with two-ray channel model is a very good alignment of the model with the experimental data on the distances up to 500 m. The absence of fading effects at a distance greater than 500 m needs further investigation. The fading can cause a significant increase in path loss at certain positions, as can be clearly seen from the measurements and simulation. Therefore, the fading effect caused by the ground reflection should be taken into account also for long-distance street backhaul links.

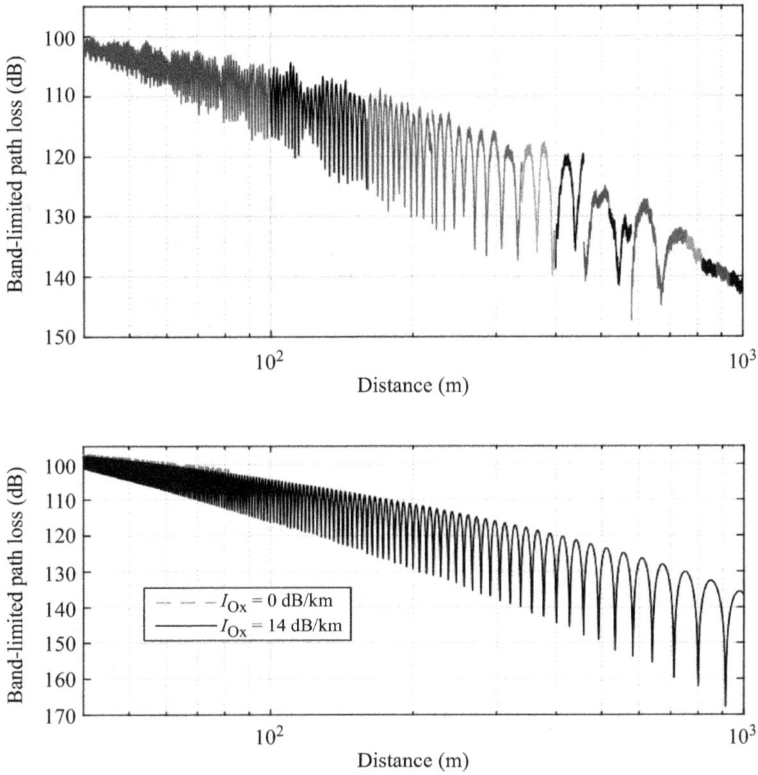

Figure 3.9 Measured vs simulated path loss ($H_{tx} = H_{rx} = 4$ m, horizontal polarization, asphalt, $\varepsilon_r = 2$, $B = 250$ MHz)

3.3.2 Experimental results interpretation

Urban street canyon measurements

In the urban street canyon path loss investigation, the measurement snapshots were taken using omnidirectional antennas as described previously. When calculating the path loss on this basis all possible propagation paths between transmitter and receiver are taken into account. In Figure 3.10, the path loss for all distances from 5 to 50 m is shown as a scatter plot. The solid line represents the linear least squares (LS) fit corresponding to the mean path loss. The parameters identified for this path loss model can be found in Table 3.4.

The least square fit was performed on a path loss model with fixed reference path loss at a reference distance:

$$L_{PL}(d_0)_{|dB} = \overline{L}_{PL}(d_0)_{|dB} + 10n\log_{10}(d/d_0) \tag{3.1}$$

The obtained path loss value is close to the free space path loss exponent of 2. This indicates that the line-of-sight propagation is dominant in this environment.

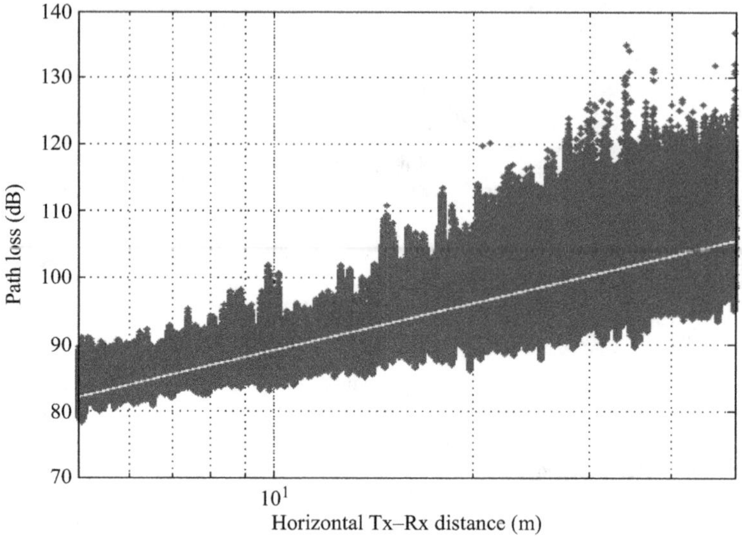

Figure 3.10 Measured path loss and mean path loss least squares fit

Table 3.4 Path loss parameters

Parameter	Value
Reference distance d_0	5 m
$\bar{L}_{PL}(d_0)_{\vert dB}$	82.02 dB
n	2.36

The high variance of the path loss however can be identified to be caused by fading from ground and sidewall reflections, as was showed in Section 3.3.1.

Time-delay bitmaps diagrams

The path loss approximation in Table 3.4 may give initial insights into the millimeter-wave channel features. Nevertheless, in-depth investigation of the channel structure requires the analysis of the channel impulse response from both the statistical and deterministic perspective. To evaluate the millimeter-wave channel properties, the measured channel impulse responses were processed by a simple threshold peak detection algorithm (3.2), according to which:

A point in the power-delay profile $P(t_k)$ is identified as a peak if:

$$\begin{cases} P(t_{k-1}) < P(t_k) > P(t_{k+1}) \\ P(t_k) > P_{noise} + 10dB \end{cases} \tag{3.2}$$

The peaks corresponding to the strongest rays or multipath components (MPCs) resolved by the sounder are shown in Figure 3.11.

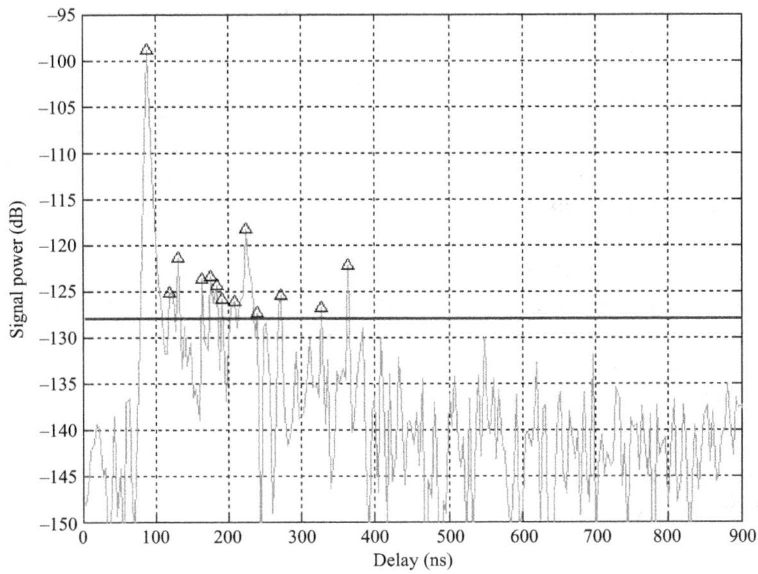

Figure 3.11 Single channel snapshot with identified peaks

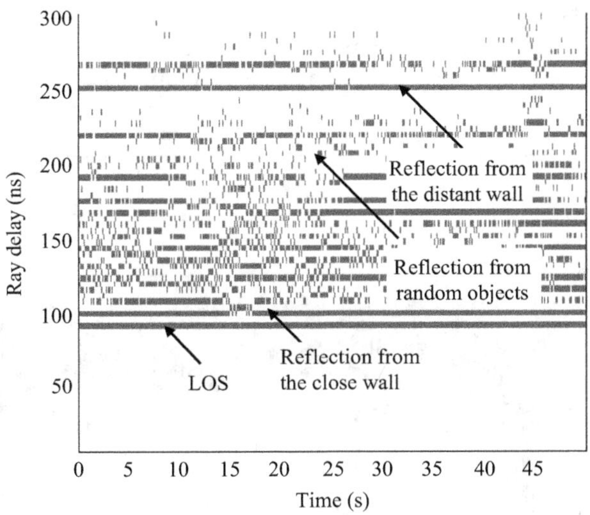

Figure 3.12 Ray delay vs. observation time bitmap

For the analysis of the channel behavior in time, the ray delay vs. observation time bitmap diagram was used. An example of such diagram is plotted in Figure 3.12, where the appearance/absence of the identified strongest rays are indicated for the whole observation time of 50 s.

It is worth noting that in every measurement the LOS ray arrives at 83 ns representing the delay caused by the 25 m separation between TX and RX. It can also be seen from Figure 3.12 that there are other very stable rays. These rays can be associated with the reflections from large static objects such as building walls and bus stop pavilions. At the same time, some rays are randomly appearing and disappearing due to small power and/or blockage. These rays can be associated with the reflections from faraway objects and closely located elements of the dynamic street environment.

For the mobile measurements with the moving RX, 40 measurement series with 62,500 snapshots each were performed. Hence, the corresponding data set comprises 2.5 million CIRs. Since the RX was moved over 25 m during each run, a fine spatial sampling (0.4 mm spacing) over the whole section of the street canyon has been achieved.

Rays classification

In the dynamic environments (city streets, squares, malls, etc.) even the direct LOS ray and the steady reflected rays are susceptible to blockage. An example of the reflected ray blockage events recorded during the experimental measurement campaign in the urban street canyon environment can be seen at the multipath components power graph in Figure 3.13.

The blockage effects may be statistically analyzed by using the time-delay bitmap diagrams of the measurements data (see Figure 3.12) obtained from multiple experiments.

The percentage of the "ray activity" may be estimated from these diagrams. Assuming ergodic properties of the blockage random process, the percentage of

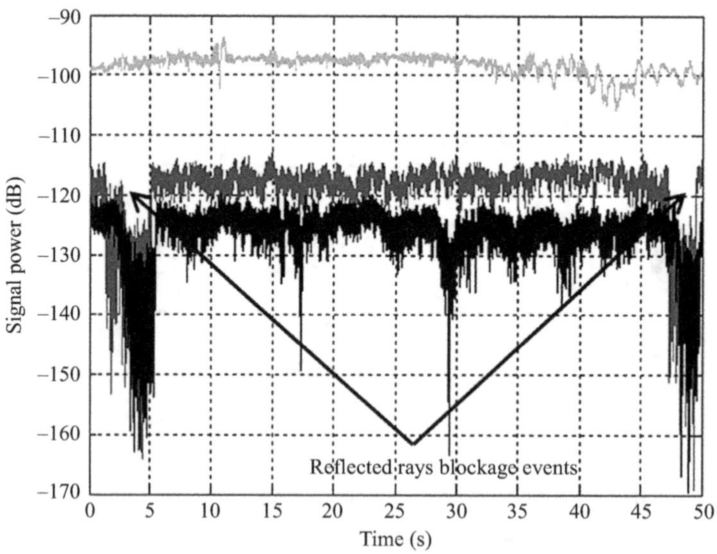

Figure 3.13 Ray blockage moments in the experimental measurements

activity in time may be used as an estimation of the blockage probability in statistical ensemble. Figure 3.14 shows the bar chart of ray activity for the Street Canyon measurements scenario (near-wall position).

Figure 3.14 allows for the classification of rays in three different groups:

- The rays with activity percentage above 80%–90% are the **D-rays**: strong and always present, if not blocked. The blockage percentage for D-rays may be estimated at around 2%–4%
- The rays with activity percentage of about 40%–70% are the **R-rays**: the reflections from faraway static objects, weaker and more susceptible to blockage due to longer travel distance.
- The rays with activity percentage below 30% are the **F-rays**: the flashing reflection from random moving objects. Such rays are not "blocked," they actually are "appearing" for a short time.

The first type of rays make the major contribution into the signal power, by being present all the time, and, usually, can be clearly identified as reflections from scenario-important macro objects. It is logical to include them into the channel model as deterministic (D-rays), explicitly calculated values. The element of randomness, important for the statistical channel modeling, may be introduced on the intra-cluster level, by adding a random exponentially decaying cluster to the main D-ray.

The second type of rays (R-rays) are reflections from random objects or objects that is not mandatory in the scenario environment. Such type of rays may be

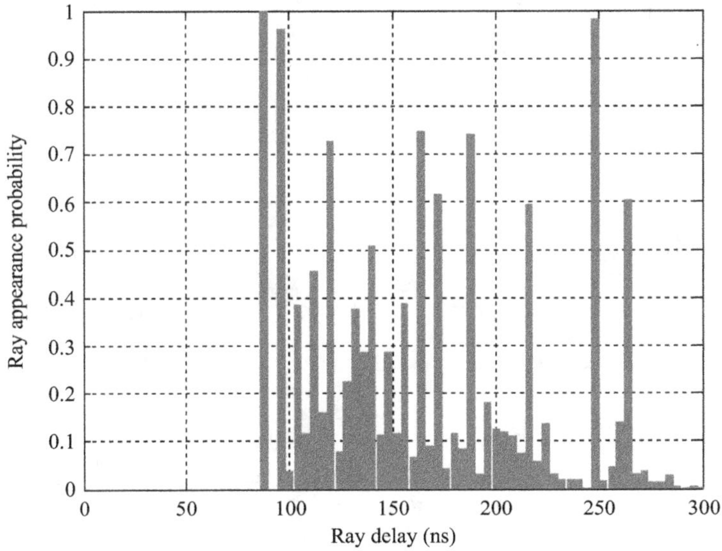

Figure 3.14 Ray appearance probability for Street Canyon measurements (near-wall position)

included in the model in a classical statistical way, as rays with parameters (power and delays) selected randomly in accordance with a predefined distributions.

The third type of rays may be introduced to the model in the same way as R-rays, but with some additional stochastic modeling to account for probability of appearance and duration.

Simplified two-ray approximation

It was demonstrated in Section 3.3.2 that two-ray channel model composed of the LOS and ground reflected rays only provides a very accurate description of the millimeter-wave channel for distances up to 500 m in "pure" open area environment (airport field). In this section, we will show that simple two-ray model may be used for approximate description of more complex and multipath environments.

The first example, which shows the predictive accuracy of two-ray channel model for the more complex open area environment (university campus), is given in Figure 3.15. The figure shows one measured channel transfer function (CTF) realization from Figure 3.7, and the corresponding calculation based on the two-ray model. The error standard deviation is about 2 dB only.

Figure 3.16 depicts the signal power of fast fading comparison for the urban street canyon experimental measurement results versus a two-ray channel model approximation. From this data it can be seen that even for this multipath environment we observe a good coincidence between the model and experimental data. It should be noted that at distances of more than 25 m the ground-reflected ray

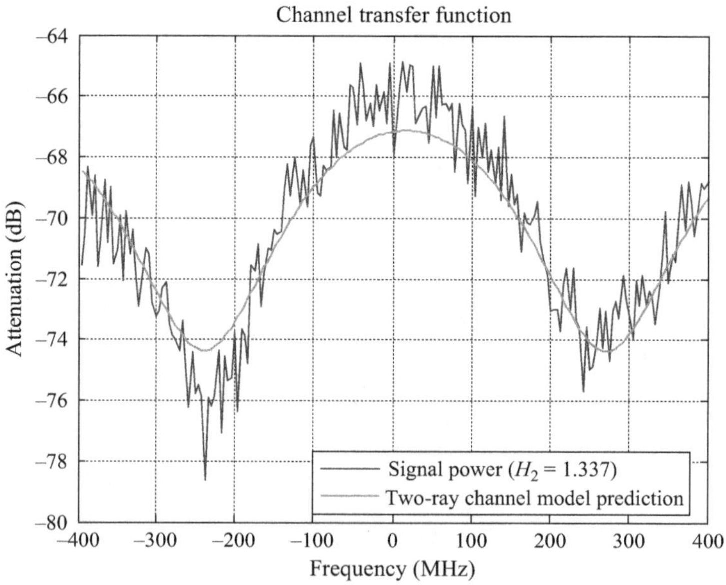

Figure 3.15 Open area experimental measurements results (thin line) versus two-ray model prediction (thick line)

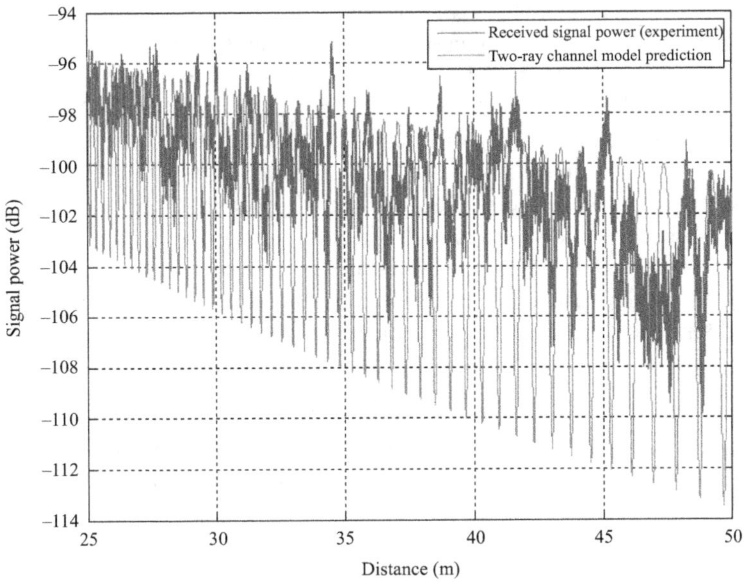

Figure 3.16 Street canyon experimental measurement results versus two-ray model results for the same parameters

is almost as strong as the direct ray due to sliding incidence with respect to the street surface. The interference between the two D-rays produces the large quasi-periodic fading gaps observed in Figure 3.16. Moreover, the fading depth can be used for a rough estimation of the reflection losses and the power of all other residual rays.

Very good agreement of the simple two-ray model with the experimental results in different measurement outdoor environments provides evidence that millimeter-wave channel can be very well described by a few dominant rays.

The joint analysis of experimental measurement data and theoretical evaluations suggests the suitability to the quasi-deterministic (Q-D) approach to the millimeter-wave channel modeling. The combination of the deterministic approach to the strongest ray parameters evaluation with the statistical description of the random components is key to meeting new millimeter-wave channel model requirements.

3.4 Quasi-deterministic approach for millimeter-wave channel modeling

The experimental results presented in the previous section confirm the quasi-optical nature of the millimeter wave propagation. The communication channel can be represented as a set of rays with different delays and propagation paths. Moreover, the analysis of the experimental data has shown that generally those rays may be divided into the three major groups: deterministic D-rays, random R-rays, and flashing F-rays. The parameters and properties of the rays in each group can be derived from the experimental measurements and ray-tracing analysis.

Deterministic rays (D-rays) in the model correspond to the map-based channel modeling approaches [19], with only one difference, that the Q-D methodology does not require exact environment map for the reconstruction because only the strongest and always present rays are considered as D-rays in this methodology. The R-rays correspond to the stochastic channel modeling approach and take into account the rest of the rays, which is difficult to count directly due to environment complexity and random factors. The F-rays represent the strong paths that are appearing for a short time due to momentary reflections from the passing cars, opening doors and other fast moving objects.

3.4.1 General structure of Q-D channel model

According to the 802.11ad channel modeling document [12] the generalized description of the clustered channel impulse response (without taking into account polarization) may be given by:

$$\begin{aligned} h(t,\varphi_{tx},\theta_{tx},\varphi_{rx},\theta_{rx}) &= \sum_i A^i C^i\left[t-T^i,\varphi_{tx}-\Phi^i_{tx},\theta_{tx}-\Theta^i_{tx},\varphi_{rx}-\Phi^i_{rx},\theta_{rx}-\Theta^i_{rx}\right] \\ C^i(t,\varphi_{tx},\theta_{tx},\varphi_{rx},\theta_{rx}) &= \sum_k \alpha^{i,k}\delta(t-\tau_k)\delta\left(\varphi_{tx}-\varphi^{i,k}_{tx}\right)\delta\left(\theta_{tx}-\theta^{i,k}_{tx}\right) \\ &\quad \times\delta\left(\varphi_{rx}-\varphi^{i,k}_{rx}\right)\delta\left(\theta_{rx}-\theta^{i,k}_{rx}\right) \end{aligned} \tag{3.3}$$

where h is a generated total channel impulse response function and t is current time; φ_{tx}, θ_{tx}, φ_{rx}, θ_{rx} are azimuth and elevation angles at the transmitter and receiver, respectively; A^i and C^i are the gain and the channel impulse response function for ith cluster, respectively; $\delta(\,)$ is the Dirac delta function; T^i, Φ^i_{tx}, Θ^i_{tx}, Φ^i_{rx}, Θ^i_{rx} are time-angular coordinates of ith cluster; $\alpha^{i,k}$ is the amplitude of the kth ray of ith cluster; $\tau^{i,k}$, $\varphi^{i,k}_{tx}$, $\theta^{i,k}_{tx}$, $\varphi^{i,k}_{rx}$, $\theta^{i,k}_{rx}$ are relative time-angular coordinates of kth ray of ith cluster.

The proposed channel model adopts the clustering approach with each cluster consisting of several rays closely spaced in time and angular domains. In a real environment, time and angular parameters of different clusters and rays are time varying functions due to non-stationarity. However, the rate of these variations is relatively slow.

The millimeter-wave channel model should also support the polarization properties of the antennas and propagation paths. Polarization of the electromagnetic waves describes the orientation of electric field **E** and magnetic intensity orientation in space and time. The magnetic field vector can always be unambiguously found if electric field vector **E** orientation and the direction of propagation are known. So the polarization properties are usually described for electric field vector **E** only. In the present channel modeling methodology we will follow the approach for the polarization effects simulation which was successfully applied in the 802.11ad channel modeling methodology [12,13].

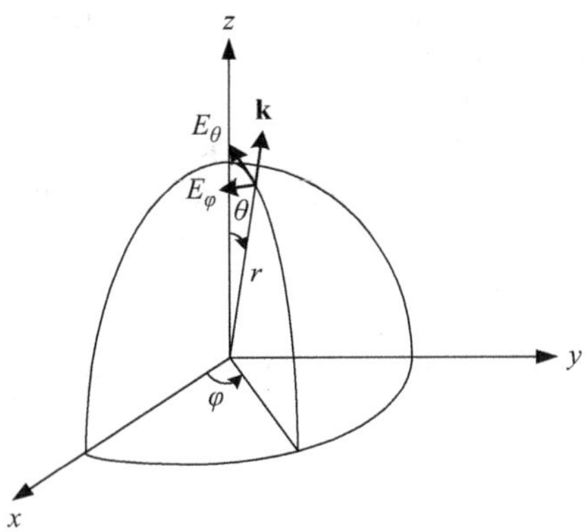

Figure 3.17 Transmitted **E** *vector in the far-field zone*

The emitted radio signal polarization is determined by the polarization properties of the antenna. In the far-field zone of the EM field radiated by the antenna, the electric vector **E** is a function of the radiation direction (defined by the azimuth angle φ and elevation angle θ in the reference coordinate system). An illustration of the transmitted **E** vector in the far-field zone is shown in Figure 3.17.

Vector **E** is perpendicular to the propagation direction **k** and can be decomposed into two orthogonal components: E_θ and E_φ that belong to the planes of constant φ and constant θ angles, respectively. Knowledge of E_θ and E_φ of the radiated signal (which may be functions of φ and θ) fully describes polarization characteristics of the antenna in the far-field zone. The polarization direction may be described by normalized vector **e**, the so-called Jones vector. Wave polarization can be described using *Jones calculus* introduced in optics for description of the polarized light. In the general case, a Jones vector is composed by two components of the electric field of the EM wave. The Jones vector **e** is defined as the normalized two-dimensional electrical field vector **E**. The first element of the Jones vector may be reduced to a real number. The second element of this vector is complex and, in the general case, defines phase difference between the orthogonal components of the **E** field.

With the selected **E** field bases (E_θ and E_φ components) for the TX and RX antennas, the polarization characteristics of each ray of the propagation channel may be described by the channel polarization matrix **H**. Based on experimental results and theoretical analysis of the phenomenon, the polarization characteristics of the model were introduced at the cluster level, assuming that all rays comprising one cluster have (approximately) the same polarization characteristics. Therefore, extending the channel structure for polarization support requires changing the scalar cluster gain coefficients $A^{(i)}$ in (3.3) by 2×2 cluster polarization matrices $\mathbf{H}^{(i)}$, and the channel impulse responses realization to be described by matrix **h**:

$$\mathbf{h}(t, \varphi_{tx}, \theta_{tx}, \varphi_{rx}, \theta_{rx}) = \sum_i \mathbf{H}^i C^i \left[t - T^i, \varphi_{tx} - \Phi^i_{tx}, \theta_{tx} - \Theta^i_{tx}, \varphi_{rx} - \Phi^i_{rx}, \theta_{rx} - \Theta^i_{rx}\right] \tag{3.4}$$

The generalized channel impulse response description (3.4) may be done in a more specific form, following the proposed hybrid quasi-deterministic, stochastic methodology. The channel impulse response can be represented as sum of clusters corresponding to D-rays (direct ray, ground reflection, etc.), clusters corresponding to R-rays and clusters corresponding to F-rays. Each ray has its own matrix **H** that characterizes the polarization, power and phases of the two polarization components, the ray delay T^i, and angular characteristics, such as angles of departure (AoD) (φ_{tx},θ_{tx}), and angles of arrival (AoA) (φ_{rx},θ_{rx}).

In this case, the transmission equation for a single ray channel may be written as:

$$y = G_{tx}(\varphi_{tx}, \theta_{tx})G_{rx}(\varphi_{rx}, \theta_{rx})\boldsymbol{e}_{rx}^{H}\mathbf{H}\boldsymbol{e}_{tx}x \tag{3.5}$$

where x and y are the transmitted and received signals, $\boldsymbol{e}_{tx}$ and $\boldsymbol{e}_{rx}$ are the polarization (Jones) vectors for the TX and RX antennas, respectively, and $G_{tx}(\varphi,\theta)$ and $G_{rx}(\varphi,\theta)$ are antenna gains at given angular coordinates. In the general case, the G_{tx} and G_{tx} are different for different polarizations and should be represented as vectors, just like $\boldsymbol{e}_{tx}$ and $\boldsymbol{e}_{rx}$.

For the LOS signal path, matrix $\mathbf{H}_{LOS}$ is close to the identity matrix (non-diagonal components may be non-zero but significantly smaller than diagonal elements) multiplied by the corresponding gain coefficient due to path loss and phase rotation due to signal propagation delay. LOS propagation does not change polarization characteristics of the signals. However, polarization characteristics of the signals are changed upon reflections. The change of the polarization characteristics upon reflection is defined by the type of the surface and the incident angle. Thus, polarization characteristics may be different for different clusters but are the same for the rays comprising one cluster.

Figure 3.18 illustrates an example for the general channel impulse response structure: a number of D-rays clusters and R-rays clusters with an exponentially decaying intra-cluster thin structure.

The parameters required for calculation of the channel model are summarized in Table 3.5. This notation will be used further for the specific scenarios description.

3.4.2 Deterministic rays (D-rays) modeling

The quasi-deterministic rays are explicitly calculated in accordance with scenario parameters, geometry and propagation conditions.

Important part of the proposed Q-D approach to the channel modeling is the calculation of the reflected ray parameters. The calculations are based on the method of images for determining the angular characteristics of the first and higher order reflections (see Figure 3.20).

The Friis equation (3.6) determines the path loss P (in dB) at the distance d:

$$P = 20\log_{10}\left(\frac{\lambda}{4\pi d}\right) \tag{3.6}$$

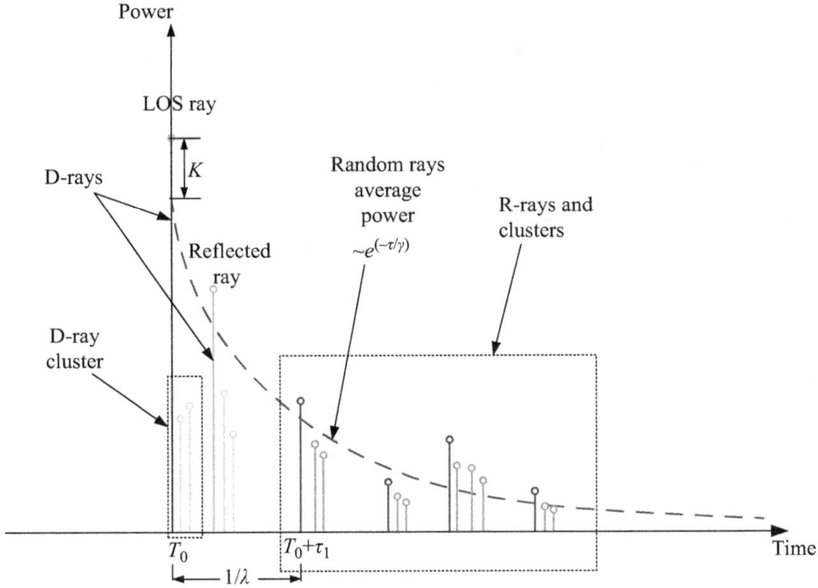

Figure 3.18 General channel impulse response structure

Table 3.5 Common input parameters of the Q-D channel model

Parameter	Description
F_c	Carrier frequency, Hz
c	Light speed, $c = 299{,}792{,}458$ m/s
A_0	Oxygen absorption, dB/km (0.015 dB/m for 60 GHz)
H_{tx}	Transmitter height above ground, m (see Figure 3.19)
H_{rx}	Receiver height above ground, m
L	Horizontal distance between TX and RX, m
$G_{tx}(\varphi,\theta)$	TX antenna gain at (φ,θ) angular coordinates
$G_{rx}(\varphi,\theta)$	RX antenna gain at (φ,θ) angular coordinates
ε_r	Relative permittivity of the ground surface at F_c,
σ	Surface roughness: standard deviation
ξ	Cross-polarization discrimination, XPD

The Fresnel equation (3.7) is used for calculation of the reflected rays power for different polarization components:

$$R = 20\log_{10}\left(\frac{\sin\phi - \sqrt{B}}{\sin\phi + \sqrt{B}}\right) \tag{3.7}$$

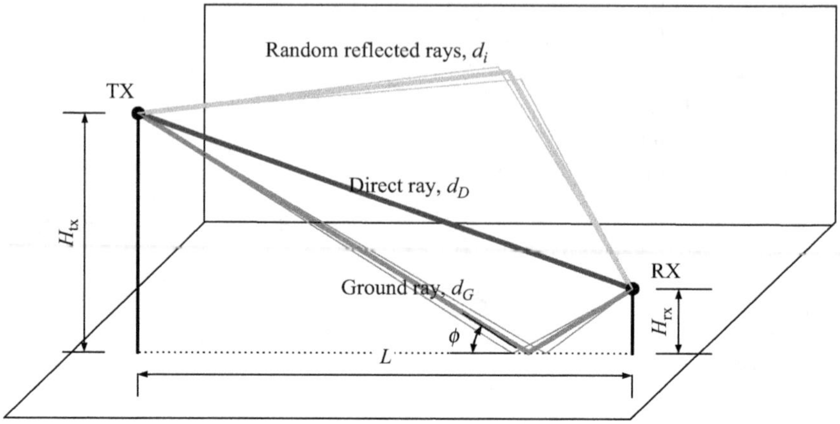

Figure 3.19 Channel model geometry

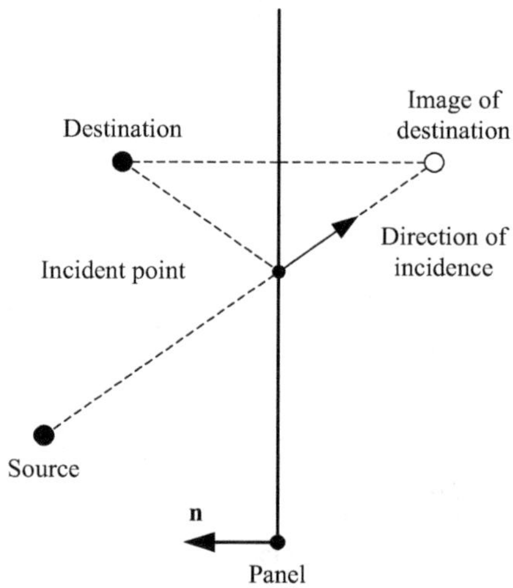

Figure 3.20 Method of images for reflected ray parameters calculation

where R is the reflection coefficient (power ratio) of the surface, ϕ is the grazing angle and

$B = \varepsilon_r - \cos^2\phi$ for horizontal polarization.
$B = (\varepsilon_r - \cos^2\phi)/\varepsilon_r^2$ for vertical polarization.

However, the Fresnel equation describes only specular reflection from the ideal surface, while in reality most surfaces can be treated as "rough" comparing to the signal wavelength in accordance with the Rayleigh roughness criterion [38]:

$$g = 4\pi\sigma_h \sin\phi/\lambda \tag{3.8}$$

where σ_h is the standard deviation of the surface height about the local mean value within first Fresnel zone, λ is the free-space wavelength, ϕ is the grazing angle measured with respect to tangent to the surface.

The surface can be considered smooth for $g < 0.3$. When the surface is rough, the reflected signal has two components: a specular component which is coherent to the incident signal, and a diffuse component which fluctuates in amplitude and phase with a Rayleigh distribution.

For example, asphalt with surface deviations of about 0.5 mm can be considered as smooth surface for grazing angle 10° ($g = 0.2$).

The specular component reflection coefficient is different from the one given by the Fresnel equations by constant multiplier: $R_s = \rho_s R$, where

$$\rho_s = e^{-g^2/2} \tag{3.9}$$

By substituting (3.8) into (3.9) in the log scale, the signal power decrease due to the reflecting surface roughness is described by factor F:

$$F = \frac{80}{\ln 10}(\pi\sigma_h \sin\phi/\lambda)^2 \tag{3.10}$$

The feasibility of the proposed approach to the prediction of the signal power is proved in [39] for outdoor microcell environments and in [40,41] for inter-vehicle communication modeling. In general, problems of the signal power prediction considered in [38].

The D-rays are strictly scenario-dependent, but in all considered scenarios two basic D-rays are present: the direct LOS ray and the ground reflected ray. The calculation of those two basic rays parameters will be the same for all scenarios.

Direct ray

Direct LOS ray is a ray between TX and RX. Its parameters are calculated using the Friis free space path loss equation (3.2) (see Table 3.6).

Ground-reflected ray

Ground-reflected rays are present in all considered scenarios. Its parameters are calculated based on Friis free space path loss equation (3.6), and the Fresnel equation (3.7) to take into account reflection and rough surface scattering factor F (equation (3.8)). Note that the horizontally and vertically polarized components of the transmitted signal will be differently reflected and thus, the channel matrix **H** should have different diagonal elements (see Table 3.7).

Polarization for D-rays

Following the proposed channel modeling methodology, all properties of quasi-deterministic rays are explicitly calculated. The channel matrix **H** contains all polarization characteristics of the ray and is calculated based on the reflection from the defined in the scenario surfaces.

Table 3.6 Direct ray parameters

Parameter	Value
Delay	Direct ray delay is calculated from the model geometry: $\tau_D = d_D/c$ $d_D = \sqrt{L^2 + (H_{tx} - H_{rx})^2}$
Power	Direct ray power calculated as free-space path loss with oxygen absorption: $P_D = 20\log_{10}\left(\frac{\lambda}{4\pi d_D}\right) - A_0 d_D$, in dB
Channel matrix	$\mathbf{R} = \begin{bmatrix} 10^{P_D/20} & 0 \\ 0 & 10^{P_D/20} \end{bmatrix} e^{\frac{j2\pi d_D}{\lambda}}$
AoD	0° azimuth and elevation
AoA	0° azimuth and elevation

It is known that reflection coefficients are different for **E** field components both parallel and perpendicular to the plane of incidence and depend on the incident angle. Theoretical coupling between parallel and perpendicular components of the reflected signal is zero for plane media interfaces (boundaries). But due to non-idealities (roughness) of the surfaces some coupling always exists in real channels.

An example of a first-order reflected signal path is shown in Figure 3.21.

The polarization matrix for the first-order reflected signal path may be found as a product of the matrix that rotates **E** vector components from the coordinate system associated with the TX antenna to the coordinate system associated with the incident plane. Next, reflection matrix **R** with reflection coefficients and cross-polarization coupling coefficients is applied, followed by a rotation to the coordinate system associated with the RX antenna. Thus, the channel propagation matrix for the case of the first-order reflected signals may be defined as:

$$\mathbf{H}_1 = \begin{bmatrix} \cos\psi_{rx} & \sin\psi_{rx} \\ -\sin\psi_{rx} & \cos\psi_{rx} \end{bmatrix} \begin{bmatrix} R_\perp(\alpha_{inc}) & \xi_1 \\ \xi_2 & R_\parallel(\alpha_{inc}) \end{bmatrix} \begin{bmatrix} \cos\psi_{tx} & \sin\psi_{tx} \\ -\sin\psi_{tx} & \cos\psi_{tx} \end{bmatrix} \tag{3.11}$$

The leftmost matrix in (3.11) is the recalculation of the polarization vector from the plane of incidence basis to RX coordinates, the rightmost matrix is the recalculation of the TX polarization vector to the plane of incidence basis (see Figure 3.21). In the center is the reflection matrix **R** that includes the reflection coefficients $R_\perp$ and $R_\parallel$ for the perpendicular and parallel components of the electric field $E_\perp$ and $E_\parallel$, respectively. Elements ξ_1 and ξ_2 in the matrix **R** are cross-polarization coupling coefficients.

Table 3.7 Ground-reflected ray parameters

Parameter	Value
Delay	Ground-reflected ray delay is calculated from the model geometry: $\tau_G = d_G/c$ $d_G = \sqrt{L^2 + (H_{tx} + H_{rx})^2}$
Power	Ground-reflected power calculated as free-space path loss with oxygen absorption, with additional reflection loss calculated on the base of Fresnel equations. Reflection loss R is different for vertical and horizontal polarizations $P_\perp = 20\log_{10}\left(\frac{\lambda}{4\pi d_G}\right) - A_0 d_G + R_\perp + F;\quad P_\parallel = 20\log_{10}\left(\frac{\lambda}{4\pi d_G}\right) - A_0 d_G + R_\parallel + F$ $F = \frac{80}{\ln 10}(\pi\sigma_h \sin\phi/\lambda)^2$ $R_\perp = 20\log_{10}\left(\frac{\sin\phi - \sqrt{B_\perp}}{\sin\phi + \sqrt{B_\perp}}\right);\quad R_\parallel = 20\log_{10}\left(\frac{\sin\phi - \sqrt{B_\parallel}}{\sin\phi + \sqrt{B_\parallel}}\right)$ $B_\parallel = \varepsilon_r - \cos^2\phi$ for horizontal polarization $B_\perp = (\varepsilon_r - \cos^2\phi)/\varepsilon_r^2$ for vertical polarization, where $\tan\phi = \frac{H_{tx} + H_{rx}}{L}$
Channel matrix	$\mathbf{R} = \begin{bmatrix} 10^{P_\perp/20} & \xi \\ \xi & 10^{P_\parallel/20} \end{bmatrix} e^{\frac{j2\pi d_G}{\lambda}}$
AoD	Azimuth: 0° Elevation: $\theta_{AoD} = \tan^{-1}\left(\frac{L}{H_{tx} - H_{rx}}\right) - \tan^{-1}\left(\frac{L}{H_{tx} + H_{rx}}\right)$
AoA	Azimuth: 0° Elevation: $\theta_{AoA} = \tan^{-1}\left(\frac{H_{tx} + H_{rx}}{L}\right) - \tan^{-1}\left(\frac{H_{tx} - H_{rx}}{L}\right)$

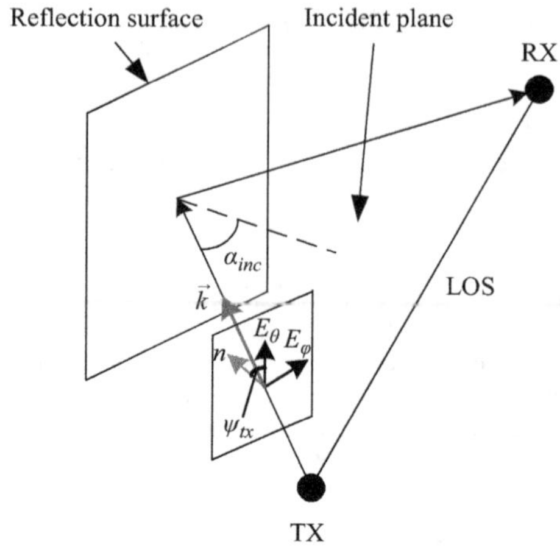

Figure 3.21 First-order reflected signal path

Note that the structure of matrix **H** given in (3.11) does not include the propagation loss along the corresponding reflected path, which should be taken into account in the final model, but does not impact polarization properties.

Similar to (3.11), the structure of polarization matrix **H** for a second-order reflection is given in (3.12) and includes additional rotation and reflection matrices:

$$\mathbf{H}_2 = \begin{bmatrix} \cos\psi_{rx} & \sin\psi_{rx} \\ -\sin\psi_{rx} & \cos\psi_{rx} \end{bmatrix} \begin{bmatrix} R_\perp(\alpha_{2inc}) & \xi_{21} \\ \xi_{22} & R_\parallel(\alpha_{2inc}) \end{bmatrix} \begin{bmatrix} \cos\psi_p & \sin\psi_p \\ -\sin\psi_p & \cos\psi_p \end{bmatrix}$$
$$\times \begin{bmatrix} R_\perp(\alpha_{1inc}) & \xi_{11} \\ \xi_{12} & R_\parallel(\alpha_{1inc}) \end{bmatrix} \begin{bmatrix} \cos\psi_{tx} & \sin\psi_{tx} \\ -\sin\psi_{tx} & \cos\psi_{tx} \end{bmatrix} \quad (3.12)$$

Note that in the general case the first incident plane does not coincide with the second incident plane. Therefore the additional recalculation from the coordinate system associated with the first incident plane to the coordinate system associated with the second incident plane is required.

The off-diagonal elements of the channel matrix **H** represent signal depolarization, i.e., the coupling between orthogonal components of the **E** vector at the TX and RX sides. There are two mechanisms for depolarization: the reflection coupling (coupling between parallel and perpendicular **E** vector components at the reflection) and geometrical coupling (coupling because of the different relative orientations of the TX and RX antennas). It can be seen that the proposed approach allows accounting for both mechanisms to create an accurate polarization impact model.

3.4.3 Statistical random rays (R-rays) modeling

For taking into account a number of rays that cannot be easily described deterministically (reflections from objects that are not fully specified in the scenario, objects with random or unknown placement, objects with complex geometry, higher-order reflections, etc.) a statistical approach is used in the Q-D channel modeling methodology.

More specifically, statistical approach to the clustered channel impulse response modeling is used [42].

Statistical approach is a basic way of R-rays generation, which is used in the Q-D channel modeling methodology. The clusters arrive at moments T^i according to Poisson process and have inter-arrival times that are exponentially distributed. The cluster amplitudes $A(T^i)$ are independent Rayleigh random variables and the corresponding phases Θ^i are independent uniform random variables over $[0,2\pi]$. The angular characteristics (AoD, AoA) of the random rays depend on the scenario and will be defined in corresponding sections with channel models parameters description.

Under such assumptions, the random rays components of the channel impulse response described in (3.3) are given by:

$$h_{cluster}(t) = \sum_{i=1}^{N_{cluster}} A(T^i)e^{j\Theta^i}C^i(t - T^i), \tag{3.13}$$

where T^i is the arrival time of the ith cluster measured from the arrival time of the LOS ray, $A(T^i)$, Θ^i are the amplitude and phase of the ith cluster. C^i is the intra-cluster impulse response defined below. The R-rays are random, with Rayleigh-distributed amplitudes and random phases, with exponentially decaying power delay profile. The total power of R-rays is determined by the K-factor with respect to the power of the direct LOS component:

$$\overline{P(T^i)} = P_0 e^{-T^i/\gamma} \tag{3.14}$$

$$\frac{P_{LOS}}{\sum P(T^i)} = K \tag{3.15}$$

where P_{LOS} is the power of the LOS ray, $P(T^i)$ is the power of cluster component with T^i arrival time, K is the ratio of LOS to all NLOS components (K-factor), and γ is power-decay constant for the clusters.

Cluster delays T^i are described by the independent inter-arrival exponential probability density function parameterized by arrival rate λ:

$$p(T^i|T^{i-1}) = \lambda e^{-\lambda(T^i - T^{i-1})} \tag{3.16}$$

The specific values of PDP parameters λ, K, and γ are defined in the channel models sections per given scenario.

Polarization for R-rays

The random rays (R-rays) are defined by the power-delay profile and angular characteristics. The R-rays model higher-order reflections, faraway reflections in the open area environment, and reflections from various random objects appeared in the area of the communication system operation. They are generated stochastically by producing those parameters as random values with the predefined probability distribution functions (PDFs).

In the 802.11ad channel model [12] the statistical approximations for different types of reflected clusters were developed based on the available experimental data [15] and ray-tracing simulations. A set of approximations were proposed for diagonal and off-diagonal elements of the first- and second-order reflections for typical indoor environments (conference room, cubicle, and living room) as a combination of log-normal and uniform distributions. In the Q-D model the ray amplitude approximated by the Rayleigh distribution (which is close to log-normal) so that the simple fixed polarization matrix $\mathbf{H}_\mathrm{p}$ may be used for introducing polarization properties to the R-rays.

The channel polarization matrix $\mathbf{H}_\mathrm{p}$ for R-rays is defined by (3.17):

$$\mathbf{H}_\mathrm{p} = \begin{bmatrix} 1 & \pm 0.1 \\ 0.1 & \pm 1 \end{bmatrix} \tag{3.17}$$

The values with sign $\pm$ are assumed to have random sign (+1 or -1, for instance) with equal probability, independently from other values.

For the cluster rays the main random ray, the polarization matrix is the same as for the random ray.

3.4.4 Random reflection R-ray generation

The synthetic aperture processing of the experimental results [43] have shown that the reflections from various environmental objects such as trees, lampposts, bus stops, etc. can be clearly identified (with exact estimation of the reflector position) from the experimental data. Such rays should be taken into account along with D-rays, which originate from large-scale objects, but the definition of the position of each reflector makes scenario description complex and specific. Thus, it is proposed to generate such type of rays (R-rays or F-rays) as reflections from the randomly placed spherical objects that (unlike the flat objects) can create specular reflection path between any two points in the 3D space.

For now, based on the experimental measurements, the R-rays as reflections from random objects are introduced for Street Canyon scenario only, in addition to statistically generated R-rays power delay profile. Also, the F-rays generates in this way, with only difference of the path existence period in the applications where the longer periods of time are analyzed.

3.4.5 Intra-cluster structure

The surface roughness and presence of the various irregular objects on the considered reflecting surfaces and inside them (bricks, windows, borders, manholes, advertisement boards on the walls, etc.) lead to the separation of the specular reflection ray to a number of additional rays with similar delays and angles: a cluster. The mechanism of the cluster formation on the surface with large-scale irregularities is illustrated in Figure 3.22. It can be seen that the reflection from the rough surfaces results in spreading the signal by angle and delay.

The intra-cluster parameters of the channel model were estimated from the measurement data [16]. The individual rays were identified in the time domain, and statistical characteristics, including average number of rays, ray arrival rate, and ray power decay time, were measured.

Based on the obtained results, the statistical model for the cluster time domain parameters is given. The structure of the model is schematically shown in Figure 3.23.

For every base ray, the intra-cluster structure is given by:

$$C(t) = \sum_{k=1}^{N_{intra}} \alpha(\tau_k) e^{j\theta_k} \delta(t - \tau_k), \tag{3.18}$$

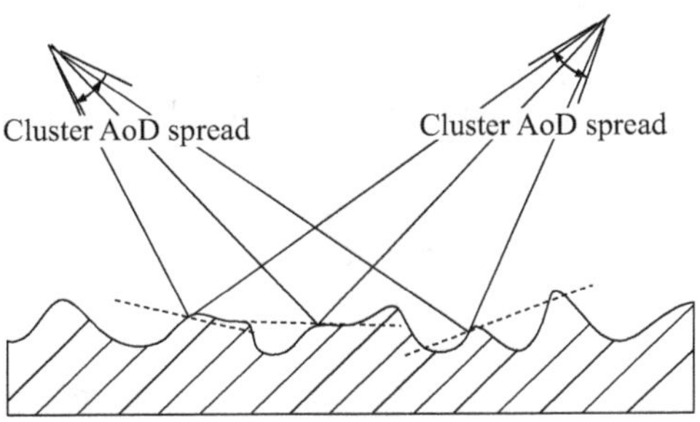

Figure 3.22 Reflection scattering mechanism

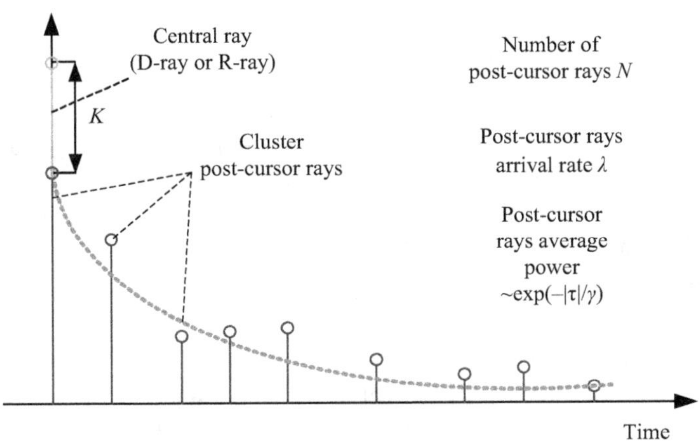

Figure 3.23 Time domain model of the intra-cluster structure

where N_{intra} is the number of intra-cluster components, τ_k is the arrival time of the kth intra-cluster component measured from the arrival time of the base D-ray or R-ray, $\alpha(\tau_k)$, $p(\tau_k)$and θ_k are the amplitude, power and phase of the kth intra-cluster component. The intra-cluster components are random, with Rayleigh-distributed amplitudes and random phases, with exponentially decaying power delay profile. The total power of intra-cluster components is determined by the K-factor with respect to the base D-ray or R-ray power:

$$\overline{p(\tau_k)} = p_0 e^{-\tau_k/\gamma} \tag{3.19}$$

$$\frac{p_{base\ ray}}{\sum P(\tau_k)} = K \tag{3.20}$$

Generally, the intra-cluster structure generation is very similar to the R-rays generation, except that for R-ray generation the LOS ray is used as a timing and power base, and for intra-cluster structure generation cluster-base D-ray or R-ray is used for that purpose.

Combining all D-rays, R-rays and their respective intra-cluster structure components will give the final channel impulse response.

The identification of rays inside the cluster in the angular domain requires an increase of the angular resolution by using directional antennas with very high gain. Another approach, the application of the "virtual antenna array" technique, where a low directional antenna element is used to perform measurements in multiple positions along the virtual antenna array to form an effective antenna aperture, was used in MEDIAN project [14,15]. These results were analyzed in [17], deriving the recommendation to model the intra-cluster angle spread for azimuth and elevation angles for both transmitter and receiver as independent normally distributed random variables with zero mean and RMS equal to 5^0: $N(0, 5^0)$.

Note that it is reasonable to assume that different types of clusters may have distinctive intra-cluster structure. For example, properties of the clusters reflected from the road surface may be different from the properties of the clusters reflected from brick walls because of the different material surface structure. Also one may assume the properties of the first- and second-order reflected clusters to be different, with the second-order reflected clusters having larger spreads in temporal and angular domains. All these effects are understood to be reasonable. However since the number of available experimental results was limited, a common intra-cluster model for all types of clusters was developed. Modifications with different intra-cluster models for different types of clusters may be a subject of future channel model enhancements.

3.4.6 Blockage modeling

Proper simulation of the blockage effects is an important part of millimeter-wave channel modeling. In all the considered scenarios that include access link simulations the signal propagation paths are subject to blockage—by humans or vehicles interrupting the rays with static positions of the AP and UE, or by UE movement in the areas where some rays are shadowed. The necessity of introduction of blockage into the 3D channel model is proven by experimental measurements (see Section 3.3.2).

The opposite, but similar, effect is the appearance of new rays for a short time—for example reflections from passing vehicles and other moving reflecting objects. Both effects can be observed in the experimental measurement results with omni-directional antennas

Figure 3.24 illustrates the mechanism of ray blockage and ray appearance. Following the picture, the average duration of blockage and duration of flashing reflections can be estimated as:

$$T_{\text{blockage}} \sim 0.5\ \text{m (human width)}/1\ \text{m/s (average speed)} \sim 0.5-1\ \text{s}$$
$$T_{\text{flash}} \sim 4.5\ \text{m (car length)}/15\ \text{m/s (average speed)} \sim 0.2-0.3\ \text{s}$$

The analysis of experimental data in ray delay-time bitmap diagrams (Figure 3.12) gives approximately the same values.

The average service period (SP) of the typical millimeter-wave communication systems is equal to 1–3 ms (802.11ad). This means that during a blockage or flashing period thousands of service periods and tens of thousands frames will pass. The system level simulations rarely include more than thousand frames, so the blockage may be modeled as a static event, instead of a dynamic process. The parameters derived from the analysis are summarized in Table 3.8. The blockage is determined once per channel snapshot and stays the same in time.

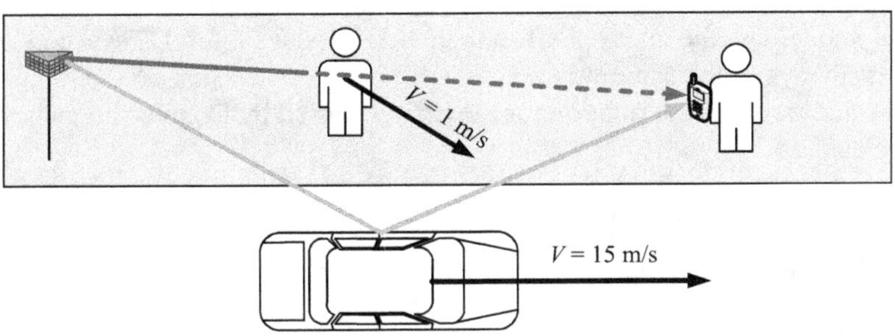

Figure 3.24 Ray blockage and random ray appearance illustration

Table 3.8 Blockage parameters for system level simulation

Parameter	Value
D-ray blockage probability, P_D	0.03
R-ray blockage probability, P_R	0.3
*Flashing R-ray appearance probability, P_F	0.2

*It should be noted that for the considered outdoor scenarios (open area and street canyon) we have developed channel models on the basis of static reflections only (D-rays and R-rays). The models with flashing R-rays require additional investigations and may be used for special studies, with the more specific vehicle/human traffic models.

Table 3.9 *Blockage parameters for VoIP and video streaming simulations*

Parameter	Value
D-ray blockage rate, λ_D	0.05 s^{-1}
R-ray blockage rate, λ_R	0.3 s^{-1}
D-ray and R-ray blockage duration, T	1 s
*Flashing R-ray appearance rate, λ_F	0.2 s^{-1}
*Flashing R-ray appearance duration, T_F	0.25 s

*It should be noted that for the considered outdoor scenarios (open area and street canyon) we have developed channel models on the basis of static reflections only (D-rays and R-rays). The models with flashing R-rays require additional investigations and may be used for special studies, with the more specific vehicle/human traffic models.

For VoIP and video streaming simulations, which require analysis of longer periods of time, the blockage moments may be introduced as a Poisson process with corresponding parameters, shown in Table 3.9.

3.4.7 Channel mobility modeling and impact

The mobility effects in the Q-D channel model are described by directly introducing the velocity vector for each UE. In multi-path channel the UE movement leads to additional phase rotation for each propagation path, but for the purposes of channel modeling, the motion effect can be introduced for D-rays and to the strongest R-rays in the same way.

The additional phase rotation for the ith ray caused by Doppler frequency shift is calculated as

$$\begin{aligned} \Delta\varphi_i(t) &= 2\pi f_i^D t \\ f_i^D &= (\boldsymbol{v}, \boldsymbol{r}_i)\frac{F_c}{c} \end{aligned} \tag{3.21}$$

where f_i^D is the frequency shift for the ith ray, $\boldsymbol{v}$ is the instantaneous vector of UE velocity (see Figure 3.25), $\boldsymbol{r}_i$ is the unity vector of the ith ray direction of arrival, and (,) denotes scalar product. The velocity vector $\boldsymbol{v}$ can be represented by the sum of its scalar components:

$$\boldsymbol{v} = v_x\boldsymbol{i} + v_y\boldsymbol{j} + v_z\boldsymbol{k}$$

The horizontal components of the velocity vector are scenario-specific. For scenarios without preferred direction of motion, such as hotel lobby or open area, the horizontal component of velocity may have uniformly distributed direction and random or fixed value. For the street canyon environment, the motion along the street may be modeled with predefined speed (3–5 km/h for pedestrians, 30–120 km/h for vehicles) and the motion across the street may be modeled as zero mean random Gaussian process.

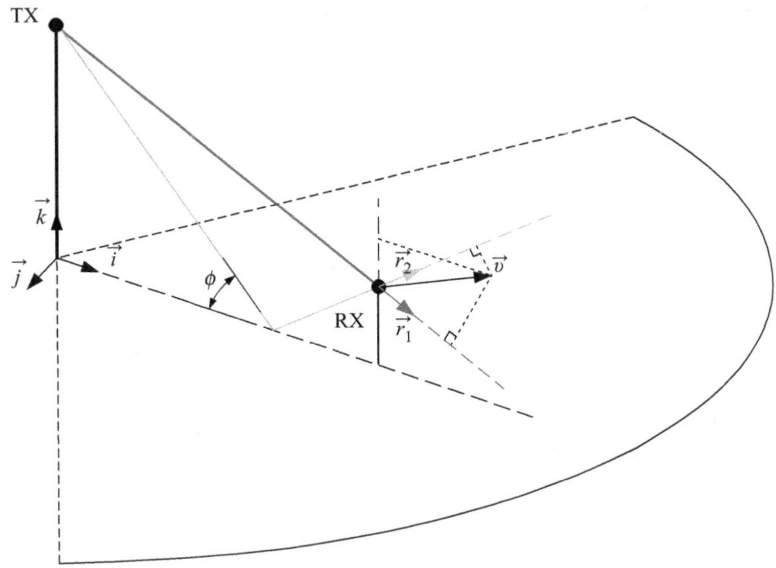

Figure 3.25 UE motion modeling illustration

As was shown in Section 3.3.1, the vertical movement of the mobile UE has more significant impact on the channel and should be considered independently. In the important case when the mobile UE is held by a human, the different models of human gait can be applied for vertical motion $z(t)$ representation [44]. In the frame of the Q-D methodology, the vertical motion is considered as a stationary Gaussian random process with correlation function equal to:

$$K_z(\tau) = \sigma_z^2 e^{-\frac{\tau^2}{\tau_z^2}}$$

The correlation function parameters should be adjusted to the real pedestrian motion with a speed of 3–5 km/h. For considered pedestrian channel models we assume $\sigma_z = 0.05$ m and $\tau_z = 0.5$ s. The vertical component v_z of velocity vector $\boldsymbol{v}$ can be defined through the vertical motion $z(t)$ of the user as the first derivative.

With the knowledge of the velocity vector and ray angles of arrival (AoAs), the values of the phase rotations can be calculated from (3.21) and added to the corresponding D-rays and R-rays phases.

The UE motion can lead to time-selective fading caused by interference of two or more rays. The fading depth can be large especially in the case of LOS and ground reflected rays interference at large distances with relatively low signal bandwidth.

For a mobile traveling at 3 km/h, from (3.21) one can calculate the maximal Doppler shift which will be about 170 Hz for the 60 GHz carrier frequency with typical channel coherence time τ_c equal to $1/f^D$= 6 ms. But for the vehicular speed 120 km/h the maximal Doppler spread will achieve 6,700 Hz with corresponding channel coherence time $\tau_c = 0.15$ ms.

Table 3.10 IEEE 802.11ad system parameters vs. millimeter-wave channel properties

IEEE 802.11ad system parameters		Millimeter-wave channel parameters	
Sample duration	0.3–0.5 ns	Delay spread	3–5 ns
Bandwidth	2,160 MHz	Coherence bandwidth	200–350 MHz
Symbol duration	242 ns/328 ns	Doppler frequency	167–6,700 Hz
Frame duration	0.1–0.7 ms	Channel coherence time	0.15–6 ms

To evaluate potential impact of the mobility effects on future 5G millimeter-wave communication systems, the parameters of IEEE 802.11ad standard for millimeter-wave Wi-Fi are compared in Table 3.10 with the typical parameters of the millimeter-wave channel.

The large available bandwidth (2,160 MHz) is compensated by a very short symbol period: 242 ns for OFDM mode and 328 ns for SC FDE mode [45]. So, the typical millimeter-wave channel with coherence time more than 0.15 ms in the worst case of high mobility will not be changed during the symbol period. At the same time, on the typical frame duration of about 0.5 ms the vehicular channel can experience significant changes which should be taken into account in future systems PHY layer design, scheduling and user time–frequency resource allocation.

Table 3.10 also allows evaluating bandwidth-related aspects of the system design. The multi-path propagation with typical delay spread of about 3–5 ns leads to frequency selectivity of the millimeter-wave channel with channel coherence bandwidth about 300 MHz. So, the millimeter-wave communication system with channel bandwidth less than this value may experience deep time-selective flat fading, which may lead to system outage.

3.5 Q-D channel models implementation

The generation of the three-dimensional millimeter-wave channel model in accordance with the Q-D methodology consists of the following steps:

- Scenario and model parameters definition.
- Calculation of the D-rays parameters in accordance with the selected scenario recommendations.
- Calculation of R-rays parameters in accordance with the selected scenario recommendations.
- Application of path blockage in accordance with the scenario requirements to the selected clusters.
- Application of TX and RX antenna patterns and beamforming algorithms.
- Conversion of the raw channel impulse response data into the discrete time required by the simulations.

The channel modeling generation flow is illustrated in Figure 3.26.

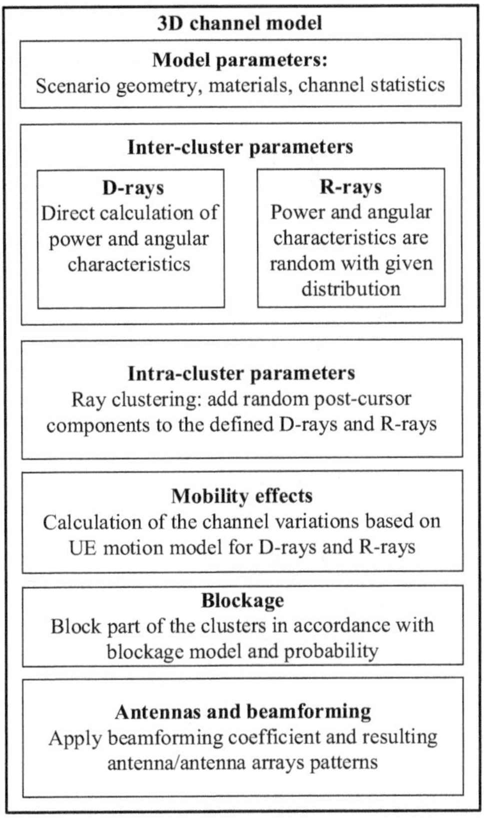

Figure 3.26 3D channel model generation

3.5.1 University campus access channel model

The university campus channel model represents the scenario with large open areas with low and spare buildings like university campus, park areas, city squares.

Modeling geometry and UE deployment

The geometry for Open area scenario is based on 3GPP Heterogeneous network scenario [46]. That scenario represents the deployment with hexagonal layout for LTE Macro nodes and a number of LTE low power nodes uniformly dropped within each macro geographical area hot zones. The scenario is extended by adding millimeter-wave BSs into the hot zones. Considering the use of directional antennas several millimeter-wave BSs should be placed with different orientations into the hot zone. The parameters of the Open Area scenario deployment are summarized in Table 3.11 and illustrated in Figure 3.27.

- Fix the total number of users, N_{users}, dropped within each macro geographical area.
- Randomly and uniformly drop the configured number of hotzone nodes, N_{HN}, within each macro geographical area (the same number N_{HN} for every macro geographical area).

Table 3.11 Open-area model parameters

Parameter	**Value**
Inter-site distance	500 m
Number of Macro cells per site	3
Number of hotzones per macro cell, N_{HN}	4
$d_{MCN\text{-}HN\_min}$	75 m
$d_{HN\text{-}HN\_min}$	40 m
Hotzone node height, H_{HN}	6 m
UE dropping	Clustered*
Number of users per macro cell, N_{users}	>100
Fraction of hotspot users, $P^{Hotspot}$	9/10
$r_{Hotzone}$	40 m
$d_{HN\text{-}UE\_min}$	5 m
$d_{MCN\text{-}UE\_min}$	35 m
UE height, H_{UE}	1.5 m
Number of millimeter-wave BS per hotzone	3
Surface material	Asphalt
Surface ε_r	4 + 0.2*j*
Surface roughness (standard deviation)	3 mm

*Clustered UE dropping:

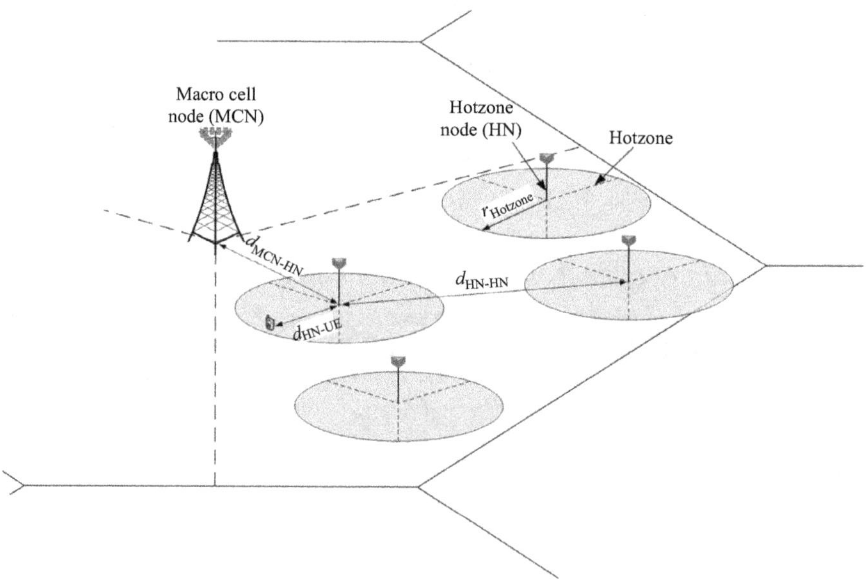

Figure 3.27 3GPP HetNet scenario with small cells

- Randomly and uniformly drop N_{users_HN} users within $r_{Hotzone}$ radius of each hotzone node, where $N_{users\_HN} = \lfloor P^{Hotspot} N_{users}/N_{HN} \rfloor$ and $P^{Hotspot}$ is a fraction of the hotspot users.
- Randomly and uniformly drop the remaining users, $N_{users} - N_{users\_HN} N_{HN}$ to the entire macro geographical area of the given macro cell (including the hotzone area).

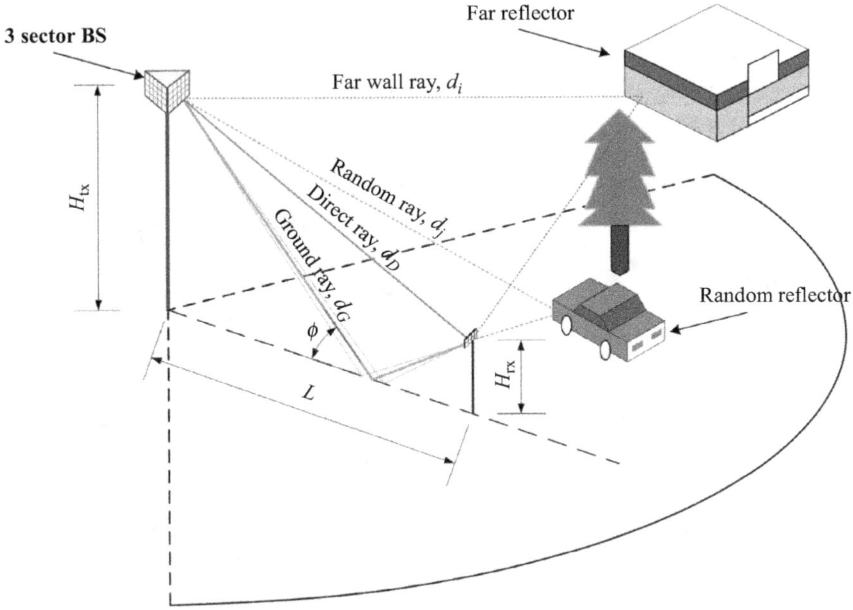

Figure 3.28 Open area millimeter-wave small cell environment

Open area D-rays and R-rays calculation

In the university campus (outdoor access larger area) scenario deployment the UEs have only two dominant rays: the direct ray and the ground reflected ray (Figure 3.28). These rays are counted as deterministic and explicitly calculated as specified in Section 3.4.2.

In addition to the main deterministic components the direct and ground rays, there are random components that represent reflection scattering. The reflection from the distant walls, random objects, and second-order reflection are taken into account as random components, and their parameters are summarized in Table 3.12.

Both D-rays and R-rays in the open-area channel model have thin cluster structure that adds post-cursor rays to the main D-ray and R-ray component. Although the direct LOS ray may also have clustered structure due to propagation path variations and partially closed by obstacles Fresnel zones, in the proposed model direct ray does not have clustering (see Table 3.13).

3.5.2 Street canyon access channel model

The street canyon (outdoor access ultra-high-rate hot-spots in a high-rise city environment) channel model represents typical urban scenario: city streets with pedestrian sidewalks along the high long buildings. The access link between the APs on the lampposts and the UEs at human hands is modeled in this scenario.

Modeling geometry and UE deployment

The geometry of the street canyon access scenario that is used for channel model parameters evaluation via ray-tracing simulations is shown in Figure 3.29. The corresponding numerical parameters are summarized in Table 3.14.

Table 3.12 Open area model R-rays parameters

Parameter	Value
Number of clusters, $N_{cluster}$	3
Cluster arrival rate, λ	0.05 ns^{-1}
Cluster power-decay constant, γ	15 ns
K-factor	6 dB
AOA	Elevation: U[AOA_{ground}:AOA_{LOS}] Azimuth: U[−60:60°]
AOD	Elevation: U[AOD_{ground}:AOD_{LOS}] Azimuth: U[−60:60°]

Table 3.13 Open area model intra-cluster parameters

Parameter	Value
Post-cursor rays K-factor, K	4 dB (NLOS only)
Post-cursor rays power decay time, γ	4.5 ns
Post-cursor arrival rate, λ	0.31 ns^{-1}
Post-cursor rays amplitude distribution	Rayleigh
Number of post-cursor rays, N	4

Street canyon D-rays and R-rays calculation

In the street canyon (outdoor access ultra-high-rate hot-spots) scenario deployment the UEs are grouped on a relatively narrow path, and with two dominant reflected rays in addition to the direct: the ground reflected ray and the wall-reflected ray. All those rays are counted as deterministic and explicitly calculated during the channel modeling.

The LOS ray calculation procedure is specified in Table 3.6, the ground-reflected ray calculation procedure is specified in Table 3.7, the wall-reflected ray parameters are determined in the same way as ground-reflected, but instead of the antenna height H_{tx} and H_{rx} the distances between the antennas and the nearest wall D_{tx} and D_{rx} are used.

In addition to deterministic components the direct and reflected rays, there are random components that represent reflection scattering. The reflection from distant walls and second-order reflections are taken into account as random components.

The random components of channel impulse response statistics are derived from the street canyon (outdoor access ultra-high-rate hot-spots) ray-tracing modeling in previous sections and from measurement data (see Table 3.15).

The cluster parameters for the street canyon model D-rays and R-rays are shown in Table 3.16.

3.5.3 Hotel lobby access channel model

The hotel lobby (indoor access large public area) channel model represents a typical indoor scenario: a large hall with multiple users within. Similar indoor

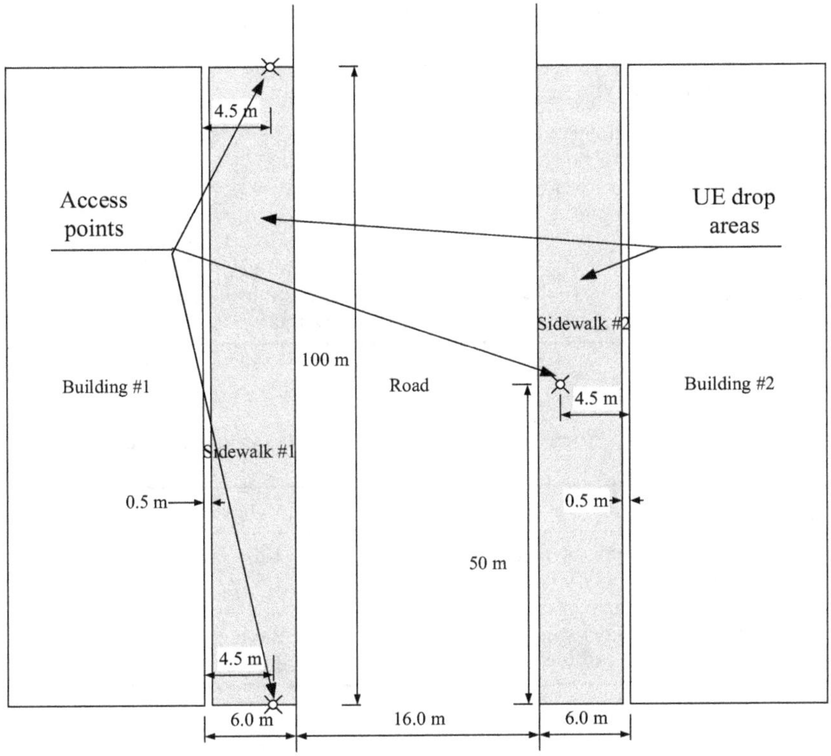

Figure 3.29 Street canyon (outdoor access ultra-high-rate hot-spots) scenario

Table 3.14 Street canyon (outdoor access ultra-high-rate hot-spots) scenario parameters

Parameter	Value
AP height, H_{tx}	6 m
UE height, H_{rx}	1.5 m
AP distance from nearest wall, D_{tx}	4.5 m
Sidewalk width	6 m
Road width	16 m
Street length	100 m
AP–AP distance, same side	100 m
AP–AP distance, different sides	50 m
Road and sidewalk material	Asphalt
Road and sidewalk ε_r	$4 + 0.2j$
Road and sidewalk roughness σ_g (standard deviation)	0.2 mm
Building walls material	Concrete
Building walls ε_r	$6.25 + 0.3j$
Building walls roughness σ_w (standard deviation)	0.5 mm

Table 3.15 Street canyon model random rays parameters

Parameter	Value
Number of clusters, $N_{cluster}$	5
Cluster arrival rate, λ	0.03 ns^{-1}
Cluster power-decay constant, γ	20 ns
Ray K-factor	10 dB
AoA	Elevation: U[−20:20°] Azimuth: U[−180:180°]
AoD	Elevation: U[−20:20°] Azimuth: U[−180:180°]

Table 3.16 Street canyon model intra-cluster parameters

Parameter	Value
Post-cursor rays K-factor, K	4 dB (NLOS only)
Post-cursor rays power decay time, γ	4.5 ns
Post-cursor arrival rate, λ	0.31 ns^{-1}
Post-cursor rays amplitude distribution	Rayleigh
Number of post-cursor rays, N	4

channel models were considered in [12,13], with statistical approach to channel modeling, suitable for link layer simulations. The proposed quasi-deterministic approach is based on the specified UE location and may be used also for system level simulations.

Modeling geometry and UE deployment

The basic parameters and geometry are summarized in Table 3.17 and illustrated in Figure 3.30.

Hotel lobby D-rays and R-rays calculation

The 3D channel model for hotel lobby scenario should include up to second-order reflection rays as D-rays, calculated based on the method of images and the Fresnel equations (Table 3.18). The exact formulas for all D-rays are too lengthy and it is recommended to use some ray-tracing algorithm for their calculation. In addition to deterministic reflections from the walls, ceiling and floor, R-rays represent reflections from other objects in the room that are not explicitly described in the scenario. The parameters of R-rays in the Hotel lobby channel model are summarized in Table 3.19.

The intra-cluster parameters for indoor access scenario are taken directly from the corresponding indoor scenario, developed in [29] and are based on the experimental measurements [14].

Table 3.17 Hotel lobby scenario parameters

Parameter	Value
AP height, H_{tx}	3.5 m
AP position	Middle of the nearest wall (see Figure 3.30)
UE height, H_{rx}	1.5 m
Room height	4 m
Room width	10 m
Room length	15 m
Floor material	Concrete
Floor ε_{rf}	$4 + 0.2j$
Floor roughness σ_f (standard deviation)	0.1 mm
Walls material	Concrete
Walls ε_{rw}	$4 + 0.2j$
Walls roughness σ_w (standard deviation)	0.2 mm
Ceiling material	Plasterboard
Ceiling ε_{rc}	$6.25 + 0.3j$
Ceiling roughness σ_c (standard deviation)	0.2 mm

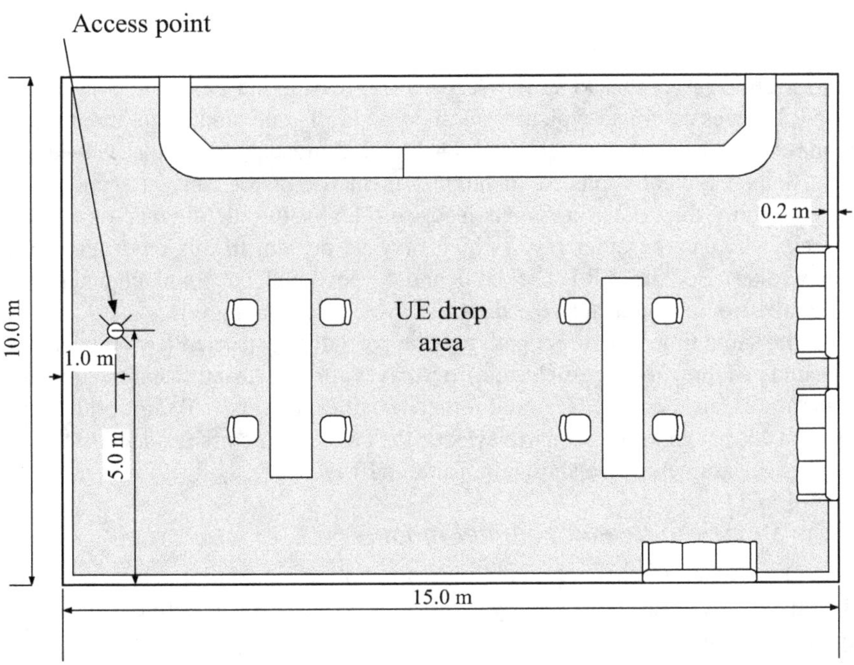

Figure 3.30 Hotel lobby (indoor access large public area) scenario

Table 3.18 Hotel lobby model random rays parameters

Parameter	Value
Number of clusters, $N_{cluster}$	5
Cluster arrival rate, λ	0.01 ns^{-1}
Cluster power-decay constant, γ	15 ns
Ray K-factor	10 dB
AoA	Elevation: U[−80:80°] Azimuth: U[−180:180°]
AoD	Elevation: U[−80:80°] Azimuth: U[−180:180°]

Table 3.19 Hotel lobby (indoor access large public area) model intra-cluster parameters

Parameter	Value
Post-cursor rays K-factor, K	10 dB
Post-cursor rays power decay time, γ	4.5 ns
Post-cursor arrival rate, λ	0.31 ns^{-1}
Post-cursor rays amplitude distribution	Rayleigh
Number of post-cursor rays, N	6

3.5.4 Backhaul channel models

Above rooftop (ART) backhauling scenario represents the backhaul link between two ART access points, typically with very high gain and high directionality antennas. Similarly, the street level backhaul represents the link between two access points, typically placed in lampposts in the street canyon scenario. The usage of highly directional antennas leads to the absolute dominance of the direct LOS ray, whereas the other rays (which may be present in this environment) are much weaker. So, the ART backhaul and Street level backhaul channel model should take into account only the direct LOS ray (see Table 3.6).

At the same time, the direct path may be partially obstructed (by objects within the bounds of the first Fresnel zone) or may vary due to atmosphere instability (temperature, air density, fog and other weather effects). To introduce those important for backhauling scenario effects, the cluster structure is added to the LOS ray. Cluster parameters are shown in Table 3.20.

3.5.5 Device to device channel models

The Q-D approach to channel modeling with explicit calculation of the major rays and statistical description of the rays with minor impact once defined for one scenario can be easily extended to scenarios with similar geometries, but slightly different parameters. For example, the model for access scenarios (with the TX height of about 6 m and RX height of about 1.5 m) can be successfully applied for

Table 3.20 Cluster parameters for ART backhaul scenario

Parameter	Value
Post-cursor rays *K*-factor, *K*	10 dB
Post-cursor rays power decay time, γ	10 ns
Post-cursor arrival rate, λ	0.03 ns^{-1}
Post-cursor rays amplitude distribution	Rayleigh
Number of post-cursor rays, *N*	4

D2D scenarios (both TX and RX have 1.5 m height) in Open area, Street canyon, or Hotel lobby scenarios.

3.6 Conclusion

In this chapter, a novel quasi-deterministic modeling methodology for millimeter-wave channels was proposed. The methodology is based on a number of previous experimental works and a few special measurement campaigns at 60 GHz performed during the MiWEBA project in different scenarios and environments. Since the 60 GHz band is exactly in the middle of the millimeter-wave frequency range of 30–90 GHz, the results presented in this work may be extrapolated in both directions, making the developed methodology applicable to this frequency range.

The street canyon measurement campaign was performed with omnidirectional antennas in combination with supporting ray tracing simulations in a real Berlin street environment. The measurement results were used for the methodology elaboration and model parameter evaluation. Other measurements were performed in the ideal open area environment (old airport) to study the LOS and ground reflected rays with the directional antennas setup. The last campaign was performed using directional antennas in a more realistic open area scenario of university campus. The parameters of ground reflection, scattering, and cross-polarization were estimated. An additional study was carried out for experimental investigation of the UE movement impact on the channel transfer function.

As a result, a new Q-D approach has been developed for modeling outdoor channels at 60 GHz. This methodology is based on the representation of the millimeter-wave channel impulse response function by a few strong scenario-specific deterministic rays (D-rays) and a certain number of relatively weak random rays (R-rays) that represent random or undefined components of the environment. The proposed methodology allows one to create the channel models for new scenarios using only basic assumptions on the environment geometry, environmental conditions, and user mobility. Flexible combination of the deterministic and statistical approaches has reasonable complexity and accuracy comparable with full site-specific deterministic models.

As an example of the Q-D methodology application, the channel models for open area (university campus), street canyon, and hotel lobby scenarios have been developed. The appropriate model parameters for access links were selected based on experimental measurements and ray tracing simulations. Explicit introduction of the deterministic D-rays and random R-rays within the Q-D channel modeling approach has allowed for proper description of the real dynamic outdoor environment, taking into account the blockage and mobility effects. The versatility of the Q-D methodology allows easy extension of the created channel models to other usage cases in the same environment. For example, the street access channel model can be modified for D2D and street-level backhaul links.

For now, the Q-D channel modeling methodology developed within the framework of the MiWEBA FP7 joint EU-Japan project is accepted as a basic approach for channel modeling in the upcoming IEEE 802.11ay standard. This standard includes not only new scenarios, but also a new usage model, such as SU-MIMO and MU-MIMO. The problem of enhancement of the Q-D model to support MIMO modes, ensuring proper spatial and frequency consistency is addressed in recent work [43,47], but require an additional studies and evaluations. Also, the channel models intra-cluster parameters should be refined for all specified scenarios and environments by additional experimental measurements using the equipment with higher angular and time resolution.

References

[1] Rappaport, T., Sun, S., Mayzus, R., *et al.*, "Millimeter wave mobile communications for 5G cellular: it will work!," *IEEE Access*, vol. 1, pp. 335–349, 2013.

[2] Boccardi, F., Heath, R. W., Lozano, A., Marzetta, T. L., and Popovski, P., "Five disruptive technology directions for 5G," *IEEE Communications Magazine*, vol. 52, no. 2, pp. 74–80, Feb. 2014

[3] Torkildson, E., Madhow, U., and Rodwell, M., "Indoor millimeter wave MIMO: feasibility and performance," *IEE Transactions on Wireless Communications*, vol. 10, no. 12, pp. 4150–4160, 2011.

[4] Maltsev, A., Sadri, A., Pudeyev, A., *et al.*, "mm-Wave small cells is a key technology for future 5G wireless communication systems," in *European Conference on Networks and Communications*, Bologna, 2014.

[5] Weiler, R. J., Peter, M., Keusgen, W., *et al.*, "Enabling 5G backhaul and access with millimeter-waves," in *European Conference on Networks and Communications*, Bologna, 2014, pp. 1–5.

[6] Azar, Y., Wong, G. N., Wang, K., *et al.*, "28 GHz propagation measurements for outdoor cellular communications using steerable beam antennas in New York city," in *IEEE International Conference on Communications (ICC)*, Budapest, 2013, pp. 5143–5147.

[7] Peter, M., Keusgen, W., and Felbecker, R., "Measurement and ray-tracing simulation of the 60 GHz indoor broadband channel: model accuracy and parameterization," in *The Second European Conference on Antennas and Propagation, 2007, EuCAP 2007*, Edinburgh, 2007.

[8] Hammoudeh, A., Sanchez, M., and Grindrod, E., "Experimental analysis of propagation at 62 GHz in suburban mobile radio microcells," *IEEE Transactions on Vehicular Technology*, vol. 48, no. 2, pp. 576–588, 1999.

[9] Hang, Z., Mayzus, R., Shu S, *et al.*, "28 GHz millimeter wave cellular communication measurements for reflection and penetration loss in and around buildings in New York city," in *IEEE International Conference on Communications (ICC)*, Budapest, 2013, pp. 5163–5167.

[10] Peter, M., Raceala-Motoc, M., Felbecker, R., Jacob, M., Priebe, S., and Kürner, T., "Analyzing human body shadowing at 60 GHz: systematic wideband MIMO measurements and modelling approaches," in *IEEE Sixth European Conference on Antennas and Propagation*, Prague, 2012, pp. 468–472.

[11] Maltsev, A., Maslennikov, R., Sevastyanov, A., *et al.* "Characteristics of indoor millimeter-wave channel at 60 GHz in application to perspective WLAN system," in *Antennas and Propagation (EuCAP)*, Barcelona, 2010, pp. 1–5.

[12] Maltsev, A., Erceg, V., Perahia, E., *et al.*, Channel Models, *IEEE* 802.11ad 09/0334r8.

[13] Maltsev, A., Maslennikov, R., Sevastyanov, A., Lomayev, A., and Khoryaev, A., "Statistical channel model for 60 GHz WLAN systems in conference room environment," in *Antennas and Propagation (EuCAP)*, Barcelona, 2010, pp. 1–5.

[14] "IMST 60 GHz indoor radio channel measurement data," ACTS MEDIAN project technical report.

[15] Sato, K., Manabe, T., Ihara, T., *et al.*, "Measurements of reflection and transmission characteristics of interior structures of office building in the 60-GHz band," *IEEE Transactions on Antennas and Propagation*, vol. 45, no. 12, pp. 1783–1792, 1997.

[16] Sawada, H., "Intra-cluster response model and parameter for channel modelling at 60 GHz (Part 3)," IEEE doc. 802.11-10/0112r1, Jan. 2010.

[17] Davydov, A., Maltsev, A., and Sadri, A., "Saleh-Valenzuela channel model parameters for library environment," IEEE document 802.15-06-0302-02-003c, Jul. 2006.

[18] METIS2020, "D1.2: initial channel models based on measurements," 2014, available at http://www.metis2020.com/documents/deliverables

[19] METIS2020, "Summary of deliverable D1.4: METIS channel models," 2015, available at http://www.metis2020.com/documents/deliverables

[20] Maltsev, A., Perahia, E., Maslennikov, R., Sevastyanov, A., Lomayev, A., and Khoryaev, A., "Impact of polarization characteristics on 60-GHz indoor radio communication systems," in *IEEE Antennas and Wireless Propagation Letters*, vol. 9, pp. 413–416, 2010.

[21] MiWEBA D1.1 "Definition of scenarios and use cases access and back-hauling and front hauling scenarios," available at http://miwiba.eu

[22] "NG 60 Use Cases," doc.: IEEE 802.11-15/0328r3, Mar. 2015.

[23] WINNER, "Spatial channel model for multiple input multiple output (MIMO) simulations," 3GPP TR 25.996 V6.1.0, September 2003. [Online]. Available at http://www.3gpp.org/ftp/Specs/html-info/25996.htm

[24] Khan, F., and Pi, Z., "Millimeter-wave mobile broadband (MMB): unleashing 3–300 GHz spectrum," in *IEEE Sarnoff Symposium*, 2011.

[25] Rappaport, T. S., Gutierrez, F., Ben-Dor, E., Murdock, J. N., Yijun, Q., and Tamir, J. I., "Broadband millimeter-wave propagation measurements and models using adaptive-beam antennas for outdoor urban cellular communications," *IEEE Transactions on Antennas and Propagation*, vol. 61, no. 4, pp. 1850–1859, 2013.

[26] Lebedew P., "Ueber die Dopplbrechung der Strahlen electrischer Kraft," *Annalen der Physik und Chemie*, vol. 56, no. 9, pp. 1–17, 1895.

[27] Bose, J., "On the determination of the wavelength of electric radiation by a diffraction grating," *Proceedings of the Royal Society*, vol. 60, pp. 167–178, 1897.

[28] Davydov, A., Maltsev, A., and Sadri, A., "Saleh-Valenzuela channel model parameters for library environment," IEEE document 802.15-06-0302-02-003c, Jul. 2006.

[29] ICT FP7 MiWEBA project #608637, "Deliverable D5.1, channel modelling and characterization," Public Deliverable, Intel Editor, Jun. 2014.

[30] Maltsev, A., Pudeyev, A., Karls, I., *et al.*, "Quasi-deterministic approach to mmWave channel modelling in a non-stationary environment," *IEEE GLOBECOM* 2014, Austin, TX, USA, pp. 966–971.

[31] Maltsev, A., Pudeyev, A., Karls, I., *et al.*, "Quasi-deterministic approach to mmWave channel modelling in the FP7 MiWEBA project," *WWRF'33*, 2014, Guildford, Great Britain.

[32] Maltsev, A., Sadri, A., Pudeyev, A., *et al.*, "Partially adaptive arrays application for MU-MIMO mode in a mmWave small cells," *IEEE PIMRC2015*, Hong Kong, 2015, pp. 315–319.

[33] Maltsev, A., Sadri, A., Pudeyev, A., *et al.*, "Performance evaluation of the isolated mmWave small cell," *IEEE PIMRC2015*, Hong Kong, 2015, pp. 2354–2358.

[34] Maltsev, A., Sadri, A., Pudeyev, A., *et al.*, "Performance evaluation of the mmWave small cells communication system in MU-MIMO mode," *EuCNC'2015*, Paris, France, 2015.

[35] Keusgen, W., Kortke, A., Peter, M., and Weiler, R., "A highly flexible digital radio testbed and 60 GHz application examples," *2013 European Microwave Conference (EuMC)*, 2013, pp. 740–743.

[36] Weiler, R. J., Peter, M., Keusgen, W., Kortke, A., and Wisotzki, M., "Millimeter-wave channel sounding of outdoor ground reflections," *2014 IEEE Radio and Wireless Symposium (RWS)*, San Diego, CA, 2015, pp. 95–97.

[37] Liebe, H. J., "MPM—an atmospheric millimeter-wave propagation model," *International Journal of Infrared and Millimeter Waves*, vol. 10, pp. 631–650, 1989.

[38] ITU Report 1008-1, Reflection from the surface of the Earth. Available at http://wwwa.itu.int/pub/R-REP-P.1008

[39] Hammoudeh, A., Sanchez, M., and Grindrod, E., "Modelling of propagation in outdoor microcells at 62.4 GHz," *Microwave Conference*, vol. 1, pp. 119–123, 1997.

[40] Sarabandi, K., Li, E., and Nashashibi, A., "Modelling and measurements of scattering from road surfaces at millimeter-wave frequencies," *IEEE Transactions on Antennas and Propagation*, vol. 45, no. 11, pp. 1679–1688, 1997.

[41] Yamamoto, A., Ogawa, K., Horimatsu, T., Kato, A., and Fujise, M., "Path-loss prediction models for intervehicle communication at 60 GHz," *IEEE Transactions on Vehicular Technology*, vol. 57, no. 1, 2008.

[42] IEEE P802.11 Wireless LANs, TGn Channel models, doc.: IEEE 802.11-03/940r4.

[43] Maltsev, A., Sadri, A. Pudeyev, A., Lomayev, A., and Bolotin, I., "Channel modeling in the next generation mmWave Wi-Fi: IEEE 802.11ay standard," European Wireless 2016, EU-JAPAN Collaboration Workshop.

[44] Alexander, R. M., "Simple models of human movement,' *Applied Mechanics Review*, vol. 48, pp. 461–470, 1995.

[45] "WiGig MAC and PHY specification, v 1.1," 2012. Available at http://www.wi-fi.org

[46] "3GPP TR 36.814 V9.0.0 (2010-03)," 3GPP, 2010. Available at http://portal.3gpp.org/desktopmodules/Specifications/SpecificationDetails.aspx?specificationId=2493

[47] Maltsev, A., Pudeyev, A., Gentile, C., *et al.*, "Q-D channel modeling application for specific IEEE 802.11ay indoor scenarios", WWRF38, Hsinchu, Taiwan, 2017.

Chapter 4
Large-scale antenna systems

Christos Masouros[1] and Adrian Garcia-Rodriguez[1,2]

In this chapter we overview one of the most prominent 5G candidate technologies, that is, Large-Scale Antenna Systems (LSAS). The stringent spectral efficiency requirements imposed by both the development of novel data-hungry communication applications and the increasing number of mobile devices have driven the design of communication systems in the past, in the present, and most likely in the foreseeable future [1–3]. The introduction of multiple-input multiple-output (MIMO) communication systems has revolutionised the approaches for enhancing the data speeds. Indeed, multi-antenna systems constitute the basis of current mobile communication standards such as Long-Term Evolution (LTE) [4]. However, 5G communication systems are not only expected to provide high data rates, but they should also do that in a power-efficient manner [1,3]. This has created an increasing research momentum on LSAS, where the cost reduction and improvements offered in both the hardware and signal processing components of base stations (BSs) motivate the addition of a larger number of antennas to further increase capacity, while reducing the overall power consumption.

LSAS, also commonly referred to as massive MIMO systems, aim at simultaneously improving the spectral and energy efficiency of current mobile standards by incorporating hundreds of antennas at the base stations (BSs) to serve dozens of users simultaneously [5–9]. Clearly, the employment of such a large number of antennas has a substantial number of implications at multiple design levels ranging from channel capacity implications, to hardware and signal processing implications such as the design of antenna arrays or the requirement of instantaneous channel state information (CSI) for attaining the promised enhancements. This wide range of implications motivate the contents of this chapter, which focuses on describing the fundamental characteristics of LSAS and places special emphasis on the most important practical implementation aspects. Specifically, Section 4.1 overviews the main pros and cons of LSAS: the central benefits promised by large antenna systems together with the challenges involved in their practical implementation and operation. Subsequently, Section 4.2 details the models conventionally employed

[1]Department of Electronic and Electrical Engineering, University College London, Torrington Place, London WC1E 7JE, UK
[2]Nokia Bell Labs, Clyde House, Business and Technology Park, Blanchardstown, Dublin 15, Ireland

for characterising massive communication systems and summarises the main results obtained in measurement campaigns. Lastly, Section 4.3 concludes this chapter by describing a number of relevant aspects that have to be considered in the optimisation and design of LSAS.

4.1 Promises and challenges of LSAS

Intuitively, increasing the number of antennas employed for communication generally improves the achievable rates due to the increased diversity and multiplexing gains provided [10–12]. However, the advantages envisaged by the employment of LSAS go beyond the above concept by exploiting the particular characteristics of the large-scale communication channels as detailed in the following. For ease of exposition, the LSAS principles are illustrated in this section by focusing on the transmission of signals from the mobile stations (MSs) towards the BS (uplink stage), since the fundamental concepts presented here are directly translatable to the downlink.

Consider a mobile communication system comprised of M cells as shown in Figure 4.1. In this generic system, each BS is equipped with N antennas and simultaneously serves K users with a single antenna each. The distinctive characteristic of LSAS with respect to systems currently deployed is that N can be in the order of several dozens or hundreds of antennas, i.e., the relationship $N \gg K$ is usually satisfied [5,7–9]. The complex narrowband baseband model of a frequency flat uncorrelated channel in the above system is typically written as

$$\mathbf{h}_{k,c,m} = \sqrt{d_{k,c,m}}\mathbf{g}_{k,c,m}, \tag{4.1}$$

where $\mathbf{h}_{k,c,m} \in \mathbb{C}^{N\times 1}$ represents the communication channel between the kth user positioned in the cth cell and the N antennas of the BS located in the mth cell. Moreover, $d_{k,c,m}$ characterises the large-scale fading component, which might include both path loss and shadow fading, whereas $\mathbf{g}_{k,c,m} \in \mathbb{C}^{N\times 1}$ denotes the small-scale fading channel component of the channel between the kth user located in the cth cell and the BS that provides coverage to the users in the m-cell [13]. The above channel model assumes that the large-scale fading component is identical for all

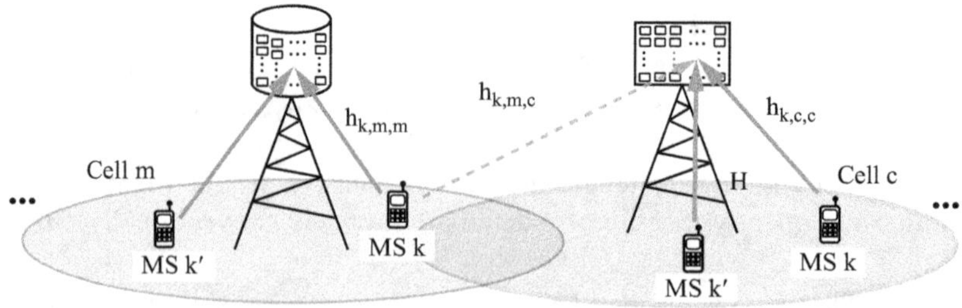

Figure 4.1 Illustration of a multicell transmission scenario

the antennas at the BS, which is a reasonable consideration when the dimensions of the antenna array are small when compared with the distance to the users. Note that, while we assume a single carrier transmission in this chapter for notational simplicity, this model can be readily extended to the widely deployed orthogonal frequency division multiplexing (OFDM). Indeed, $\mathbf{h}_{k,c,m}$ accurately characterises the narrowband propagation channel in a given subcarrier of an OFDM transmission [14].

It is often useful to employ the composite communication channel $\mathbf{H}_{c,m} \in \mathbb{C}^{N\times K}$ between the K MSs in cell c and the mth BS, defined as

$$\mathbf{H}_{c,m} = \left(\mathbf{h}_{1,c,m}, \mathbf{h}_{2,c,m}, \ldots, \mathbf{h}_{K,c,m}\right), \quad \forall\, c, m \in \{1, \ldots, M\}. \tag{4.2}$$

Similarly, in the following $\mathbf{D}_{c,m} = \mathrm{diag}\left(d_{1,c,m}, d_{2,c,m}, \ldots, d_{K,c,m}\right), \ \forall\, c, m \in \{1, \ldots, M\}$ and $\mathbf{G}_{c,m} = \left(\mathbf{g}_{1,c,m}, \mathbf{g}_{2,c,m}, \ldots, \mathbf{g}_{K,c,m}\right), \ \forall\, c, m \in \{1, \ldots, M\}$. Throughout this section, the entries of $\mathbf{G}_{c,m}$ are considered to be independent and identically distributed (i.i.d.) random variables satisfying $\mathbb{E}[\mathbf{G}_{c,m}\mathbf{G}_{c,m}^H] = \mathbf{I}_N, \ \forall\, c, m \in \{1, \ldots, M\}$. In the previous expression, $\mathbb{E}[\cdot]$ is the expectation operator and $\mathbf{I}_N$ represents the $N \times N$ identity matrix. Note that, at this stage, there is no assumption on a particular distribution from which the channel coefficients are derived.

The signal received by the cth BS, $\mathbf{y}_c \in \mathbb{C}^{N\times 1}$, at a given channel use for uplink symbol transmission is typically expressed as

$$\mathbf{y}_c = \sum_{m=1}^{M} \sqrt{\rho}\mathbf{H}_{c,m}\mathbf{x}_m + \mathbf{n}, \tag{4.3}$$

where $\mathbf{x}_m \in \mathbb{C}^{K\times 1}$ denotes the composite signal transmitted by the MSs in the mth cell and $\mathbf{n} \in \mathbb{C}^{N\times 1}$ represents the standard additive white Gaussian noise (AWGN) vector with entries satisfying $n_i \sim \mathcal{CN}(0, 1)$. The signal transmitted by the kth user is constrained to satisfy an average power constraint $\mathbb{E}[|x_{m,k}|^2] = 1$, although more complex power allocation schemes are typically implemented in practice [15]. Moreover, ρ characterises the ratio between the average transmission power and average noise power. In the previous expressions $\sim$ can be read as "distributed as" and $\mathcal{CN}(a, b)$ represents a random variable following a complex normal distribution with mean a and variance b [16]. The concepts presented in this section, unless otherwise stated, are introduced by considering a single-cell ($M = 1$) system when it suffices to reveal the particular insights of LSAS and the subindices c and m are omitted for simplicity. Below we discuss the key promises from the theoretical study of LSAS, along with major challenges in their practical deployment.

4.1.1 LSAS promises

4.1.1.1 Favourable propagation and channel hardening

The spectral efficiency gains offered by LSAS essentially rely on two propagation phenomena that occur when N grows large: the so-called favourable propagation and channel hardening effects. It has been shown that in the asymptotic domain, as

N grows large with respect to K, an LSAS with $N \gg K$ experiences favourable propagation conditions [5,7]; that is

$$\mathbf{h}_k^H \mathbf{h}_{k'} = \sqrt{d_k} \mathbf{g}_k^H \mathbf{g}_{k'} \sqrt{d_{k'}} \underset{N \gg K}{\approx} 0, \ \forall k, k' \in \{1, \ldots, K\} \quad \text{and} \quad k \neq k'. \tag{4.4}$$

and

$$\mathbf{h}_k^H \mathbf{h}_k = \sqrt{d_k} \mathbf{g}_k^H \mathbf{g}_k \sqrt{d_k} \underset{N \gg K}{\approx} N d_k, \forall k \in \{1, \ldots, K\}. \tag{4.5}$$

In the above expressions $(\cdot)^H$ denotes the Hermitian conjugate vector or matrix. Note that the above two expressions entail that:

$$\mathbf{H}^H \mathbf{H} = \mathbf{D}^{1/2} \mathbf{G}^H \mathbf{G} \mathbf{D}^{1/2} \approx N\mathbf{D}. \tag{4.6}$$

In plain words, favourable propagation conditions are experienced when the communication channels of different users become orthogonal as per (4.4) and the small-scale fading component vanishes for $\mathbf{h}_k^H \mathbf{h}_k$ as shown in (4.5). Physically, channel orthogonality appears as a result of having a large number of antennas at the BS and can be interpreted as follows: the closer to orthogonal the user channels become, the larger the disparity in the communication channels of different users will be. This facilitates the spatial separation of the data streams transmitted by different users at the BS side, since their channels become distinct. Indeed, the favourable propagation conditions have a direct impact on the uplink capacity of the multi-user MIMO system, defined as C_u. Specifically, the uplink capacity measured in bits per second (bits/s) is given by [10]

$$\begin{aligned} C_u &= B \log_2 \det\left(\mathbf{I}_K + \rho \mathbf{H}^H \mathbf{H}\right) \approx B \log_2 \det(\mathbf{I}_K + N\rho \mathbf{D}) \\ &= B \sum_{k=1}^{K} \log_2(1 + N\rho d_k) \text{bits/s}, \end{aligned} \tag{4.7}$$

where B denotes the system bandwidth in Hertz (Hz) and $\det(\cdot)$ is the determinant operator [17]. The above expression is obtained by assuming that the LSAS BS has a perfect knowledge of the channel matrix $\mathbf{H}$. Remarkably, the last equality in (4.7) does not depend on the small-scale fading coefficients, g_k, which occurs due to the favourable propagation conditions shown in (4.6). In plain words, the effects of the small-scale fading are averaged out, which is one of the main motivations behind the deployment of LSAS in 5G systems.

The last equality of the uplink capacity (4.7) also reveals its direct dependence on the large-scale fading coefficients, d_k, which typically vary at a much slower rate than the small-scale fading parameters [18]. Indeed, the variations of the achievable rates over time are conventionally regarded as a sole function of the small-scale fading coefficients [19,20]. Since these parameters do not influence the expression in (4.7), the achievable rates experience reduced temporal variations as the number of BS antennas grows, a phenomenon referred to as *channel hardening*. In other words, the instantaneous achievable rates approximate the ergodic

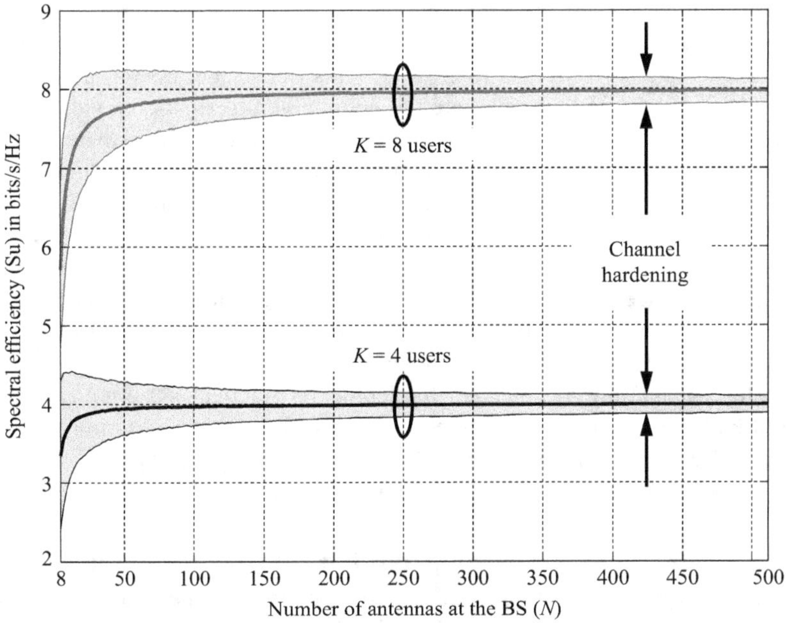

Figure 4.2 Illustration of the channel hardening effect. Spectral efficiency (S_u) vs. number of BS antennas (N) for K = {4,8} and $\rho = 1/N$. Marked lines denote the average spectral efficiency; shaded areas denote the 90% confidence intervals of the instantaneous spectral efficiency

achievable rates [21]. Overall, the channel hardening phenomenon indicates that the deployment of LSAS at the BS of cellular systems may facilitate coherent user data rates throughout long periods of time by removing the variations introduced by the small-scale fading effects, hence improving the users' quality of experience (QoE).

Let us study the above observation by using an example shown in Figure 4.2. Here, the evolution of the average spectral efficiency, $S_u = C_u/B$, is shown for an increasing number of BS antennas for the cases of $K = 4$ and $K = 8$. This is denoted by the bold solid lines in the figure. Figure 4.2 also shows the 90% confidence intervals, which bound the region containing 90% of the instantaneous spectral efficiency values. These are denoted by the shaded areas around the solid lines for each of the $K = \{4, 8\}$ cases. For reasons of illustration, here we consider a Rayleigh fading channel, i.e., the ith entry of the channel vector follows $h_{k,i} \sim \mathcal{CN}(0, 1)$, and omit the large-scale fading effect ($\mathbf{D} = \mathbf{I}_K$). Moreover, $\rho = 1/N$, i.e., the transmission power is independent on the number of transmit antennas N, but grows proportionally with the number of users K. The results of Figure 4.2 show that the 90% confidence intervals, i.e., the thickness of the shaded areas, are reduced as N grows for both $K = \{4, 8\}$, which entails that the user rates vary less with respect to their mean, as expected by the channel hardening effect.

Additionally, it can be observed that the maximum average rates are approximately achieved for the reasonable finite values of $N \approx 50$ for the case of $K = 4$, whereas $N \approx 100$ antennas are required for $K = 8$. This occurs because a larger number of antennas is required to approximate the favourable propagation conditions when K grows, a trend that is also consistent when the downlink is considered.

The above results show that both the favourable propagation and channel hardening effects promised by LSAS have a clear positive impact on the communication rates. However, the system complexity in terms of both hardware and software should also be considered when characterising the benefits attainable by LSAS. For instance, the significant increase in the number of antennas could enormously complicate the signal processing required at the BS for symbol transmission (precoding) and detection, which are essential for experiencing both favourable propagation and channel hardening in practical systems. However, it turns out that the simplest linear precoders and detectors become optimal with an asymptotically large number of BS antennas, which simplifies the operation of LSAS. We discuss this aspect in the following section.

4.1.1.2 Linear signal processing and ideal power scaling law

The baseband signal processing in the communication link is typically carried out in the form of precoding and signal detection [12,20,22–24]. Specifically, precoding schemes are employed at the BS during downlink transmission. They aim to shape the transmitted symbols at the BS according to the channel coefficients, such that the information symbols can be accurately detected by simple processing at the MSs [19,22]. Similarly, detection schemes are employed for the BS during uplink transmission to determine the original symbols transmitted by the user terminals [12]. The above procedure offers the beneficial effect of shifting the signal processing complexity to the BSs, thus simplifying the complexity at the user terminals, where the computational capabilities and the energy supply are more constrained. Precoding and detection schemes can be classified into linear and non-linear depending on the signal processing operations used, which further impacts their signal processing complexity [22]. In general, a fundamental trade-off exists between these, i.e., non-linear precoding and detection schemes typically offer an improved performance when compared with the linear ones at the cost of requiring a more complicated signal processing. However, things change in the realm of LSAS, where even the simplest linear precoding and detection schemes become asymptotically optimal. We show this explicitly in the following, for the representative case of uplink transmission.

By applying a linear detection operation on the received signal of (4.3), expressed by the detection matrix $\mathbf{W} \in \mathbb{C}^{K\times N}$, the output signal can be expressed as

$$\mathbf{r} = \mathbf{W}\mathbf{y} = \mathbf{W}\left(\sqrt{\rho}\,\mathbf{H}\mathbf{x} + \mathbf{n}\right), \tag{4.8}$$

where $\mathbf{r} \in \mathbb{C}^{K\times 1}$ is a column vector containing estimates of the K symbols conveyed by the single-antenna user terminals, i.e., r_k represents the estimate of the symbol transmitted by the kth user terminal in a given channel use. Note that for notational simplicity we have assumed a single cell scenario, i.e., $m = 1$ in (4.3). Let us consider the simplest case where $\mathbf{W} = \mathbf{H}^H$, which is commonly referred to as matched filtering (MF) or maximum ratio combining (MRC) detection [20,23]. Then, the post-processed signal, under the favourable propagation conditions of (4.6), can be expanded as

$$\mathbf{r} = \mathbf{H}^H \left(\sqrt{\rho}\,\mathbf{H}\mathbf{x} + \mathbf{n}\right) \approx N\sqrt{\rho}\,\mathbf{D}\mathbf{x} + \mathbf{H}^H\mathbf{n}. \tag{4.9}$$

The first term in the final summation of (4.9), $N\sqrt{\rho}\mathbf{D}\mathbf{x}$, corresponds to the signal component, whereas the second term, $\mathbf{H}^H\mathbf{n}$, represents the noise component resulting after linear detection. Note that $\mathbf{H}^H$ is statistically independent of $\mathbf{n}$. Therefore, the signal-to-noise ratio (SNR) of the kth user can be expressed as

$$\mathrm{SNR}_k = \frac{\mathbb{E}\left[|N\sqrt{\rho}d_k x_k|^2\right]}{\mathbb{E}\left[|\mathbf{h}_k^H\mathbf{n}|^2\right]} = \frac{N^2\rho d_k^2\mathbb{E}\left[|x_k|^2\right]}{d_k\mathbb{E}\left[|\mathbf{g}_k^H\mathbf{n}|^2\right]} = N\rho d_k, \tag{4.10}$$

where the expectation is taken over the small-scale fading components $\mathbf{g}_k$, the transmitted symbols $\mathbf{x}_k$ and the noise component $\mathbf{n}$. We remark that this is justified by the fact that the spatial channel statistics, which include the large-scale fading components, vary at a slower rate (several channel coherence periods [18]) when compared with the previous factors, which change at a rate comparable to the symbol rate. Note that the last equality in (4.10) follows from the assumption that $\mathbb{E}[|x_k|^2] = 1$ and from the favourable propagation characteristics, discussed in the previous section. Indeed, for favourable propagation conditions, we have $\mathbb{E}[|\mathbf{g}_k^H\mathbf{n}|^2] = N$ for the unit variance noise assumed in the model. The above SNR produces an uplink rate per user of:

$$R_k = B\log_2(1 + \mathrm{SNR}_k) = B\log_2(1 + N\rho d_k)\mathrm{bits/s}. \tag{4.11}$$

Note that the uplink rate per user R_k with MF detection is identical to the maximum achievable rate that the kth user experiences under favourable propagation conditions shown in (4.7). In other words, the application of the simplest signal processing strategy can theoretically achieve the maximum rates under favourable propagation conditions, which constitute one of the most remarkable aspects of LSAS. From the physical perspective, the resulting intuition is that the large number of antennas at the BS enables concentrating the energy on the users' particular spatial directions [7].

The result in (4.10) reveals another notable aspect of LSAS: the linear increase of the effective SNR with the number of antennas at the BS N, when an MF with perfect CSI is employed for detection. In other words, the required transmission power to achieve a certain SNR scales inversely proportionally with the number of antennas N at the BS, under favourable propagation conditions. This is referred to as the ideal *power scaling law* of LSAS that can only be achieved when perfect channel state information (CSI) is available at the BS and it straightforwardly generalises to other precoding and detection schemes [20,23]. The power scaling law also has an important implication in the hardware design of LSAS: since the power employed for transmission can be reduced to achieve the performance of a system with a smaller number of antennas, the transceivers can be constructed with radio frequency (RF) components of a lower dynamic range and can therefore be cheaper and more energy efficient. Furthermore, the reliability of LSAS is also enhanced when compared with their small-scale counterparts, since the failure of one of the multiple RF transceivers does not significantly affect the performance of the overall system.

The above results are, however, obtained under the idealistic assumptions of favourable propagation conditions and perfect CSI at the BS. In this context, the required number of antennas for which the favourable propagation conditions take effect have not been clearly specified up to this point. Moreover, it is well known that different precoding and detection strategies provide different attainable rates when a finite number of antennas is considered at the BS [19,20,23]. This aspect can be explicitly observed in Figure 4.3, where the achievable rates of different linear detection schemes are illustrated in comparison with the channel's achievable spectral efficiency, for increasing numbers of BS antennas N, $K = 8$ users and $\rho = 10/N$ in (4.3). The detection schemes considered are the already introduced MRC ($\mathbf{W} = \mathbf{H}^H$), the zero-forcing (ZF) detector ($\mathbf{W} = \mathbf{H}^\dagger = (\mathbf{H}^H\mathbf{H})^{-1}\mathbf{H}$, [12,23]) and the minimum mean squared error (MMSE) detector ($\mathbf{W} = \left(\mathbf{H}^H\mathbf{H} + \xi\mathbf{I}_K\right)^{-1}\mathbf{H}^H$, [12,23]). In the previous expressions, $(\cdot)^\dagger$ and $(\cdot)^{-1}$ denote the pseudoinverse and

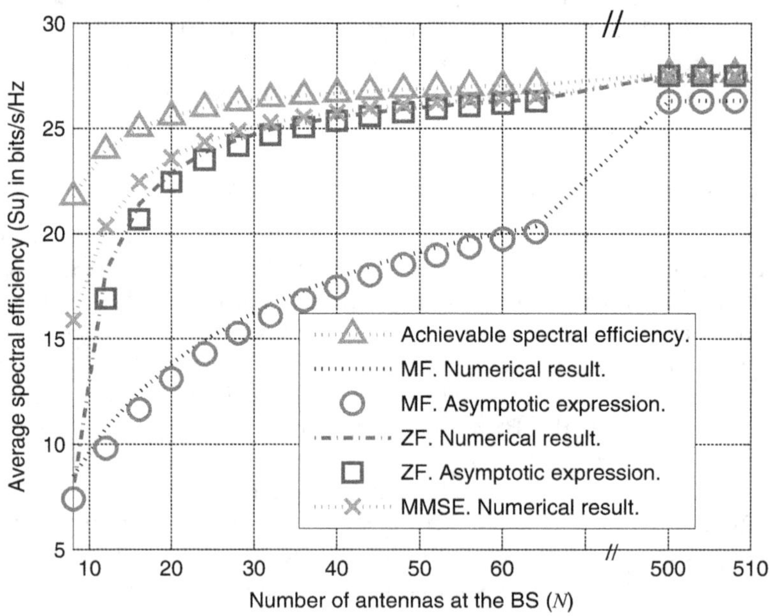

Figure 4.3 Average spectral efficiency (S_u) for different linear detection schemes and increasing number of BS antennas (N). $K = 8$ users and $\rho = 10/N$

inverse of a matrix, respectively, and $\xi > 0$ is a constant that generally depends on the system SNR [20,23,25]. Additionally, the numerical results are compared with the simple asymptotic expressions developed in [23].

The results shown in Figure 4.3 illustrate that linear precoding schemes perform close to the maximum sum spectral efficiency as N grows large. Specifically, it can be seen that ZF and MMSE detection schemes closely approximate the attainable sum spectral efficiency at lower values of N antennas at the BS, while MF detection becomes close to optimal at higher values of N. Figure 4.3 also shows that the asymptotic approximations accurately characterise the average spectral efficiency for a reasonably small number of BS antennas N; a constant in the theoretical results derived for LSAS [19,20,23].

Notably, the results derived in this section assume that perfect knowledge of the channel coefficients is available at the BS. Since the collection of accurate CSI is one of the most important aspects of LSAS, in the following we describe the CSI acquisition procedure as well as its impact on the resulting performance.

4.1.2 LSAS challenges

4.1.2.1 Acquisition of channel state information in LSAS

The acquisition of instantaneous CSI is essential in LSAS for exploiting the excessive number of antennas and reaching the favourable propagation conditions with linear precoding and detection techniques [26]. This procedure has not been considered in the previous sections where, for illustrative reasons, it was assumed that accurate CSI is available at the large BS at no cost. However, this idealistic assumption does not hold in real systems, since part of the channel coherence time must be employed for CSI acquisition and imperfect channel estimates are inevitably produced as a result of this procedure. Indeed, the large number of antennas deployed in LSAS promotes a change of philosophy in the CSI acquisition. We detail this in the following.

Throughout this chapter we consider a block-fading channel model, where the channel remains constant during a number of symbol periods that coincide with the frame length and varies independently between frames [20]. The limited number of symbols over which the channel can be considered constant, $\tau = B_c T_c$, determines the time that the BS can spend for CSI acquisition before data transmission and reception. In the previous expression, B_c denotes the channel coherence bandwidth and T_c denotes the channel coherence time [27]. The limited channel coherence time is the reason why the majority of LSAS research has focused on time division duplex (TDD) transmission of Figure 4.4(a), where both the data transmission and reception are carried out in the same frequency band [5,7,28]. TDD is favoured for LSAS because the training interval length τ_t is proportional to the number of active terminals K and is instead independent of the large number of antennas deployed at the BSs. This contrasts with frequency division duplex (FDD) systems of Figure 4.4(b) where, under rich scattering conditions, the training and feedback lengths grow with N [29]. Owing to the uplink–downlink channel reciprocity in TDD transmission [30], the downlink channel can be accurately estimated during uplink

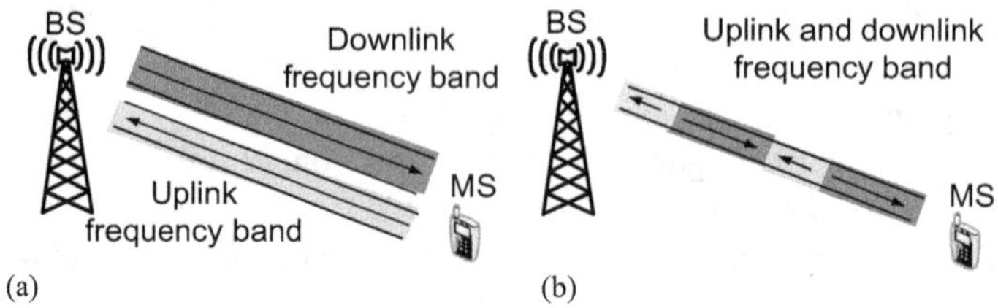

Figure 4.4 Uplink–downlink duplexing approaches. (a) Frequency division duplex (FDD) transmission vs. (b) time division duplex (TDD) transmission

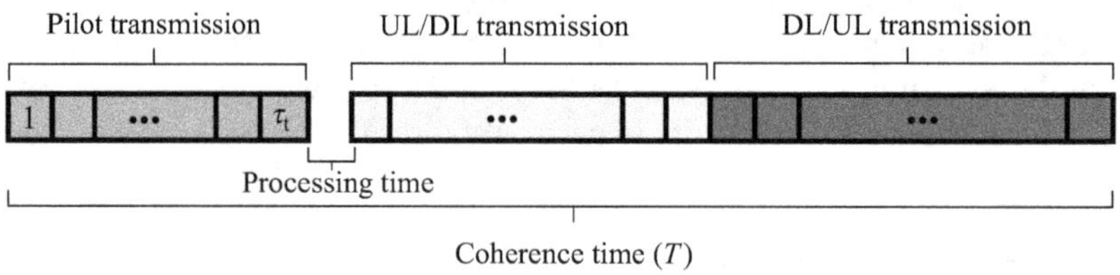

Figure 4.5 Basic structure of the TDD operation protocol

CSI acquisition, provided that the transmit and receive communication ends are accurately calibrated [5,31,32]. Explicitly, under the latter assumption it follows that $\mathbf{H}^d = (\mathbf{H}^u)^H$, where the superindices 'd' and 'u' denote downlink and uplink, respectively.[1] Accordingly, the TDD operation facilitates an *open loop* CSI acquisition, which makes it particularly attractive for LSAS deployments. While FDD operation has also attracted considerable research interest in the recent LSAS literature, owing to its above advantage we focus on TDD transmission in the following.

Under TDD operation, the channel coherence interval T_c, which generally depends on the speed at which the user terminals are moving [10], is divided into three phases: pilot training, data transmission and data reception. A simplified illustration of the operation protocol of TDD systems is shown in Figure 4.5. Initially, the BS acquires the CSI via uplink pilots transmitted from the user terminals in a process that occupies $\tau_t \geq K$ consecutive symbol periods. Once the pilot information is retrieved, the BS then requires a short processing time in which the CSI is computed from the pilot signals received in the previous phase. The condition $\tau_t \geq K$ in the CSI acquisition process allows for orthogonality between the pilot sequence of the kth user, $\theta_k = \sqrt{\rho_{\text{tr}}}[\theta_k(1), \ldots, \theta_k(\tau_t)]^T$ and the pilot signals of the rest of the intra-cell users. In the previous expression, $\rho_{\text{tr}} = \rho\tau_t$ represents the effective training SNR that depends on both the time spent for uplink

[1] $\mathbf{H}^d = (\mathbf{H}^u)^T$ actually holds in realistic propagation scenarios, since there is no physical phenomenon that influences the phase of the complex frequency coefficients depending on the direction of communication. However, the consideration $\mathbf{H}^d = (\mathbf{H}^u)^H$ adopted in this chapter has been commonly employed and does not have any impact on the results [20].

pilot transmission and the normalised transmit SNR of the MSs. Mathematically, the training pilots should ideally satisfy:

$$\theta_k^H \cdot \theta_{k'} = \begin{cases} 1, & \text{if } k = k', \\ 0, & \text{if } k \neq k'. \end{cases} \tag{4.12}$$

The signal received at the BS after the pilot training phase $\mathbf{Y}_{\text{tr}} \in \mathbb{C}^{N \times \tau_t}$ in a single cell is given by

$$\mathbf{Y}_{\text{tr}} = \sqrt{\rho_{\text{tr}}}\mathbf{H}\Theta + \mathbf{N}, \tag{4.13}$$

where $\Theta \in \mathbb{C}^{K \times \tau_t}$ is defined as $\Theta = [\theta_1, \ldots, \theta_K]^T$ and $\mathbf{N} \in \mathbb{C}^{N \times \tau_t}$ is an AWGN matrix formed by independent and identically distributed (i.i.d.) standard Gaussian entries. The received pilots are then processed by correlating them with the pilots that were known at the BS. The resultant channel estimate $\hat{\mathbf{H}} \in \mathbb{C}^{N \times K}$ can then be expressed as [33]

$$\hat{\mathbf{H}} = \mathbf{Y}_{\text{tr}}\Theta^H = \left(\sqrt{\rho_{\text{tr}}}\mathbf{H}\Theta + \mathbf{N}\right)\Theta^H = \sqrt{\rho_{\text{tr}}}\mathbf{H} + \mathbf{N}\Theta^H. \tag{4.14}$$

where the final equality follows from $\Theta\Theta^H = \mathbf{I}_K$ as per (4.12). Note that the noise distribution and its statistical properties are not affected by the above operation, due to Θ being unitary. The above decision statistic, $\hat{\mathbf{H}}$, can be subsequently employed to estimate the communication channel via traditional estimation methods. For instance, the commonly employed MMSE estimate of (4.14) is given by [20,23,34,35]

$$\tilde{\mathbf{H}} = \left(\mathbf{H} + \frac{1}{\sqrt{\rho_{\text{tr}}}}\mathbf{N}\right)\hat{\mathbf{D}}, \tag{4.15}$$

where $\hat{\mathbf{D}} = \left(\frac{1}{\sqrt{\rho_{\text{tr}}}}\mathbf{D}^{-1} + \mathbf{I}_K\right)^{-1}$ is a diagonal matrix. It should be noted that the employment of the realistic imperfect CSI estimates harms the ideal behaviour of LSAS with perfect CSI described in Section 4.1.1, where it was highlighted that the effective SNR experienced a linear growth with N as shown in (4.10). Instead, it can be shown that the effective SNR grows proportionally to $\sqrt{N}$ when the resultant MMSE estimate (4.15) is employed for linear precoding and detection [20,23].

The channel acquisition process can limit the global performance of practical LSAS due to the necessity of satisfying $\tau_t \geq K$ for obtaining precise intra-cell CSI. For instance, a coherence period comprised of $\tau \approx 200$ symbols can be expected for scenarios with high-mobility users or high frequencies [26]. This reduced coherence period might prevent the assignment of orthogonal pilot sequences to all intra-cell users, hence inducing a significant error in the estimated channel, which in turn affects the achievable rates for transmission and reception. However, when orthogonality between every pair of user pilot sequences is preserved, the employment of the MMSE estimator allows deriving one of the most remarkable theoretical results of LSAS: Both the thermal noise and the imperfect CSI due to channel estimation asymptotically vanish ($N \rightarrow \infty$) when linear precoders and detectors are employed [5,23]. Intuitively, this occurs because both the noise and the channel estimation error are independent of the available channel estimate $\tilde{\mathbf{H}}$ [23]. In other words, the performance of a single-cell LSAS with simple linear precoding and detection schemes is asymptotically optimal, even under imperfect CSI. Naturally, this condition does not occur in practice, since the number of antennas is finite and the BSs are distributed in different cells. Therefore, it is natural to question whether this ideal condition would still be satisfied in multi-cell scenarios with an asymptotically large number of antennas. We study this aspect in the following.

4.1.2.2 LSAS in multi-cell systems: pilot contamination

The coherence time is generally not long enough to ensure orthogonality between the pilot sequences of users located in different cells. This forces to employ non-orthogonal pilot sequences that generate the so-called pilot contamination effect, which is conventionally regarded as the ultimate performance limit of LSAS [7]. Pilot contamination appears when a given BS m—which would only ideally receive channel information for the intra-cell users—acquires inaccurate CSI during the training phase, because it cannot fully decorrelate the pilot signals generated from the users of another cells that share the same pilot sequence. Indeed, the worst-case scenario where identical pilot sequences are simultaneously re-used in all cells is usually considered in the literature [5,20,36]. Physically, the pilot contamination effect results in that the affected BSs partially beamform their signals to the mobile users placed in other cells that share the same pilot sequence. In other words, an inaccurate estimation of the intra-cell channel generates inter-cell interference as illustrated in Figure 4.6(b) and described in the following.

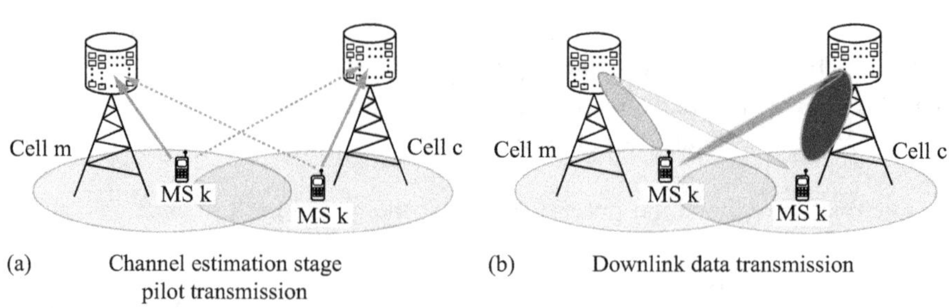

Figure 4.6 Illustration of the pilot contamination effect. (a) The user terminals sharing the same pilot sequence interference with each other in the CSI acquisition stage. (b) The BSs partially beamform their signals towards the user terminals that created pilot contamination during the CSI acquisition stage

Following a procedure similar to the derivation of (4.14), the decision metrics obtained at the cth BS after pilot correlation in a multi-cell system can be expressed as [23,33]

$$\hat{\mathbf{H}}_{c,c} = \sqrt{\rho_{\mathrm{tr}}}\,\mathbf{H}_{c,c} + \sqrt{\rho_{\mathrm{tr}}}\sum_{m\neq c}\mathbf{H}_{m,c} + \mathbf{N}\Theta^H, \tag{4.16}$$

where $\mathbf{H}_{m,c}$ denotes the global uplink channel matrix as defined in (4.2). The resulting MMSE estimate is given by [20,23,35]

$$\tilde{\mathbf{H}}_{c,c} = \left(\sum_{m=1}^{M}\mathbf{H}_{m,c} + \frac{1}{\sqrt{\rho_{\mathrm{tr}}}}\mathbf{N}\Theta^H\right)\hat{\mathbf{D}}_c, \tag{4.17}$$

where $\hat{\mathbf{D}}_c \in \mathbb{C}^{K\times K}$ is a diagonal matrix with kth diagonal element given by $\hat{d}_k = d_{k,c,c}\left(\sum_{m=1}^{M} d_{k,m,c} + \frac{1}{\rho_{\mathrm{tr}}}\right)^{-1}$. Here, $d_{k,m,c}$ represents the large-scale fading component between the kth user positioned in the cth cell and the N antennas of the BS located in the mth cell. Note that (4.17) explicitly includes the communication channels from other cells $m \neq c$.

To illustrate the insights of the pilot contamination in the uplink we employ the simplest detection strategy: the MF. This linear detector is able to reach the asymptotic sum rates in the uplink of LSAS under ideal conditions as shown in Section 4.1.1 and its simplicity contributes to understand the effect of the pilot contamination in multi-cell scenarios. Following (4.3), consider a multi-cell system comprised of M cells. The signal received at the cth BS in such a system is given by

$$\mathbf{y}_c = \sqrt{\rho}\,\mathbf{H}_{c,c}\mathbf{x}_c + \sqrt{\rho}\sum_{m\neq c}\mathbf{H}_{m,c}\mathbf{x}_m + \mathbf{n}_c. \tag{4.18}$$

Note that the effect of the intercell interference is explicitly shown in (4.18) as each BS receives information from the user terminals of all M cells. When MF is applied at reception, the resultant decision vector at the BS of the cth cell can be expressed as

$$\begin{aligned}
\mathbf{r}_c &= \tilde{\mathbf{H}}_{c,c}^H\mathbf{y}_c = \tilde{\mathbf{H}}_{c,c}^H\left[\sum_{m_2=1}^{M}\sqrt{\rho}\,\mathbf{H}_{m_2,c}\mathbf{x}_{m_2} + \mathbf{n}_c\right] \\
&= \hat{\mathbf{D}}_c\left(\sum_{m_1=1}^{M}\mathbf{H}_{m_1,c} + \frac{1}{\sqrt{\rho_{\mathrm{tr}}}}\mathbf{N}_c\right)^H\left(\sum_{m_2=1}^{M}\sqrt{\rho}\,\mathbf{H}_{m_2,c}\mathbf{x}_{m_2} + \mathbf{n}_c\right) \\
&= \hat{\mathbf{D}}_c\Bigg[\sqrt{\rho}\sum_{m_1=m_2}\mathbf{H}_{m_1,c}^H\mathbf{H}_{m_2,c}\mathbf{x}_{m_2} + \sum_{m_1=1}^{M}\mathbf{H}_{m_1,c}^H\mathbf{n}_c + \frac{1}{\sqrt{\rho_{\mathrm{tr}}}}\mathbf{N}_c^H\mathbf{n}_c \\
&\quad + \frac{1}{\sqrt{\rho_{\mathrm{tr}}}}\mathbf{N}_c^H\sum_{m_2=1}^{M}\sqrt{\rho}\mathbf{H}_{m_2,c}\mathbf{x}_{m_2} + \sqrt{\rho}\sum_{m_1\neq m_2}\mathbf{H}_{m_1,c}^H\mathbf{H}_{m_2,c}\mathbf{x}_{m_2}\Bigg],
\end{aligned} \tag{4.19}$$

where $m_1, m_2 \in \{1, \ldots, M\}$ enumerate the cell indices and the MF matrix is constructed using the channel estimate obtained after MMSE estimation in (4.17). As N grows large, products of the form $\mathbf{H}^H_{m_1,c}\mathbf{n}_c$ in (4.19) can be considered negligible since the noise is uncorrelated to the communication channel [5]. A similar argument applies for the products $\mathbf{N}^H_c\mathbf{H}_{m_2,c}\mathbf{x}_{m_2}$. Moreover, products of the form $\mathbf{H}^H_{m_1,c}\mathbf{H}_{m_2,c}, \ \forall\, m_1 \neq m_1$ are much smaller than the correlated products $(m_1 = m_2)$ for large N, which instead remain significant. The pilot contamination effect is caused by the interference terms of the form $\mathbf{H}^H_{m_1,c}\mathbf{H}_{m_2,c}, m_1 = m_2 \neq c$ that do not vanish for large N and are caused by the mobile stations that share the same pilot sequences as shown in Figure 4.6(a) [5,23]. These provoke a partial beamforming of the signals towards the mobile stations located in other cells as shown in Figure 4.6(b).

The study of the pilot contamination effect has attracted considerable attention from the research community since it was first identified as the main limiting performance phenomenon in LSAS [5,20,28,33]. Indeed, a variety of strategies for mitigating pilot contamination capitalising on different aspects for discriminating the interfering inter-cell signals from the intra-cell have also emerged under the LSAS framework. For instance, the differences between the spatial covariance matrices of the intra-cell users and the rest of the interfering mobile stations sharing identical pilot sequences can be exploited for reducing the pilot contamination [37–39]. Moreover, the pilot contamination effect can also be reduced by identifying the path loss differences between the interfering signals from other cells and the intra-cell signals of interest [40,41], by unsynchronising the pilot and uplink phases of adjacent cells [36] or by designing sophisticated strategies encompassing both downlink and uplink training [42]. While the detailed description of the above category of techniques falls out of the scope of this chapter, the reader is referred to the above mentioned works for further discussion.

The above works demonstrate that there are a number of strategies to mitigate the harmful pilot contamination effect. However, there exist other aspects that limit the ultimate performance of LSAS, which we aim to describe in the following.

4.1.2.3 Hardware limitations in LSAS

Clearly, the large number of antennas required in LSAS entail a substantial increase in the RF chains required and the overall hardware complexity of these systems when compared with state-of-the-art standards. This burden promotes the employment of cheaper hardware components that might introduce signal distortions and imperfections, hence diminishing the returns offered by practical LSAS [46]. The effect that the use of non-ideal hardware has over LSAS has been incorporated to their analysis in the literature [47]. The imperfections are typically modelled as additive distortions in the baseband signal model. These distortions are introduced by incorporating additional Gaussian noise sources when compared to the conventional MIMO system model: one at the transmitter and an additional one in each receiver. The noise sources aim at modelling the combined effect of harmful effects such as the *I*/*Q* imbalance of non-ideal mixers, the amplifiers' non-linearities, or the imperfect analogue filters [46,47].

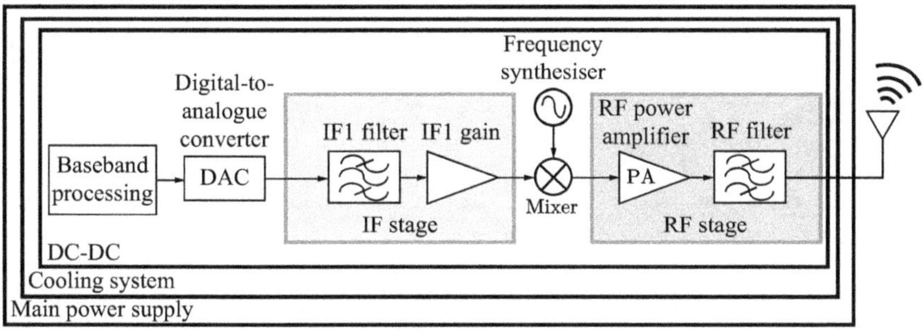

Figure 4.7 Simplified block diagram of a standard transmit RF chain and the supplementary system components that contribute towards the overall power consumption [43–45]

A simplified block diagram of a transmit RF chain with the typical hardware components is shown in Figure 4.7.

Similarly to (4.3), the model that incorporates the above hardware non-idealities for the uplink data transmission in a single-user system is given by [47]

$$\mathbf{y} = \mathbf{h}\left(s + \eta_t^{\mathrm{UT}}\right) + \eta_r^{\mathrm{BS}} + \mathbf{n}', \tag{4.20}$$

where both the hardware impairments at the user terminal, η_t^{UT}, and the BS, η_r^{BS}, are comprised of Gaussian random variables, which is motivated by the additive effect of the impairments and the central limit theorem [16,47]. In the above expression, $s \in \mathbb{C}$ denotes the symbol transmitted by the user and $\mathbf{n}'$ is the AWGN vector with independent entries satisfying $n_i' \sim \mathcal{CN}(0, \sigma^2)$. Interestingly, the more complete model (4.20) allows concluding the ultimate capacity of LSAS is more likely limited by the hardware impairments at the user terminals rather than those of the BS, which instead vanish for the large number of antennas [47]. This motivates the employment of cheaper components in LSAS, which is a crucial aspect due to the large number of RF components that have to be implemented [26]. For more information about the characterisation and effect of the hardware impairments in LSAS, we refer the interested reader to the relevant works [48–50] and references therein.

4.1.2.4 Antenna array topologies and limited physical space

Typically, the antenna array geometry has a significant impact on the physical beamforming gains of the array and the overall array performance. Both the impact of the array geometry and the limitations of having a constrained physical space have been ignored up to this point. These are important aspects that have to be considered in the design of realistic LSAS, since the physical space where the antennas must be allocated may indeed be constrained. Consider, for example, an array with $N = 160$ elements and central transmission frequency of $f = 2.6$ GHz, a standard frequency for LTE systems currently deployed [4]. This results in a wavelength of $\lambda = c/f \approx 11.5$ cm, where $c = 3 \times 10^8$ m/s refers to the speed of light.

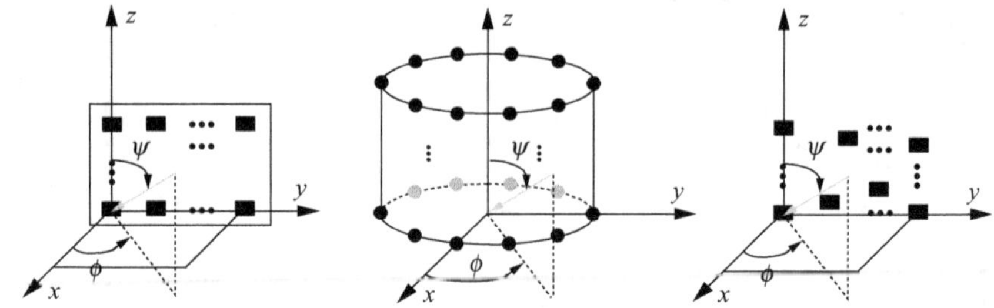

Figure 4.8 Illustration of planar, cylindrical and irregular antenna array topologies. ϕ and ψ denote the azimuth and elevation angles, respectively

The elements in antenna arrays are usually placed at a distance $d \geq 0.5\lambda$, which is commonly referred to as the critical spacing [51]. Clearly, placing the antennas along a straight line as in traditional uniform linear arrays (ULAs) seems impractical in the above example, since the size of the resultant array would be $d_t = N\lambda/2 \approx 9.2$ m, which could be difficult to accommodate in the antenna structures currently deployed. Moreover, ULAs do not allow performing beam steering in the elevation domain, which might be necessary for discriminating the user positions and for accurately focusing the transmitted energy [52].

The employment of other antenna topologies such as planar, cylindrical, spherical or even irregular arrays seems a more suitable choice for LSAS as illustrated in Figure 4.8. [9,53,54]. Indeed, most of the antenna prototypes currently implemented for LSAS adopt one of these geometries with dual-polarised antennas [53,55–57]. The employment of dual-polarised antennas seems essential for LSAS, since they allow doubling the number of antenna ports without increasing the mutual coupling and the physical space required for antenna arrays [58]. For instance, the dimensions of the above antenna array with $N = 160$ dual-polarised antenna elements distributed in a rectangular arrangement with $N_{hor} = 20$ antennas and $N_{ver} = 8$ antennas would occupy an approximate area of 115×46 cm^2. Note that placing a larger number of antennas horizontally allows obtaining a finer resolution in the azimuth domain than in the elevation [52,57].

Nevertheless, for a number of LSAS deployment scenarios, the colossal number of antennas envisaged may have to be fitted in the limited physical space of today's BSs. Therefore, there might be scenarios where the antennas must be placed at a distance $d \leq 0.5\lambda$ to satisfy specific physical constraints. Accordingly as the number of antennas increases in a fixed physical space, this requires a diminishing antenna separation. This further results in an increase in the spatial correlation of the antennas as signal sources and in the mutual coupling of the antennas as electrical components [59]. In this context, it is interesting to study the trade-off that arises in these physically constrained scenarios [59–61]. That is, by increasing N, the number of antennas as information sources increases, while their spatial separation decreases aggravating the effects of mutual coupling and spatial correlation [61,62].

4.1.2.5 Signal processing complexity in LSAS

Despite favouring linear signal processing techniques for data transmission and reception, the increased system dimensions envisaged by LSAS drastically increase the complexity of the signal processing operations [26,63]. Moreover, the power consumption of digital signal processors (DSPs) and data interfaces might be significant in LSAS [64]. In this section, we discuss the complexity of linear processing in LSAS, highlighting the most complex operations involved in the process.

The initial communication stage of an LSAS consists in acquiring CSI for the subsequent data transmission and detection processes, as shown in Figure 4.5. However, the employment of channel estimation strategies such as the MMSE conventionally considered in the LSAS literature might impose a significant computational burden due to the need to invert matrices of large dimensions [26,34,63,65]. In this context, the employment of polynomial approximations to the matrix inverse pose a relevant trade-off between complexity and estimation quality depending on the polynomial order selected [34]. Once the training stage has been completed, the BS moves onto the data transmission and reception phases. Here, the number of fast Fourier transforms (FFTs) and inverse FFTs that have to be computed on a symbol-by-symbol basis is also severely increased with respect to current standards. This constitutes one of the main computational burdens since they have to be computed at the symbol rate involving [26,65]:

$$C_{\text{FFT}} = \mathcal{O}(N\tau N_{\text{sub}} \log_2(N_{\text{sub}}))\text{flops}, \tag{4.21}$$

where $\mathcal{O}$ represents the Big-O operator, τ denotes the number of symbols contained in the channel coherence time T and N_{sub} denotes the number of subcarriers in the system bandwidth B.

Regarding the complexity of the linear precoders and detectors, the main drawback of the ZF and MMSE strategies can be found in the matrix inversions required to build the precoding matrix [66,67]. Note that the channel inversions only have to be performed at the start of the channel coherence time and not on a symbol-by-symbol basis, since the result remains valid while the channel remains unchanged [26,65,68,69]. Specifically, assuming $N \geq K$, the complexity of the pre-computations required for data reception in a specific coherence time is given by [65]

$$C_{\text{inv}} = \mathcal{O}\left(NK^2\tau_u N_{\text{blocks}}\right)\text{flops}, \tag{4.22}$$

where $\tau_u < \tau$ denotes the number of symbols dedicated for symbol reception (uplink), and N_{blocks} denotes the total number of frequency blocks with a similar channel in the total bandwidth B [65]. The polynomial complexity of the above operations with the number of users K has promoted the introduction of signal processing methods for reducing the complexity of this stage. These include strategies where the matrix inverse is approximated via the Neumann series expansion [67] and polynomial decompositions [63,66]. More importantly however, matrix-vector multiplications of dimension N must be performed at a symbol rate between the received (information) symbol vector and the detection (precoding) matrix for producing the resulting decision (transmitted) symbols. These significantly contribute towards the overall computational complexity in LSAS [26,65].

Overall, it can be concluded that LSAS impose stringent real-time signal processing requirements. Still, performing the above operations seems feasible with technologies currently available [26]. Moreover, it is envisaged that the rapid development of the digital signal processing technology will simultaneously alleviate both the strict requirements and the economical costs involved in the computational capabilities of LSAS.

4.2 LSAS channel models and propagation aspects

The majority of the theoretical results described in Section 4.1 has been derived by analysing the performance of LSAS under a set of restricted uncorrelated channel models. As explained above however, in practice there are a number of scenarios where typically the spatial streams are correlated. Accordingly, a number of theoretical and semi-empirical channel models are available in the literature, for studying such effects. This section overviews the models commonly employed to accurately characterise the behaviour of LSAS for facilitating the interested reader to model these systems. Moreover, the conclusions extracted from measurement campaigns and the specific channel models derived for LSAS are also described.

4.2.1 Theoretical LSAS channel models

The correlation-based stochastic channel model has been widely employed for characterising theoretically the behaviour of LSAS due to its simplicity [9,19,20]. As a first step, in this section we concentrate on the correlation-based channel models and highlight their main characteristics and limitations for modelling the communication channels of LSAS. For coherence with the previous sections, we focus on the uplink system model of (4.3), particularised to the single cell scenario for notational simplicity. We further focus on the one-side correlation model, with the correlation drawn at the BS, while the models can be extended to address the antenna correlation at the MS.

Consider a single-cell LSAS comprised of a BS with N transmit antennas and K single-antenna users. The correlation-based model expresses the uplink channel between the kth user and the BS as [9,19,20,29]

$$\mathbf{h}_k = \mathbf{R}_k^{\frac{1}{2}} \mathbf{g}_k, \tag{4.23}$$

where $\mathbf{R}_k \in \mathbb{C}^{N \times N}$ is an Hermitian deterministic matrix that characterises the channel correlation matrix of the kth user and, similarly to (4.1), $\mathbf{g}_k \in \mathbb{C}^{N \times 1} \sim \mathcal{CN}(\mathbf{0}, \mathbf{I}_N)$ is a random vector whose complex coefficients $g_{k,n}$ represent the channel gain between the nth antenna at the BS and the single-antenna kth user. Throughout this section, $\mathbf{g} \sim \mathcal{CN}(\mathbf{a}, \mathbf{B})$ denotes that the random vector $\mathbf{g}$ obeys a complex Gaussian distribution with mean $\mathbf{a}$ and covariance matrix $\mathbf{B}$ [16]. Therefore, as per (4.23), $\mathbf{h}_k \sim \mathcal{CN}(\mathbf{0}, \mathbf{R}_k)$. The above model considers that there is no spatial correlation amongst users, which is a reasonable assumption since they are usually separated by distances larger than λ [29]. Moreover, note that the covariance matrices $\mathbf{R}_k, \forall k \in \{1, \ldots, K\}$ typically vary at a rate slower than the channel coherence time [18,19].

The above channel model allows characterising a variety of channel effects that impact the performance of LSAS. For instance, the simple Rayleigh fading model with path loss introduced in Section 4.1 can be expressed as a particularisation of (4.23) by letting $\mathbf{R}_k = \mathbf{D}$. Note that in general, $\text{rank}(\mathbf{R}_k) \neq N$, where $\text{rank}(\cdot)$ refers to the rank of a matrix [17]. In other words, the model introduced in (4.23) can be employed to characterise sparse communication channels [70,71]. Equivalently, the Karhunen–Loève (KL) representation of the communication channel can be expressed as [72]

$$\mathbf{h}_k = \mathbf{U}_k \sum\nolimits_k^{\frac{1}{2}} \mathbf{g}_k, \tag{4.24}$$

where $\mathbf{U}_k \in \mathbb{C}^{N\times N}$ is a unitary matrix with columns given the eigenvectors of the channel covariance matrix $\mathbf{R}_k = \mathbf{U}_k \sum_k \mathbf{U}_k^H$ and $\sum_k \in \mathbb{C}^{N\times N}$ is a diagonal matrix containing the eigenvalues of $\mathbf{R}_k$. The equivalent representation introduced in (4.24) has been recurrently employed to analyse the behaviour of LSAS communication channels [29,72], since it allows explicitly showing channel sparsity, which relates to the number of zero eigenvalues in $\sum_k^{\frac{1}{2}}$, and it facilitates the development of reduced-RF antenna array designs as detailed in Section 4.3.

Physically, a reduced rank in the covariance matrix $\mathbf{R}_k$ might be produced by different effects such as significant inter-antenna correlation or insufficient scattering, which are particularly important for the characterisation of LSAS. These effects can be characterised by employing definitions of the channel covariance matrix in (4.23) with different degrees of complexity, which are described in the following. Amongst a number of existing correlation models, here we focus on the most relevant channel models for LSAS to date.

4.2.1.1 Exponential correlation model

The exponential correlation model aims at characterising the impact of the inter-antenna correlation, which induces a similarity in the communication channels of neighbouring antennas. The channel covariance matrix of the kth user in the exponential correlation model is given by [73]

$$\mathbf{R}_k = \begin{pmatrix} 1 & t_k & t_k^2 & \cdots & t_k^{N-1} \\ (t_k)^H & 1 & t_k & \cdots & t_k^{N-2} \\ \vdots & \vdots & \vdots & \cdots & \vdots \\ (t_k^H)^{N-1} & \cdots & (t_k^H)^2 & (t_k)^H & 1 \end{pmatrix}, \tag{4.25}$$

where $t_k = |t_k|e^{j\phi_k} \in \mathbb{C}$ is the correlation coefficient for user k and it has been assumed that the massive BS follows a linear arrangement (ULA) for simplicity, although this model can be immediately generalised to other topologies [73,74]. The phase ϕ_k of t_k is usually set as zero or identical [34,63]. However, more recent works define the phases ϕ_k as independent random variables to characterise the differences between the channels of different users [73–75].

The exponential correlation model is particularly relevant for constrained-size LSAS as there exists a direct dependence between the antenna distance and the entries of the channel correlation matrix. However, since this simple model is not able to accurately capture other important communication effects such as the limited channel scattering, their employment is conventionally restricted to the support of conceptual proofs [34,63,75].

4.2.1.2 One-ring channel model

The one-ring model has been conventionally employed for characterising mobile communication systems in which the BS does not experience significant local scattering, while it is the user terminals that are surrounded by scatterers [13,76]. This corresponds to the typical scenario in which the mobile stations are at ground level whereas the BS is located on an elevated structure, as is the case for typical cellular BSs today. For this reason, this model has been adopted in a variety of LSAS works such as [19,72,77].

The one-ring local scattering model can be physically interpreted as shown in Figure 4.9 for the particular case of a ULA, although this model can be readily extended to encompass more intricate antenna topologies. Let a mobile user located at azimuth angle κ at distance D from the BS be surrounded by a ring of scatterers over a circular area of radius R. Then, the correlation coefficient between the antennas i and j of a ULA at the BS is given for the kth user by

$$\{\mathbf{R}_k\}_{i,j} = \frac{1}{2\Delta}\int_{-\Delta}^{\Delta} e^{j\frac{2\pi}{\lambda}d(i-j)\cos(\vartheta+\kappa)}d\vartheta, \tag{4.26}$$

where $\{\mathbf{X}\}_{i,j}$ denotes the i,jth element of matrix $\mathbf{X}$ and d is the separation between adjacent antenna elements at the BS. The angular spread is defined as

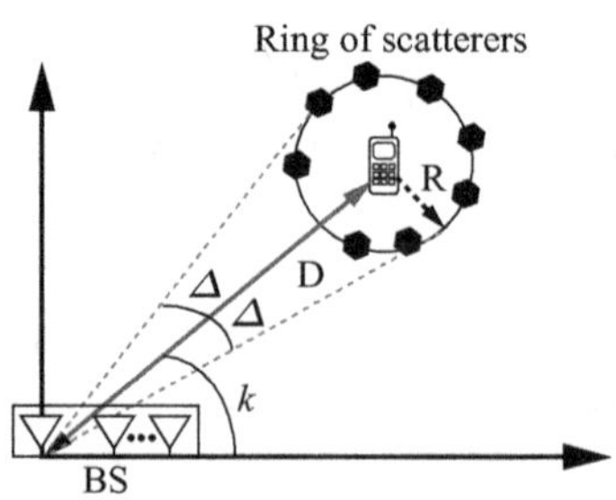

Figure 4.9 Illustration of the one-ring channel model [72,76]

$\Delta \approx \arctan(R/D)$ and $\kappa \in [0, 2\pi)$ is the angle between the BS and the kth MS as shown in Figure 4.9. Intuitively, a reduced Δ induces a larger channel correlation, since the range of angles of arrival of the impinging signals is reduced. A uniform distribution of the received power from planar waves impinging on the BS array is considered in (4.26). Note that these assumptions are only approximations for LSAS, where the spherical shape of the real impinging waves might impact the performance [78,79].

4.2.1.3 Physical or finite scatterer channel model

This channel model divides the angular domain into a finite number of directions (dimensions) $P < N$ that might correspond to the number of active scatterers [80]. Each of the P directions from user k to the BS is characterised by a pair of azimuth and elevation Angles of Arrival (AoA), $\{\phi_p, \psi_p\}$, $p \in \{1, \ldots, P\}$. The array steering vector, $\mathbf{a}(\phi_p) \in \mathbb{C}^{N\times 1}$, can be expressed as

$$\mathbf{a}(\phi_p) = \frac{1}{\sqrt{P}} \left[e^{-jf_0(\phi_p, \psi_p)}, \ldots, e^{-jf_N(\phi_p, \psi_p)} \right]^T. \tag{4.27}$$

Here, $f_n(\phi_p, \psi_p)$ is an arbitrary function that depends on the geometry of the antenna array at the BS [52]. For instance, a linear array with a uniform distance between the radiating elements follows $f_n(\phi_p) = \frac{2\pi(n-1)D}{\lambda} \cos(\phi_p)$ and the steering vector is given by [52]

$$\mathbf{a}_{\mathrm{ULA}}(\phi_p) = \frac{1}{\sqrt{P}} \left[1, e^{-j2\pi\frac{d}{\lambda}\cos(\phi_p)}, \ldots, e^{-j2\pi\frac{(N-1)d}{\lambda}\cos(\phi_p)} \right]^T. \tag{4.28}$$

In the above, d is the spacing between the antenna elements at the transmitter. Incorporating the physical scattering model to the standard model, the resulting channel vector from the kth mobile station to the BS, $\mathbf{h}_k \in \mathbb{C}^{N\times 1}$, is given by

$$\mathbf{h}_k = \sum_{p=1}^{P} \mathbf{a}(\phi_p, \psi_p)\alpha_{k,p}, \tag{4.29}$$

where $\alpha_{k,p}$ is the channel gain between the kth user and the BS in dimension p and it usually comprises the fast fading, shadowing and path loss effects. For simplicity, it is usually considered that the number of transmission paths P is identical for all users $k \in \{1, \ldots, K\}$. Another common assumption relies in considering that the fast fading coefficients depend on the physical direction, whereas the large-scale fading factors, such as the path loss, are independent of the direction p.

Similarly to (4.23), the physical channel model can also be expressed in a matrix form [20,33]. Specifically, let the composite vector $\alpha_k \in \mathbb{C}^{P\times 1}$ be defined $\alpha_k = [\alpha_{k,1}, \ldots, \alpha_{k,P}]^T$. Then, the channel of the kth user in the physical channel model can be expressed as [33]

$$\mathbf{h}_k = \mathbf{A}_k \alpha_k, \tag{4.30}$$

where $\mathbf{A} \in \mathbb{C}^{N\times P}$ is a full-rank matrix encompassing the steering vectors $\mathbf{A} = [\mathbf{a}(\phi_1), \ldots, \mathbf{a}(\phi_P)]$. Note that this model falls within the realm of correlation-based stochastic models (4.23) with $\tilde{\mathbf{R}}_k = \left[\sqrt{d_k}\mathbf{A}_k, \mathbf{0}_{N\times N-P}\right]$, where d_k refers to the large-scale fading component of the kth user [20]. This entails that $\text{rank}(\mathbf{R}_k) \leq P$, which intuitively can be interpreted as follows: the number of independent transmission paths between the BS and the kth user, which is directly related to the number of scattering clusters, determine the maximum number of independent data streams that the LSAS supports [29,72]. Furthermore, as explicitly shown in (4.26) and (4.29), both the physical and the one-ring channel models allow characterising the angles of arrival to or departure from the antenna array. This aspect combined with the possibility of characterising limited scattering environments have promoted the employment of the physical channel model for characterising the behaviour of LSAS in a variety of works [20,33,37,80,81].

4.2.2 Mutual coupling characterisation

The large increase in the number of antennas envisaged in LSAS implies that the space dedicated to allocate the antennas must also grow to secure uncorrelated operation [7,60]. However, physical antenna structures may be subject to limited space that might affect the attainable inter-antenna separation. Moreover, future LSAS implementations are likely to incorporate hundreds of small dipole antennas [82] and other low-cost components. The above may promote the appearance of mutual coupling between the radiating elements. The mutual coupling effect characterises the energy absorbed by antennas physically close to the reference operating antenna as illustrated in Figure 4.10. In the transmission stage, this entails that part of the signal energy intended for a specific antenna excites the neighbouring antennas at the transmitter, which consequently emit part of this unintended energy.

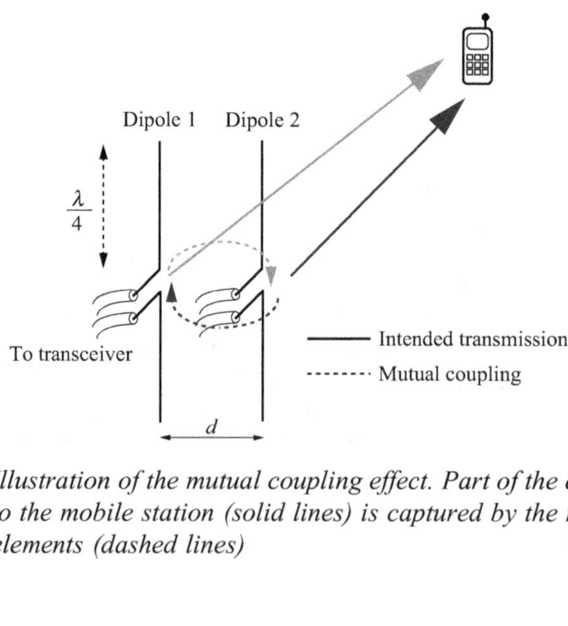

Figure 4.10 Illustration of the mutual coupling effect. Part of the energy intended to the mobile station (solid lines) is captured by the nearby antenna elements (dashed lines)

An accurate characterisation of the mutual coupling effects can be provided by analysing the S-parameters of the communication system [13,83,84]. This is the approach adopted in [58,85] for modelling the mutual coupling effect, where the S-parameters are obtained via full-wave electromagnetic simulations. Instead, a simpler approach employed in [60,86,87] is based on the use of impedance matrices to characterise mutual coupling effects. Specifically, this model introduces an additional matrix to the channel model of (4.23) to account for the effect of the mutual coupling. The main drawback of this alternative is that the results are only applicable to antennas that are minimum scatterers[2] such as the half-wave length dipole [13]. However, its direct relationship with the models currently employed in LSAS and the possibility of using analytic expressions to obtain the impedance matrices of specific types of antennas make the use of this model convenient [60,87,88].

In the mutual coupling matrix model, the channel between kth single-antenna user and the BS, $\mathbf{h}_k \in \mathbb{C}^{N\times 1}$, is typically expressed as [60]

$$\mathbf{h}_k = \mathbf{Z}\tilde{\mathbf{R}}_k\mathbf{g}_k, \tag{4.31}$$

where $\tilde{\mathbf{R}}_k$ and $\mathbf{g}_k$ correspond to those introduced in (4.23) and $\mathbf{Z} \in \mathbb{C}^{N\times N}$ is the matrix introduced for modelling the mutual coupling effect. The mutual coupling matrix is routinely expressed as

$$\mathbf{Z} = (Z_A + Z_L)(\Xi + Z_L\mathbf{I}_N)^{-1}, \tag{4.32}$$

where Z_A is the antenna impedance, Z_L is the load impedance and Ξ is a matrix given by [52]

$$\Xi = \begin{pmatrix} Z_A & Z_m^{1,2} & Z_m^{1,3} & Z_m^{1,4} & \cdots & Z_m^{1,N} \\ Z_m^{2,1} & Z_A & Z_m^{2,3} & Z_m^{2,4} & \cdots & Z_m^{2,N} \\ \vdots & \vdots & \vdots & \vdots & \cdots & \vdots \\ Z_m^{N,1} & Z_m^{N,2} & \cdots & Z_m^{N,N-2} & Z_m^{N,N-1} & Z_A \end{pmatrix}, \tag{4.33}$$

where $Z_m^{i,j}, i,j \in \{1,\ldots,N\}$, is the mutual impedance between antenna elements i and j. Typical values of the load impedance satisfy $Z_L = 50\Omega$, in the mutual coupling model (4.32), while the value of $Z_A = 73 + j42.5\Omega$ is often used for the impedance of a dipole antenna. The mutual impedances can be obtained by employing the closed-form expressions available in [52] for dipole antennas with different relative positions.

[2]Antennas that are minimum scatterers are those that act as if they were not present when open-circuited.

4.2.3 Measurement campaigns and LSAS-specific channel models

The particular characteristics of LSAS greatly influence the associated communication channels when compared to conventional small-scale MIMO setups. For instance, instead of the traditional planar wavefront often assumed in small-scale systems, a spherical wavefront may have to be considered in LSAS arrays motivated by the large physical size of the massive antenna structures. Additionally, the scattering clusters can be considered non-stationary both in the physical and time domains for similar reasons [78,79]. These aspects motivate the development of novel channel models that consider the specific characteristics of LSAS [78,79,89].

The above particular characteristics of LSAS communication channels have been observed in a variety of measurement campaigns performed (see, e.g., [53,56,89–92] and references therein). In this context, the measurement results to-date seem to corroborate the enhancements in the user spatial separability that can be achieved with LSAS [53]. In other words, the channel orthogonality among the users described in (4.4) might also be achieved in real propagation scenarios. Overall, the measurement results seem to indicate that the ideal achievable rates theoretically promised over Rayleigh fading conditions might be approximated in particular measurement conditions [53]. However, this is not a general characteristic of LSAS, since some scenarios, such as those formed by closely located users experiencing a LOS propagation, might incur significant losses with respect to the ideal achievable rates.

4.3 LSAS: system design and optimisation

4.3.1 Reduced RF-chain implementations

Despite being tolerant to cheaper RF components, LSAS still require a large number of hardware components, since their theoretically promised benefits rely on the implementation of one active RF chain per antenna. However, each RF chain has a substantial number of components that contribute towards the overall power consumption as depicted in Figure 4.7. For example, the transmission stage is comprised of digital signal processors, digital-to-analogue-converters, mixers, filters, frequency synthesisers and power amplifiers that dictate the total power consumed by the BS [93]. The study and design of conventional MIMO transceivers most commonly assumes that the number of RF chains is identical to the number of BS antennas, i.e., each antenna of the MIMO array employs its dedicated RF chain. While this high-performance approach offers a complete flexibility in the processing of the signals transmitted and received by the antenna array, for the particular case of LSAS it entails that a significant number of hardware components must be incorporated into the BS. In this section, we introduce a number of solutions and the practical designs discussed in the literature to-date for simplifying the hardware and overall operation of LSAS.

4.3.1.1 Antenna selection

An intuitive approach for reducing the hardware complexity consists in reducing the number of antennas simultaneously employed for transmission and reception at the BS. Such schemes typically involve antenna selection strategies and they essentially rely on switching off the antennas that do not significantly contribute towards the communication, based on the available CSI [94]. Similarly to (4.3), the received signal in an arbitrary antenna selection scheme with L antennas simultaneously active per BS can be expressed as

$$\tilde{\mathbf{y}}_c = \sum_{m=1}^{M} \sqrt{\rho}\,\mathbf{H}_{c,m}^{\mathcal{L}_m}\mathbf{x}_m + \mathbf{n}, \tag{4.34}$$

where $\mathbf{H}_{c,m}^{\mathcal{L}_m} \in \mathbb{C}^{L\times K}$ denotes the resulting channel after choosing $L < N$ rows from $\mathbf{H}_{c,m}$, indexed by the sets $\mathcal{L}_m, m \in \{1, \ldots, M\}$. Clearly, both the number of active antennas L and the specific antenna elements chosen by the indices $\mathcal{L}_m$ influence the ultimate system performance and its careful design is essential in antenna selection schemes [94]. In this context, the deployment of LSAS might favour performing this selection due to the great number of antennas available.

Two different hardware implementations for antenna selection are shown in Figure 4.11(a) and (b). Figure 4.11(a) illustrates a flexible antenna selection scheme where N RF chains are implemented and a subset of them might be disconnected. Instead, Figure 4.11(b) shows a hardware implementation where a reduced number $N_{RF} < N$ of RF chains are implemented and an RF switching matrix determines the N_{RF} active antennas. Note that one architecture can be more beneficial than the other depending on the actual system requirements. The first approach has the disadvantage that it still requires the deployment of large numbers of RF chains, while the RF switching approach can distort the output signals and introduce additional losses [90]. The employment of antenna selection strategies for improving the energy efficiency of LSAS has been considered in both theoretical and measured channels in [90,95], while a large number of antenna selection approaches exist [90,96,97].

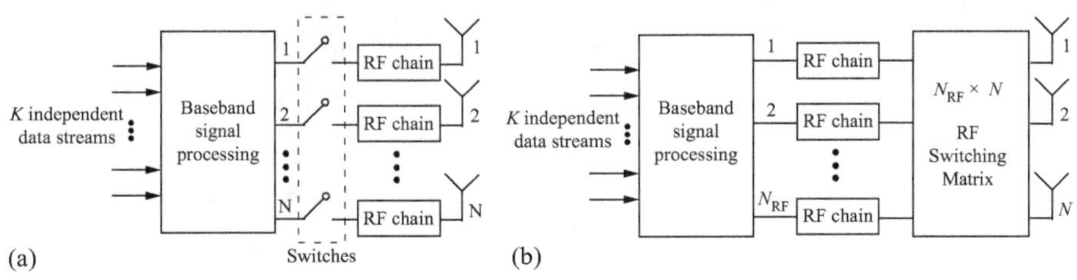

Figure 4.11 Block diagram of antenna selection schemes with (a) N RF chains and (b) $N_{RF} < N$ RF chains and an RF switching matrix

4.3.1.2 Hybrid analogue and digital beamforming: reducing the number of RF chains

An alternative strategy for reducing the number of RF chains without affecting performance in LSAS that has attracted recent interest is exploiting channel sparsity [29,70–72,98]. In this context, channel sparsity can be interpreted as a set of specific properties that results in the communication channel having a reduced number of degrees of freedom (DoF) [10]. For instance, channel sparsity might be promoted by the appearance of a severe spatial correlation originated by an insufficient inter-antenna spacing in the array, or a limited number of physical scatterers in the propagation channel as it occurs in the millimetre wave band [33,59,99–101]. In these cases, channel sparsity can be exploited for reducing the number of RF chains by incorporating analogue beamforming networks comprised of analogue phase shifters and signal splitters and combiners adjacent to the antenna front-ends [98]. These approaches are commonly referred to as hybrid analogue and digital beamforming designs and their analysis constitutes our focus in the following. While this has been a particularly vibrant research area at the time of the production of this chapter, here we overview the most prominent methodologies to date.

The architecture of a generic hybrid beamformer is shown in Figure 4.12. The hybrid beamformer is divided into two different blocks: a digital baseband beamformer and an analogue RF beamformer. The main advantage of hybrid beamforming designs when compared with traditional MIMO schemes is that the baseband beamforming can be designed to operate on signals of a reduced dimensionality $N_{\mathrm{RF}} < N$, hence allowing a reduced number of RF chains N_{RF} in the array architecture, as illustrated in Figure 4.12. This comes at the cost of employing additional RF components for splitting the $N_{\mathrm{RF}} \geq K$ input signals to the RF beamformer, applying arbitrary phase shifts that depend on the available CSI, and combining the resultant phase-shifted signals for producing the required N outputs.

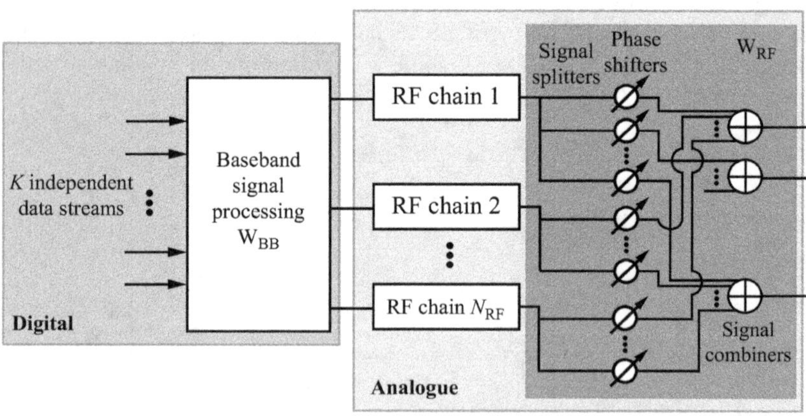

Figure 4.12 Block diagram of a generic hybrid beamforming scheme comprised of a digital baseband precoder, $\boldsymbol{W}_{BB}$, and an analogue phase shifting network, $\boldsymbol{W}_{RF}$

Similarly to (4.8), in a single-cell uplink scenario, the output when applying hybrid beamforming at the receiver can be expressed as [72,98,99]

$$\mathbf{r} = \mathbf{W}\mathbf{y} = \mathbf{W}_{\mathrm{BB}}\mathbf{W}_{\mathrm{RF}}\left(\sqrt{\rho}\,\mathbf{H}\mathbf{x} + \mathbf{n}\right), \tag{4.35}$$

where $\mathbf{W}_{\mathrm{BB}} \in \mathbb{C}^{N_{\mathrm{RF}}\times K}$ and $\mathbf{W}_{\mathrm{RF}} \in \mathbb{C}^{N\times N_{\mathrm{RF}}}$ represent the baseband and analogue precoding matrices, respectively. Note the reduced dimensions of $\mathbf{W}_{\mathrm{BB}}$ with respect to $\mathbf{W}$ in (4.8). Since the analogue beamforming matrix is a phase shifting network, the *i*,*l*th element of $\mathbf{W}_{\mathrm{RF}}$ adopts the form $(\mathbf{W}_{\mathrm{RF}})_{i,l} = \frac{1}{\sqrt{N}}e^{j\varphi(i,l)}$, where the specific phase shifts $\varphi(i,l)$ are determined according to the available instantaneous or long-term (statistical) CSI [72,98,99,102].

The design of both $\mathbf{W}_{\mathrm{RF}}$ and $\mathbf{W}_{\mathrm{BB}}$ for LSAS is considered in [72,98,99,102]. Although some works assume rich scattering channels [102], a large number of hybrid approaches rely on channel sparsity for reducing the number of RF chains without significantly affecting performance. For instance, consider the channel representation introduced in (4.24) and a single user for simplicity. A sparse communication channel can essentially be described by a subset of $v_k \leq N$ eigenvalues and eigenvectors, and satisfies [72]:

$$\mathbf{h}_k = \mathbf{U}_k \sum_k^{1/2} \mathbf{g}_k \approx \mathbf{U}_k^{(v_k)} \left(\sum_k^{1/2}\right)^{(v_k)} \mathbf{g}_k^{(v_k)}, \tag{4.36}$$

where $\sum_k^{(v_k)} \in \mathbb{C}^{v_k\times v_k}$ represents a diagonal matrix containing the v_k non-negligible eigenvalues of the channel covariance matrix $\mathbf{R}_k$ and the columns of $\mathbf{U}_k^{(v_k)} \in \mathbb{C}^{N\times v_k}$ are formed by the corresponding eigenvectors. Here, $\mathbf{g}_k \in \mathbb{C}^{v_k\times 1} \sim \mathcal{CN}(\mathbf{0}, \mathbf{I}_{v_k})$. In this particular case, only $N_{\mathrm{RF}} = v_k$ RF chains would be required for approximating the maximum attainable rates provided that $\mathbf{W}_{\mathrm{RF}} = \mathbf{U}_k^{(v_k)}$, which only depends on the channel statistics $\mathbf{R}_k$ [72]. Interestingly, it can be demonstrated that, in the specific case of an asymptotically large ULA with critical spacing, the *i*th column of the optimal RF beamforming matrix $\mathbf{W}_{\mathrm{RF}}^{\mathrm{opt}}$ corresponds to a column of the discrete Fourier transform (DFT) matrix [72,98,103]:

$$\mathbf{w}_{\mathrm{RF},i} \approx \frac{1}{\sqrt{N}}\left[1, e^{\frac{-2\pi(i-1)j}{N}}, e^{\frac{-4\pi(i-1)j}{N}}, \ldots, e^{\frac{-2\pi(i-1)(N-1)j}{N}}\right]^T. \tag{4.37}$$

Physically, this can be interpreted as a division of the physical space into N orthogonal beams by means of the RF beamforming matrix in (4.37), where, however, only a subset v_k of them can be used for data transmission due to the sparse nature of the channel. Accordingly a selection of the v_k useful directions that contribute towards propagating the signals between the BS and the users is performed by means of the hybrid precoding. Note that a similar concept was employed in [37] for reducing the pilot contamination effect.

Research has also been focused on reducing the complexity of the fully connected analogue beamforming strategy presented in this section. In this context, viable approaches for reducing the hardware and signal processing complexities are the exploitation of the inter-channel correlation between closely spaced antennas [104] and the design of sub-optimal approaches for reducing the number of connections and hardware components in the analogue beamforming network [105,106]. However, it should be noted that sparsity is not to-date regarded in general as a given characteristic of communication channels at microwave frequencies and further investigations are required for validating sparsity-based approaches.

4.3.1.3 Spatial modulation

An alternative strategy which has gained interest for reducing the number of RF chains in LSAS is spatial modulation (SM), where the transmission involves only a few (down to one) RF chains being simultaneously active [107–109]. The block diagram of an SM transmitter with a single RF chain is illustrated in Figure 4.13(a). The fundamental idea behind SM consists in, apart from using conventional modulation methods such as quadrature amplitude modulation (QAM) to convey information, encoding some additional information on the antenna indices by only activating a subset of antennas, based on the instantaneous data. This generates a three-dimensional constellation, as illustrated in Figure 4.13(b), where the first two dimensions correspond to the conventional QAM symbols and the third one is given by the antenna indices employed to transmit the chosen modulation symbols. Mathematically, the signal $\mathbf{x}_u \in \mathbb{C}^{n_t \times 1}$ conveyed by an SM transmitter with n_a active antennas out of a total of n_t antennas adopts the form [109]:

$$\mathbf{x}_u = \begin{bmatrix} 0 & \cdots & s^q_{l_1} & \cdots & s^q_{l_{n_a}} & \cdots & 0 \end{bmatrix}^T, \tag{4.38}$$

where $l_i \in \{1, \ldots, n_t\}, i \in \{1, \ldots, n_a\}$ denotes the active antenna index, s^q represents the qth symbol of the transmit constellation and $(\cdot)^T$ denotes the transpose of a matrix. Note that both the active antenna indices and the conventional constellation symbols have to be estimated at reception.

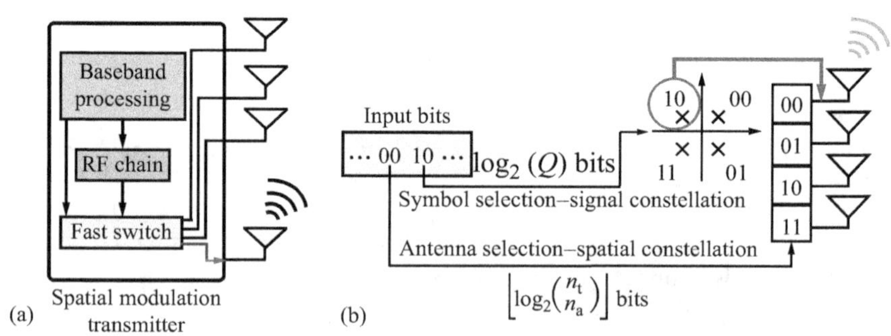

Figure 4.13 Illustration of the spatial modulation principle. (a) Block diagram of a spatial modulation transmitter and (b) symbol and antenna selection procedure

SM allows transmitting a total of $B_{\text{SM}} = \lfloor \log_2(\binom{n_t}{n_a}) \rfloor + \log_2(Q)$ bits per channel use, where $\lfloor \cdot \rfloor$ denotes the floor operator and Q is the conventional signal constellation order [109]. This results in a reduction of spectral efficiency with respect to traditional MIMO systems with $B_{\text{MIMO}} = n_t \cdot \log_2(Q) > B_{\text{SM}}$ bits per channel use simultaneously conveyed, with a dedicated RF chain for each of the n_t transmit antennas. Therefore, SM poses a trade-off between spectral efficiency, which is sacrificed due to the reduced number of active antennas, and hardware complexity, which is reduced due to the need of only requiring a reduced number of RF chains [107–109]. While SM strategies could be applied to the downlink for reducing the hardware complexity [110], they are also particularly suitable for enhancing the uplink data transmission rates without significantly increasing the complexity of the user handsets in LSAS [111].

4.3.2 Energy efficiency optimisation

In addition to the above, and in line with other disciplines in the general ICT sector, there has been an increasing concern to make wireless networks more energy efficient. Accordingly, the analysis and optimisation of the energy efficiency in LSAS has attracted significant research interest [1,112–114]. Accordingly, the analysis and optimisation of the energy efficiency in LSAS has attracted significant research interest [68,82,95,114–117]. The energy efficiency, ε, measured in bits per Joule (bits/J) is defined as

$$\varepsilon = \frac{R}{P_{\text{tot}}}, \tag{4.39}$$

where R refers to a rate metric measured in bits per second (bits/s), typically the achievable sum rate, and P_{tot} is the power consumed by the hardware components involved in the communication link.

While initial energy efficiency metrics relied on only accounting for the power consumed by the power amplifiers, the large number of additional components required in LSAS have made this approach insufficient, since N transmit RF chains such as the one represented in Figure 4.7 should be implemented. Accordingly, a number of energy efficiency metrics that consider the power consumption of both the hardware and signal processing components have emerged [45,68,93,95]. Here we adopt one of the more complete metrics in the literature to date, proposed in [45,93]. Concentrating on data transmission, the energy efficiency metric introduced in (4.39) is further expanded into:

$$\varepsilon = \frac{R}{P_{\text{tot}}} = \frac{R(1-\sigma_{\text{DC}})(1-\sigma_{\text{MS}})(1-\sigma_{\text{cool}})}{P_{\text{PA}} + P_{\text{RF}} + P_{\text{BB}}}, \tag{4.40}$$

where σ_{DC}, σ_{MS} and σ_{cool} refer to the losses incurred by the DC–DC power supply, the main power supply and active cooling losses respectively. Moreover, with reference to the RF structure of Figure 4.7, $P_{\text{PA}} = \frac{P_t}{\varsigma}$ refers to the average power consumed by power amplifiers with efficiency ς to produce a signal with an output power of P_t Watts, whereas P_{BB} denotes the power consumed by the baseband digital signal processors and the digital data interfaces. P_{RF} represents the power consumption of the electronic components of the RF chains excluding the power amplifiers, which can be further expressed as

$$P_{\text{RF}} = N_{\text{RF}}(P_{\text{DAC}} + P_{\text{mix}} + P_{\text{filt}}) + P_{\text{syn}}, \tag{4.41}$$

where P_{DAC} and P_{mix} denote the power required by the digital-to-analogue converters (DACs) and signal mixers per RF chain, respectively. The number of RF chains is N_{RF}. Furthermore, P_{filt} corresponds to the power consumption of the filters and P_{syn} is the power consumed by the frequency synthesiser. Additionally, $P_{\text{BB}} = p_c C$ denotes the power consumption of the digital signal processor (DSP). Here, C represents the average number of real floating point operations (flops) per second and p_c refers to the Joules of energy required per real flop. Representative values for the above parameters can be found in [45,68,93].

Particular focus has been placed on the energy efficiency of LSAS in the literature, explicitly considering the power consumption of the RF components [68,82,95,114,116,117]. Specific problems investigated in the context of energy efficiency include the optimisation of the energy efficiency as a function of both the number of users K and BS antennas N [68,82], the development of energy-efficient antenna selection strategies [95] and comparisons between the energy efficiency of small cells and LSAS, as both are compelling strategies for 5G [114,117]. Note that, although directly compared in [114,117], both strategies might actually coexist in future heterogeneous networks (HetNets) as described in Section 4.3.3 [115].

4.3.3 Network architectures and distributed LSAS

The principles and practical considerations of LSAS have been introduced throughout this chapter by considering a classical homogeneous multi-cell scenario, where each LSAS cell is comprised of a large antenna array located in its centre and serves the users that fall into its assigned coverage region [7,19,20]. However, LSAS are envisaged to be compatible with a variety of network architectures for further enhancing their strengths, which are briefly mentioned in the following [9,54,115,118].

One of the prominent 5G scenarios involves a multi-tier approach where multiple small-cell access points comprise the lower tier and a macro BS with a large number of antennas forms the higher tier [115,118–120]. An example of such a HetNet architecture is shown in Figure 4.14(a), where it can be explicitly seen that some users are served by closely located small cells with reduced path loss and small-scale antenna arrays. Figure 4.14(a) also demonstrates that the excess number of antennas available at the macro BS can then be employed to form a wireless backhaul to these small cells, although wired backhaul solutions are also strong candidates for 5G implementations. Particularly relevant in this context is

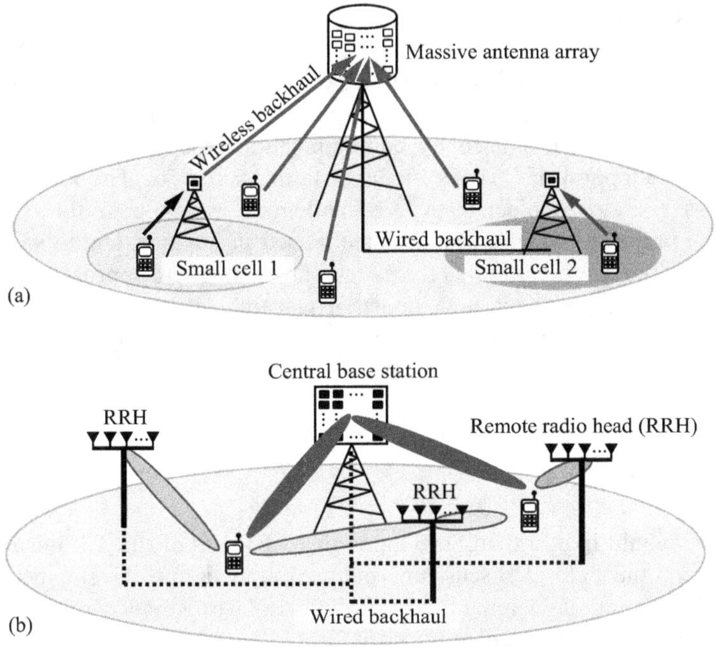

Figure 4.14 Illustration of (a) a multi-tier heterogeneous network comprised of an LSAS BS in one tier and small cells in a lower tier and (b) a distributed LSAS

the development of strategies that exploit the large number of antennas for suppressing the inter-tier interference and improving the network attainable rates [118,120]. Indeed, LSAS are also a flexible alternative for establishing wireless backhauls to remote radio heads in HetNets [121].

Another flourishing area of research in this context studies the impact of allocating the LSAS antennas in a distributed manner [54,122–124]. This network architecture, referred to as distributed LSAS, falls into the broader category of multi-cell MIMO cooperative networks [125]. An illustrative example of a distributed LSAS is shown in Figure 4.14(b), where a central BS is directly connected to a multiplicity of remote radio heads that are physically separated. As explicitly shown in Figure 4.14(b) these systems rely on sharing information such as the user data or the CSI between the BSs and remote radio heads for enhancing the performance and diversity of the network, at the expense of added backhauling overheads [54,122–124].

Overall, the discussion over the role of large-scale BSs in the network deployments is still in its early stages and further research is necessary to assess their potential and their performance with the techniques available in the current LTE standard, such as cell range expansion or almost-blank subframes [9,126].

4.4 Concluding remarks

This chapter has overviewed the principles of LSAS systems, as one of the dominant candidates for 5G communication networks and beyond. It is the remarkable advantages of LSAS, promised by communications theory that form the main driver for the adoption of LSAS systems in future networks. The implementation of LSAS is however subject to severe challenges, which pose the question of whether the promised benefits can be materialised in practice, and to what extent. This has been the central stimulus for the pursuit of practical approaches for LSAS deployment. While academic and industrial research on LSAS has been gaining momentum globally, a number of research problems remain open to this day, as identified throughout this chapter, promising a fruitful research field in the years to come.

Acknowledgements

The authors would like to thank the following Members of the Technical Staff in Bell Labs for the helpful discussions held throughout the development of this chapter: Dr Efstratios Doumanis, Dr Lei Guan, Mr Ray Kearney, Dr Pawel Rulikowski and Dr Vijay Venkateswaran.

References

[1] G. Li, Z. Xu, C. Xiong, *et al.*, "Energy-efficient wireless communications: Tutorial, survey, and open issues," *IEEE Wireless Communications*, vol. 18, no. 6, pp. 28–35, Dec. 2011.

[2] T. Chen, Y. Yang, H. Zhang, H. Kim, and K. Horneman, "Network energy saving technologies for green wireless access networks," *IEEE Wireless Communications*, vol. 18, no. 5, pp. 30–38, Oct. 2011.

[3] J. Andrews, S. Buzzi, W. Choi, *et al.*, "What will 5G be?" *IEEE Journal on Selected Areas in Communications*, vol. 32, no. 6, pp. 1065–1082, Jun. 2014.

[4] A. Ghosh, R. Ratasuk, B. Mondal, N. Mangalvedhe, and T. Thomas, "LTE-advanced: Next-generation wireless broadband technology [Invited Paper]," *IEEE Wireless Communications*, vol. 17, no. 3, pp. 10–22, Jun. 2010.

[5] T. Marzetta, "Noncooperative cellular wireless with unlimited numbers of base station antennas," *IEEE Transactions on Wireless Communications*, vol. 9, no. 11, pp. 3590–3600, Nov. 2010.

[6] E. Larsson, O. Edfors, F. Tufvesson, and T. Marzetta, "Massive MIMO for next generation wireless systems," *IEEE Communications Magazine*, vol. 52, no. 2, pp. 186–195, Feb. 2014.

[7] F. Rusek, D. Persson, B. K. Lau, *et al.*, "Scaling up MIMO: Opportunities and challenges with very large arrays," *IEEE Signal Processing Magazine*, vol. 30, no. 1, pp. 40–60, Jan. 2013.

[8] L. Lu, G. Li, A. Swindlehurst, A. Ashikhmin, and R. Zhang, "An overview of massive MIMO: Benefits and challenges," *IEEE Journal of Selected Topics in Signal Processing*, vol. 8, no. 5, pp. 742–758, Oct. 2014.

[9] K. Zheng, L. Zhao, J. Mei, B. Shao, W. Xiang, and L. Hanzo, "Survey of large-scale MIMO systems," *IEEE Communications Surveys Tutorials*, vol. 17, no. 3, pp. 1738–1760, third quarter 2015.

[10] D. Tse, and P. Viswanath, *Fundamentals of Wireless Communication*. Cambridge University Press, 2005.

[11] G. Foschini, and M. Gans, "On limits of wireless communications in a fading environment when using multiple antennas," *Wireless Personal Communications*, vol. 6, no. 3, pp. 311–335, 1998.

[12] Y. Jiang, M. Varanasi, and J. Li, "Performance analysis of ZF and MMSE equalizers for MIMO systems: An in-depth study of the high SNR regime," *IEEE Transactions on Information Theory*, vol. 57, no. 4, pp. 2008–2026, Apr. 2011.

[13] C. Oestges, and B. Clerckx, *MIMO Wireless Communications: From Real-World Propagation to Space-Time Code Design*. Access Online via Elsevier, 2010.

[14] L. Cimini, "Analysis and simulation of a digital mobile channel using orthogonal frequency division multiplexing," *IEEE Transactions on Communications*, vol. 33, no. 7, pp. 665–675, Jul. 1985.

[15] E. Björnson, E. G. Larsson, and M. Debbah, "Massive MIMO for Maximal Spectral Efficiency: How Many Users and Pilots Should Be Allocated?," *IEEE Transactions on Wireless Communications*, vol. 15, no. 2, pp. 1293–1308, Feb. 2016.

[16] P. Z. Peebles, J. Read, and P. Read, *Probability, Random Variables, and Random Signal Principles*. McGraw-Hill, New York, 2001, vol. 3.

[17] C. D. Meyer, *Matrix Analysis and Applied Linear Algebra*, vol. 2. SIAM, 2000.

[18] I. Viering, H. Hofstetter, and W. Utschick, "Validity of spatial covariance matrices over time and frequency," in *IEEE Global Telecommunications Conference (GLOBECOM),* vol. 1, November 2002, pp. 851–855.

[19] S. Wagner, R. Couillet, M. Debbah, and D. T. M. Slock, "Large system analysis of linear precoding in correlated MISO broadcast channels under limited feedback," *IEEE Transactions on Information Theory*, vol. 58, no. 7, pp. 4509–4537, Jul. 2012.

[20] J. Hoydis, S. ten Brink, and M. Debbah, "Massive MIMO in the UL/DL of cellular networks: How many antennas do we need?" *IEEE Journal on Selected Areas in Communications*, vol. 31, no. 2, pp. 160–171, Feb. 2013.

[21] T. M. Cover, and J. A. Thomas, *Elements of Information Theory*. John Wiley & Sons, 2012.

[22] C. Windpassinger, R. F. H. Fischer, T. Vencel, and J. Huber, "Precoding in multiantenna and multiuser communications," *IEEE Transactions on Wireless Communications*, vol. 3, no. 4, pp. 1305–1316, Jul. 2004.

[23] H. Q. Ngo, E. Larsson, and T. Marzetta, "Energy and spectral efficiency of very large multiuser MIMO systems," *IEEE Transactions on Communications*, vol. 61, no. 4, pp. 1436–1449, Apr. 2013.

[24] C. Masouros, "Correlation rotation linear precoding for MIMO broadcast communications," *IEEE Transactions on Signal Processing*, vol. 59, no. 1, pp. 252–262, Jan. 2011.

[25] C. Peel, B. Hochwald, and A. Swindlehurst, "A vector-perturbation technique for near-capacity multiantenna multiuser communication – Part I: Channel inversion and regularization," *IEEE Transactions on Communications*, vol. 53, no. 1, pp. 195–202, Jan. 2005.

[26] E. Björnson, E. G. Larsson, and T. L. Marzetta, "Massive MIMO: Ten myths and one critical question," *IEEE Communications Magazine*, vol. 54, no. 2, pp. 114–123, Feb. 2016.

[27] J. G. Proakis, *Digital Communications*. McGraw-Hill, 1995.

[28] J. Jose, A. Ashikhmin, T. Marzetta, and S. Vishwanath, "Pilot contamination and precoding in multi-cell TDD systems," *IEEE Transactions on Wireless Communications*, vol. 10, no. 8, pp. 2640–2651, Aug. 2011.

[29] Z. Jiang, A. Molisch, G. Caire, and Z. Niu, "Achievable rates of FDD massive MIMO systems with spatial channel correlation," *IEEE Transactions on Wireless Communications*, vol. 14, no. 5, pp. 2868–2882, May 2015.

[30] G. Smith, "A direct derivation of a single-antenna reciprocity relation for the time domain," *IEEE Transactions on Antennas and Propagation*, vol. 52, no. 6, pp. 1568–1577, Jun. 2004.

[31] J. Vieira, F. Rusek, and F. Tufvesson, "Reciprocity calibration methods for massive MIMO based on antenna coupling," in *IEEE Global Communications Conference (GLOBECOM)*, December 2014, pp. 3708–3712.

[32] H. Wei, D. Wang, H. Zhu, J. Wang, S. Sun, and X. You, "Mutual coupling calibration for multiuser massive MIMO systems," *IEEE Transactions on Wireless Communications*, vol. 15, no. 1, pp. 606–619, Jan. 2016.

[33] H. Ngo, E. Larsson, and T. Marzetta, "The multicell multiuser MIMO uplink with very large antenna arrays and a finite-dimensional channel," *IEEE Transactions on Communications*, vol. 61, no. 6, pp. 2350–2361, Jun. 2013.

[34] N. Shariati, E. Björnson, M. Bengtsson, and M. Debbah, "Low-complexity polynomial channel estimation in large-scale MIMO with arbitrary statistics," *IEEE Journal of Selected Topics in Signal Processing*, vol. 8, no. 5, pp. 815–830, Oct. 2014.

[35] S. M. Kay, *Fundamentals of Statistical Signal Processing, Volume I: Estimation Theory*. Prentice-Hall, 1993.

[36] F. Fernandes, A. Ashikhmin, and T. Marzetta, "Inter-cell interference in noncooperative TDD large scale antenna systems," *IEEE Journal on Selected Areas in Communications*, vol. 31, no. 2, pp. 192–201, Feb. 2013.

[37] H. Yin, D. Gesbert, M. Filippou, and Y. Liu, "A coordinated approach to channel estimation in large-scale multiple-antenna systems," *IEEE Journal on Selected Areas in Communications*, vol. 31, no. 2, pp. 264–273, Feb. 2013.

[38] M. Li, S. Jin, and X. Gao, "Spatial orthogonality-based pilot reuse for multi-cell massive MIMO transmission," in *International Conference on Wireless Communications Signal Processing (WCSP)*, October 2013, pp. 1–6.
[39] H. Huh, G. Caire, H. Papadopoulos, and S. Ramprashad, "Achieving 'massive MIMO' spectral efficiency with a not-so-large number of antennas," *IEEE Transactions on Wireless Communications*, vol. 11, no. 9, pp. 3226–3239, Sep. 2012.
[40] R. R. Mueller, M. Vehkaperae, and L. Cottatellucci, "Blind pilot decontamination," in *17th International ITG Workshop on Smart Antennas (WSA)*, March 2013, pp. 1–6.
[41] Y. Li, Y.-H. Nam, B. L. Ng, and J. Zhang, "A non-asymptotic throughput for massive MIMO cellular uplink with pilot reuse," in *IEEE Global Communications Conference (GLOBECOM)*, December 2012, pp. 4500–4504.
[42] J. Zhang, B. Zhang, S. Chen, X. Mu, M. El-Hajjar, and L. Hanzo, "Pilot contamination elimination for large-scale multiple-antenna aided OFDM systems," *IEEE Journal of Selected Topics in Signal Processing*, vol. 8, no. 5, pp. 759–772, Oct. 2014.
[43] L. Guan, R. Kearney, C. Yu, and A. Zhu, "High-performance digital predistortion test platform development for wideband RF power amplifiers," *International Journal of Microwave and Wireless Technologies*, vol. 5, no. 2, pp. 149–162, 2013.
[44] X. Wu, S. Sinanovic, M. Di Renzo, and H. Haas, "Base station energy consumption for transmission optimised spatial modulation (TOSM) in correlated channels," in *IEEE 17th International Workshop on Computer Aided Modeling and Design of Communication Links and Networks (CAMAD)*, 2012, pp. 261–265.
[45] G. Auer, V. Giannini, I. Godor, *et al.*, "Cellular energy efficiency evaluation framework," in *IEEE 73rd Vehicular Technology Conference (VTC Spring)*, 2011, pp. 1–6.
[46] T. Schenk, *RF Imperfections in High-Rate Wireless Systems: Impact and Digital Compensation*. Springer Science & Business Media, 2008.
[47] E. Björnson, J. Hoydis, M. Kountouris, and M. Debbah, "Massive MIMO systems with non-ideal hardware: Energy efficiency, estimation, and capacity limits," *IEEE Transactions on Information Theory*, vol. 60, no. 11, pp. 7112–7139, Nov. 2014.
[48] A. Pitarokoilis, S. Mohammed, and E. Larsson, "Uplink performance of time-reversal MRC in massive MIMO systems subject to phase noise," *IEEE Transactions on Wireless Communications*, vol. 14, no. 2, pp. 711–723, Feb. 2015.
[49] U. Gustavsson, C. Sanchez-Perez, T. Eriksson, *et al.*, "On the impact of hardware impairments on massive MIMO," in *GLOBECOM Workshops*, December 2014, pp. 294–300.
[50] X. Zhang, M. Matthaiou, M. Coldrey, and E. Björnson, "Impact of residual transmit RF impairments on training-based MIMO systems," *IEEE Transactions on Communications*, vol. 63, no. 8, pp. 2899–2911, Aug. 2015.

[51] J. Brady, N. Behdad, and A. Sayeed, "Beamspace MIMO for millimeter-wave communications: System architecture, modeling, analysis, and measurements," *IEEE Transactions on Antennas and Propagation*, vol. 61, no. 7, pp. 3814–3827, Jul. 2013.

[52] C. A. Balanis, *Antenna Theory: Analysis and Design*. John Wiley & Sons, 2012.

[53] X. Gao, O. Edfors, F. Rusek, and F. Tufvesson, "Massive MIMO performance evaluation based on measured propagation data," *IEEE Transactions on Wireless Communications*, vol. 14, no. 7, pp. 3899–3911, Jul. 2015.

[54] H. Q. Ngo, A. Ashikhmin, H. Yang, E. G. Larsson, and T. L. Marzetta, "Cell-free massive MIMO: Uniformly great service for everyone," in *IEEE 16th International Workshop on Signal Processing Advances in Wireless Communications (SPAWC)*, June 2015, pp. 201–205.

[55] C. Shepard, N. Anand, and L. Zhong, "Practical performance of MU-MIMO precoding in many-antenna base stations," in *Proceedings of the 2013 Workshop on Cellular Networks: Operations, Challenges, and Future Design*. ACM, 2013, pp. 13–18.

[56] J. Hoydis, C. Hoek, T. Wild, and S. ten Brink, "Channel measurements for large antenna arrays," in *International Symposium on Wireless Communication Systems (ISWCS)*, August 2012, pp. 811–815.

[57] J. Vieira, S. Malkowsky, K. Nieman, *et al.*, "A flexible 100-antenna testbed for massive MIMO," in *GLOBECOM Workshops*, December 2014, pp. 287–293.

[58] X. Artiga, J. Perruisseau-Carrier, and A. Perez-Neira, "Antenna array configurations for massive MIMO outdoor base stations," in *IEEE Sensor Array and Multichannel Signal Processing Workshop (SAM)*, June 2014, pp. 281–284.

[59] C. Masouros, and M. Matthaiou, "Space-constrained massive MIMO: Hitting the wall of favorable propagation," *IEEE Communications Letters*, vol. 19, no. 5, pp. 771–774, May 2015.

[60] C. Masouros, M. Sellathurai, and T. Ratnarajah, "Large-scale MIMO transmitters in fixed physical spaces: The effect of transmit correlation and mutual coupling," *IEEE Transactions on Communications*, vol. 61, no. 7, pp. 2794–2804, Jul. 2013.

[61] S. Biswas, C. Masouros, and T. Ratnarajah, "On the effect of antenna correlation and coupling on the energy-efficiency of massive MIMO systems," *IEEE Transactions on Wireless Communications*, vol. 15, no. 5, pp. 3492–3505, May 2016.

[62] G. Dahman, J. Flordelis, and F. Tufvesson, "Experimental evaluation of the effect of BS antenna inter-element spacing on MU-MIMO separation," in *IEEE International Conference on Communications (ICC)*, June 2015, pp. 1685–1690.

[63] A. Kammoun, A. Muller, E. Björnson, and M. Debbah, "Linear precoding based on polynomial expansion: Large-scale multi-cell MIMO systems," *IEEE Journal of Selected Topics in Signal Processing*, vol. 8, no. 5, pp. 861–875, Oct. 2014.

[64] C. Erdmann, D. Lowney, A. Lynam, *et al.*, "A heterogeneous 3D-IC consisting of two 28 nm FPGA die and 32 reconfigurable high-performance data converters," in *IEEE International Solid-State Circuits Conference Digest of Technical Papers (ISSCC)*, February 2014, pp. 120–121.

[65] H. Yang, and T. Marzetta, "Performance of conjugate and zero-forcing beamforming in large-scale antenna systems," *IEEE Journal on Selected Areas in Communications*, vol. 31, no. 2, pp. 172–179, Feb. 2013.

[66] S. Zarei, W. Gerstacker, R. R. Muller, and R. Schober, "Low-complexity linear precoding for downlink large-scale MIMO systems," in *IEEE 24th International Symposium on Personal Indoor and Mobile Radio Communications (PIMRC)*, September 2013, pp. 1119–1124.

[67] H. Prabhu, J. Rodrigues, O. Edfors, and F. Rusek, "Approximative matrix inverse computations for very-large MIMO and applications to linear pre-coding systems," in *IEEE Wireless Communications and Networking Conference (WCNC)*, 2013, pp. 2710–2715.

[68] E. Björnson, L. Sanguinetti, J. Hoydis, and M. Debbah, "Optimal design of energy-efficient multi-user MIMO systems: Is massive MIMO the answer?" *IEEE Transactions on Wireless Communications*, vol. 14, no. 6, pp. 3059–3075, Jun. 2015.

[69] L. Vandenberghe, "Applied numerical computing," 2013. [Online]. Available at www.seas.ucla.edu/vandenbe/103/lectures/lu.pdf

[70] A. Tulino, A. Lozano, and S. Verdu, "Impact of antenna correlation on the capacity of multiantenna channels," *IEEE Transactions on Information Theory*, vol. 51, no. 7, pp. 2491–2509, Jul. 2005.

[71] V. Raghavan, and A. Sayeed, "Sublinear capacity scaling laws for sparse MIMO channels," *IEEE Transactions on Information Theory*, vol. 57, no. 1, pp. 345–364, Jan. 2011.

[72] A. Adhikary, J. Nam, J.-Y. Ahn, and G. Caire, "Joint spatial division and multiplexing – The large-scale array regime," *IEEE Transactions on Information Theory*, vol. 59, no. 10, pp. 6441–6463, Oct. 2013.

[73] B. Clerckx, G. Kim, and S. Kim, "Correlated fading in broadcast MIMO channels: Curse or blessing?" in *IEEE Global Telecommunications Conference (GLOBECOM)*, November 2008, pp. 1–5.

[74] J. Choi, and D. Love, "Bounds on eigenvalues of a spatial correlation matrix," *IEEE Communications Letters*, vol. 18, no. 8, pp. 1391–1394, Aug. 2014.

[75] B. Lee, J. Choi, J. Seol, D. Love, and B. Shim, "Antenna grouping based feedback compression for FDD-based massive MIMO systems," *IEEE Transactions on Communications*, vol. 63, no. 9, pp. 3261–3274, Sep. 2015.

[76] W. C. Jakes, and D. C. Cox, *Microwave Mobile Communications*. Wiley-IEEE Press, 1994.

[77] S. Noh, M. Zoltowski, Y. Sung, and D. Love, "Pilot beam pattern design for channel estimation in massive MIMO systems," *IEEE Journal of Selected Topics in Signal Processing*, vol. 8, no. 5, pp. 787–801, Oct. 2014.

[78] S. Wu, C.-X. Wang, E.-H. Aggoune, M. Alwakeel, and Y. He, "A non-stationary 3-D wideband twin-cluster model for 5G massive MIMO

channels," *IEEE Journal on Selected Areas in Communications*, vol. 32, no. 6, pp. 1207–1218, Jun. 2014.

[79] S. Wu, C.-X. Wang, H. Haas, E.-H. Aggoune, M. Alwakeel, and B. Ai, "A non-stationary wideband channel model for massive MIMO communication systems," *IEEE Transactions on Wireless Communications*, vol. 14, no. 3, pp. 1434–1446, Mar. 2015.

[80] S. L. H. Nguyen, and A. Ghrayeb, "Precoding for multicell massive MIMO systems with compressive rank-*q* channel approximation," in *IEEE 24th International Symposium on Personal Indoor and Mobile Radio Communications (PIMRC)*, September 2013, pp. 1227–1232.

[81] S. L. H. Nguyen, and A. Ghrayeb, "Compressive sensing-based channel estimation for massive multiuser MIMO systems," in *IEEE Wireless Communications and Networking Conference (WCNC)*, 2013, pp. 2890–2895.

[82] E. Björnson, L. Sanguinetti, and M. Kountouris, "Energy-efficient future wireless networks: A marriage between massive MIMO and small cells," in *IEEE 16th International Workshop on Signal Processing Advances in Wireless Communications (SPAWC)*, June 2015, pp. 211–215.

[83] J. Wallace, and M. Jensen, "Mutual coupling in MIMO wireless systems: A rigorous network theory analysis," *IEEE Transactions on Wireless Communications*, vol. 3, no. 4, pp. 1317–1325, Jul. 2004.

[84] M. Morris, and M. Jensen, "Network model for MIMO systems with coupled antennas and noisy amplifiers," *IEEE Transactions on Antennas and Propagation*, vol. 53, no. 1, pp. 545–552, Jan. 2005.

[85] X. Artiga, B. Devillers, and J. Perruisseau-Carrier, "On the selection of radiating elements for compact indoor massive-multiple input multiple output base stations," *IET Microwaves, Antennas Propagation*, vol. 8, no. 1, pp. 1–9, Jan. 2014.

[86] N. Maleki, E. Karami, and M. Shiva, "Optimization of antenna array structures in mobile handsets," *IEEE Transactions on Vehicular Technology*, vol. 54, no. 4, pp. 1346–1351, Jul. 2005.

[87] R. Janaswamy, "Effect of element mutual coupling on the capacity of fixed length linear arrays," *IEEE Antennas and Wireless Propagation Letters*, vol. 1, no. 1, pp. 157–160, 2002.

[88] R. Zi, X. Ge, H. Wang, J. Zhang, and C.-X. Wang, "Multiuser massive MIMO uplink performance with mutual coupling effects," in *IEEE Global Communications Conference (GLOBECOM)*, December 2014, pp. 3296–3301.

[89] X. Gao, F. Tufvesson, and O. Edfors, "Massive MIMO channels measurements and models," in *IEEE Asilomar Conference on Signals, Systems and Computers*, November 2013, pp. 280–284.

[90] X. Gao, O. Edfors, F. Tufvesson, and E. G. Larsson, "Massive MIMO in real propagation environments: Do all antennas contribute equally?" *IEEE Transactions on Communications*, vol. 63, no. 11, pp. 3917–3928, Nov. 2015.

[91] C. Shepard, A. Javed, and L. Zhong, "Control channel design for many-antenna MU-MIMO," in *Proceedings of the 21st Annual International Conference on Mobile Computing and Networking*. ACM, 2015, pp. 578–591.

[92] S. Payami, and F. Tufvesson, "Channel measurements and analysis for very large array systems at 2.6 GHz," in *Sixth European Conference on Antennas and Propagation (EUCAP)*, March 2012, pp. 433–437.

[93] G. Auer, O. Blume, V. Giannini, *et al*., "D2.3: Energy efficiency analysis of the reference systems, areas of improvements and target breakdown," *INFSOICT-247733 EARTH (Energy Aware Radio and NeTwork TecHnologies), Tech. Rep*, 2010.

[94] S. Sanayei, and A. Nosratinia, "Antenna selection in MIMO systems," *IEEE Communications Magazine*, vol. 42, no. 10, pp. 68–73, Oct. 2004.

[95] H. Li, L. Song, and M. Debbah, "Energy efficiency of large-scale multiple antenna systems with transmit antenna selection," *IEEE Transactions on Communications*, vol. 62, no. 2, pp. 638–647, Feb. 2014.

[96] P. Amadori, and C. Masouros, "Interference driven antenna selection for massive multi-user MIMO," *IEEE Transactions on Vehicular Technology*, vol. 65, no. 8, pp. 5944–5958, Aug. 2016.

[97] G. Zheng, I. Krikidis, C. Masouros, S. Timotheou, D.-A. Toumpakaris, and Z. Ding, "Rethinking the role of interference in wireless networks," *IEEE Communications Magazine*, vol. 52, no. 11, pp. 152–158, Nov. 2014.

[98] A. Liu and V. Lau, "Phase only RF precoding for massive MIMO systems with limited RF chains," *IEEE Transactions on Signal Processing*, vol. 62, no. 17, pp. 4505–4515, Sep. 2014.

[99] O. El Ayach, S. Rajagopal, S. Abu-Surra, Z. Pi, and R. Heath, "Spatially sparse precoding in millimeter wave MIMO systems," *IEEE Transactions on Wireless Communications*, vol. 13, no. 3, pp. 1499–1513, Mar. 2014.

[100] T. Rappaport, G. MacCartney, M. Samimi, and S. Sun, "Wideband millimeter-wave propagation measurements and channel models for future wireless communication system design," *IEEE Transactions on Communications*, vol. 63, no. 9, pp. 3029–3056, Sep. 2015.

[101] P. Amadori and C. Masouros, "Low RF-complexity millimeter-wave beamspace-MIMO systems by beam selection," *IEEE Transactions on Communications*, vol. 63, no. 6, pp. 2212–2223, Jun. 2015.

[102] D. Ying, F. W. Vook, T. A. Thomas, and D. J. Love, "Hybrid structure in massive MIMO: Achieving large sum rate with fewer RF chains," in *IEEE International Conference on Communications (ICC)*, June 2015, pp. 2344–2349.

[103] A. Sayeed, "Deconstructing multiantenna fading channels," *IEEE Transactions on Signal Processing*, vol. 50, no. 10, pp. 2563–2579, Oct. 2002.

[104] A. Garcia-Rodriguez, and C. Masouros, "Exploiting the increasing correlation of space constrained massive MIMO for CSI relaxation," *IEEE Transactions on Communications*, vol. 64, no. 4, pp. 1572–1587, Apr. 2016.

[105] X. Gao, L. Dai, J. Zhang, S. Han, and C.-L. I, "Capacity-approaching linear precoding with low-complexity for large-scale MIMO systems," in *IEEE International Conference on Communications (ICC)*, June 2015, pp. 1577–1582.

[106] W. Roh, J.-Y. Seol, J. Park, *et al.*, "Millimeter-wave beamforming as an enabling technology for 5G cellular communications: Theoretical feasibility and prototype results," *IEEE Communications Magazine*, vol. 52, no. 2, pp. 106–113, Feb. 2014.

[107] R. Mesleh, H. Haas, S. Sinanovic, C. W. Ahn, and S. Yun, "Spatial modulation," *IEEE Transactions on Vehicular Technology*, vol. 57, no. 4, pp. 2228–2241, Jul. 2008.

[108] M. Di Renzo, H. Haas, and P. M. Grant, "Spatial modulation for multiple-antenna wireless systems: A survey," *IEEE Communications Magazine*, vol. 49, no. 12, pp. 182–191, Dec. 2011.

[109] M. Di Renzo, H. Haas, A. Ghrayeb, S. Sugiura, and L. Hanzo, "Spatial modulation for generalized MIMO: Challenges, opportunities, and implementation," *Proceedings of the IEEE*, vol. 102, no. 1, pp. 56–103, 2014.

[110] P. Yang, Y. Xiao, Y. L. Guan, *et al.*, "Single-carrier spatial modulation: A promising design for large-scale broadband antenna systems," *IEEE Communications Surveys Tutorials*, vol. PP, no. 99, pp. 1–1, 2016.

[111] A. Garcia-Rodriguez, and C. Masouros, "Low-complexity compressive sensing detection for spatial modulation in large-scale multiple access channels," *IEEE Transactions on Communications*, vol. 63, no. 7, pp. 2565–2579, Jul. 2015.

[112] J. Xu, and L. Qiu, "Energy efficiency optimization for MIMO broadcast channels," *IEEE Transactions on Wireless Communications*, vol. 12, no. 2, pp. 690–701, Feb. 2013.

[113] F. Heliot, M. Imran, and R. Tafazolli, "On the energy efficiency-spectral efficiency trade-off over the MIMO Rayleigh fading channel," *IEEE Transactions on Communications*, vol. 60, no. 5, pp. 1345–1356, May 2012.

[114] W. Liu, S. Han, C. Yang, and C. Sun, "Massive MIMO or small cell network: Who is more energy efficient?" in *IEEE Wireless Communications and Networking Conference Workshops (WCNCW)*, April 2013, pp. 24–29.

[115] E. Björnson, M. Kountouris, and M. Debbah, "Massive MIMO and small cells: Improving energy efficiency by optimal soft-cell coordination," in *20th International Conference on Telecommunications (ICT)*, May 2013, pp. 1–5.

[116] D. Ha, K. Lee, and J. Kang, "Energy efficiency analysis with circuit power consumption in massive MIMO systems," in *IEEE 24th International Symposium on Personal Indoor and Mobile Radio Communications (PIMRC)*, September 2013, pp. 938–942.

[117] W. Liu, S. Han, and C. Yang, "Energy efficiency comparison of massive MIMO and small cell network," in *IEEE Global Conference on Signal and Information Processing (GlobalSIP)*, December 2014, pp. 617–621.

[118] K. Hosseini, J. Hoydis, S. ten Brink, and M. Debbah, "Massive MIMO and small cells: How to densify heterogeneous networks?" in *IEEE International Conference on Communications (ICC)*, June 2013, pp. 5442–5447.

[119] M. Kountouris, and N. Pappas, "HetNets and massive MIMO: Modeling, potential gains, and performance analysis," in *IEEE-APS Topical*

Conference on Antennas and Propagation in Wireless Communications (APWC), September 2013, pp. 1319–1322.

[120] J. Hoydis, K. Hosseini, S. ten Brink, and M. Debbah, "Making smart use of excess antennas: Massive MIMO, small cells, and TDD," *Bell Labs Technical Journal*, vol. 18, no. 2, pp. 5–21, 2013.

[121] T. Marzetta, and H. Yang, "Dedicated LSAS for macro-cell wireless backhaul – Part I: Downlink," *Bell Laboratories, Alcatel-Lucent*, Tech. Rep., 2012.

[122] Z. Liu, and L. Dai, "A comparative study of downlink MIMO cellular networks with co-located and distributed base-station antennas," *IEEE Transactions on Wireless Communications*, vol. 13, no. 11, pp. 6259–6274, Nov. 2014.

[123] K. Hosseini, W. Yu, and R. Adve, "Large-scale MIMO versus network MIMO for multicell interference mitigation," *IEEE Journal of Selected Topics in Signal Processing*, vol. 8, no. 5, pp. 930–941, Oct. 2014.

[124] K. Truong, and R. Heath, "The viability of distributed antennas for massive MIMO systems," in *Asilomar Conference on Signals, Systems and Computers*, November 2013, pp. 1318–1323.

[125] D. Gesbert, S. Hanly, H. Huang, S. Shamai Shitz, O. Simeone, and W. Yu, "Multi-cell MIMO cooperative networks: A new look at interference," *IEEE Journal on Selected Areas in Communications*, vol. 28, no. 9, pp. 1380–1408, Dec. 2010.

[126] D. López-Pérez, M. Ding, H. Claussen, and A. H. Jafari, "Towards 1 Gbps/UE in cellular systems: Understanding ultra-dense small cell deployments," *IEEE Communications Surveys Tutorials*, vol. 17, no. 4, pp. 2078–2101, fourth quarter 2015.

Chapter 5

Effects of densification and randomness of infrastructure deployment in cellular networks

Stelios Stefanatos[1], Antonis G. Gotsis[1] and Angeliki Alexiou[1]

5.1 Introduction

5.1.1 Infrastructure densification in future cellular networks

The exponential increase of wireless data traffic has made clear that current cellular networks will not be able to provide acceptable service quality in the near future. This observation has shifted the focus of the wireless industry and research community toward defining and eventually implementing a new (fifth) generation of cellular networks that is able to support extremely large data rates per (wireless) link as well as an extremely large number of concurrent links [1]. Toward this end, various technological enablers have been identified [2,3] which can be roughly partitioned into three major categories:

- Advanced transmission techniques achieving high spectral efficiency, e.g., massive MIMO.
- Increase of transmission bandwidth, e.g., exploitation of mmWave frequencies.
- Re-design of system architecture.

The first two categories have been present in the evolution of cellular networks since their beginning and may be viewed as "standard" approaches toward performance enhancement (of course, this does not imply that there are no significant technical challenges associated with them). The third category on the other hand, constitutes a revolutionary (rather than evolutionary) change of a fundamentally stable attribute of the cellular network, namely, its architecture. In particular, the traditional concept of the cellular network consisting of base stations distributed on a, more or less, sparse and regular pattern is becoming outdated. The most important new architectural concept, already partially implemented by current cellular networks, is that of "small cells," generated by deploying small-size, small-range

[1]Department of Digital Systems, University of Piraeus, Androutsou 150, Pireaus, 18532, Greece

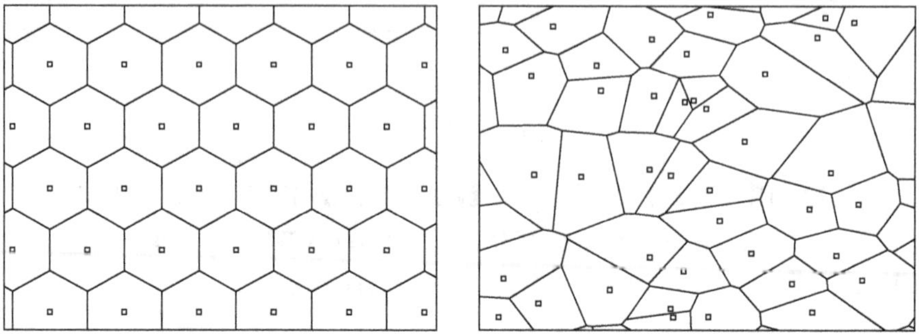

(a) (b)

Figure 5.1 Comparison of infrastructure deployments: (a) regular distribution of APs, modeling transitional cellular networks and (b) random distribution of APs, modeling future dense networks. Marks depict APs (base stations) and lines depict the (geometrical) boundaries of the cells

access points (APs), essentially providing "capacity injections" to the traditional network, thus introducing the notion of heterogeneous network (HetNet) [4].

Clearly, the notion of infrastructure *densification* [5,6] is directly related to small cells networking. Infrastructure densification has been naturally present throughout the evolution of cellular networks, however, its role in future networks is expected to be much more prevalent, not only limited to "fixed" APs but also including "opportunistic" APs corresponding to new envisioned access methods such as device-to-device communications. Terms like "ultra dense" or "hyper dense" network(s) are becoming increasingly popular in the literature [6,7], referring to extremely high infrastructure density levels (in APs per unit area), including scenarios where infrastructure density is of the same order as (or even larger than) the density of user equipment (UEs). This extreme densification of infrastructure brings about a new system attribute, namely, (quasi-) *randomness* of AP positions, that was not present in previous cellular network generations (see Figure 5.1). This randomness is the result of practical difficulties in operators deploying infrastructure in arbitrary positions, use of user-deployed equipment, and exploitation of UEs as content providers.

5.1.2 Analytical modeling of future cellular networks

It is clear that densification and randomness of infrastructure deployment results in a new operational environment, where interference levels are no longer controllable by careful AP placement and can potentially become extremely large under certain (random) deployment configurations. It is therefore of significant interest to understand the technical implications of this new environment as well as the potential of extreme infrastructure densification as a means to satisfy the future network requirements. Obviously, the well proven engineering methods of system analysis/design by formulation and solution of appropriate optimization problems can be applied for this purpose, and this has indeed been done extensively in the literature. However, although valuable, these methods suffer from a severe limitation: they are inherently limited to a finite (and typically small) number of transmitters/receivers, and, therefore, cannot reliably capture the effects of *large-scale* and dense infrastructure deployments, critical attributes of the future network. In addition, the outcomes of these studies are typically in algorithmic form, with limited analytical insights.

Toward an analytical investigation of the performance of future ultra-dense networks taking explicitly into account randomness of AP positions and large-scale deployments, it is natural to consider tools from the field of (two-dimensional) random point processes, commonly referred to as stochastic geometry [8]. Stochastic geometry tools have been applied for decades for modeling various systems, however, they were first employed as a tool to investigate communication network problems around 1996 [9,10]. Concerning wireless/mobile networks, stochastic geometry was initially (and still is) employed heavily in ad hoc network studies [11,12], which is only natural, as the fundamental characteristic of these networks is the random positions of the wireless nodes. Application to cellular networks appeared only recently in the seminal work of [13], whose elegant results and insights initiated a huge interest within the cellular research community to stochastic geometry methods, resulting in a vast amount of related papers [14] significantly contributing to the understanding of cellular system performance analysis and design. The recently published excellent books and monographs on applications of stochastic geometry for wireless networks [15–19] indicate that the field has already reached a high level of maturity, and (partial) acquaintance with related techniques and/or results appears to be a necessity for the wireless/cellular system engineer/researcher.

5.1.3 Purpose and organization of the chapter

The purpose of this chapter is two-fold. First, to characterize system performance under dense and random infrastructure deployments that will aid in obtaining design insights and, second, to demonstrate how tools from stochastic geometry can be applied to investigate these issues. Numerous representative operational scenarios and performance metrics are considered and some of the most important state-of-the-art results are presented in sufficient detail to be considered as a (technical) tutorial to newcomers in this relatively new research field. It is noted that the contents of this chapter are by no means an exhaustive summary of the results and techniques that have appeared in the field in recent years. However, it is hoped that it provides a good overview of fundamental analysis methods that are common in most related works.

The chapter is organized as follows. Section 5.2 provides an overview/summary of the most important tools from stochastic geometry that are useful for wireless network analysis. Section 5.3 presents in detail the baseline system model that will be considered throughout this chapter and relevant system performance metrics. It also includes a discussion on the important topic of AP–UE association criteria.

Specific operational scenarios are examined in detail considering the signal-to-interference ratio and the user rate as the metrics of interest in Sections 5.4 and 5.5, respectively. Finally, Section 5.6 concludes the chapter.

Notation: $\mathbb{P}(\cdot)$ and $\mathbb{P}(\cdot|\cdot)$ denote the probability and conditional probability measures, respectively. $\mathbb{E}(\cdot)$ and $\mathbb{E}(\cdot|\cdot)$ denote expectation and conditional expectation operators, respectively, and $\mathbb{I}(\cdot)$ is the indicator function. The Euclidean norm of $x \in \mathbb{R}^2$ is denoted as $|x|$ and the area of a bounded set $\mathscr{A} \subset \mathbb{R}^2$ is denoted as $|\mathscr{A}|$.

5.2 Review of homogeneous Poisson point processes

This section provides a short review of fundamental properties of spatial, homogeneous Poisson Point Processes (PPPs), as well as some of the most important and common tools employed in the analysis of cellular systems based on the homogeneous PPP model. The presentation is informal in the sense that a mathematically rigorous description/derivation of results, which requires knowledge of measure theory, is not followed here. However, for the engineering applications considered in this chapter, this level of rigorousness is unnecessary, and the reader is referred to the excellent texts [8,15] for a more advanced and in-depth treatment of the topic.

5.2.1 Fundamental properties

The two-dimensional Euclidean space $\mathbb{R}^2$ will be considered throughout this chapter. A point process $\mathbf{\Phi} = \{x_i\}_{i=1,2,\ldots} \subset \mathbb{R}^2$ is a countable set of randomly placed points in $\mathbb{R}^2$ with no accumulation points, i.e., the number of points contained within any bounded subset of $\mathbb{R}^2$ is finite. Out of the many types of point processes studied in the literature, the most common is by far the *homogeneous* PPP.[1] A PPP is uniquely described by a single parameter $\lambda > 0$, referred to as *density* (or intensity) of the process, and has the following fundamental statistical properties [16]:

1. The number of points $N(\mathscr{A})$ in any bounded set $\mathscr{A} \subset \mathbb{R}^2$ is a Poisson random variable with mean $\lambda|\mathscr{A}|$, i.e., $\mathbb{P}(N(\mathscr{A}) = k) = \frac{(\lambda|\mathscr{A}|)^k}{k!}e^{-\lambda|\mathscr{A}|}$, $k = 0, 1, \ldots$, where $|\mathscr{A}|$ is the area (Lebesgue measure) of $\mathscr{A}$. Note that this property implies that λ is in units of *average* number of points per unit area, i.e., λ provides an indication of how densely the points of the PPP are distributed on the plane.
2. Given the number of points N of a PPP in any bounded set $\mathscr{A} \subset \mathbb{R}^2$, the positions of these N points are independently and uniformly distributed in $\mathscr{A}$.
3. For any two disjoint bounded sets $\mathscr{A} \subset \mathbb{R}^2$ and $\mathscr{B} \subset \mathbb{R}^2$, $\mathscr{A} \cap \mathscr{B} = \emptyset$, the number of points of a PPP in these regions, $N(\mathscr{A})$ and $N(\mathscr{B})$, respectively, are independent.

It can be shown that the above properties imply that the PPP is simple, i.e., no two points of $\mathbf{\Phi}$ are the same almost surely (although they can be arbitrarily close in Euclidean distance). Properties 2 and 3 are referred to as complete (or maximum) randomness [16].

[1] In the following, the abbreviation PPP will refer to the homogeneous Poisson point process.

Two important properties of the PPP that follow from of its definition but are of great importance that deserve explicit mentioning are the following.

4. Superposition of PPPs: The point process $\Phi_1 \cup \Phi_2$, generated by the superposition of two independent PPPs $\Phi_1$and Φ_2 of densities $\lambda_1$and λ_2, respectively, is also a PPP of density $\lambda_1 + \lambda_2$.
5. Thinning of PPP: Let $\Phi = \{x_k\}$ denote a PPP of density λ. The point process $\Phi' \subseteq \Phi$ resulting by independently removing from Φ the point x_k with a probability $1 - p$ for each $x_k \in \Phi$, is also a PPP of density $p\lambda$.

An important extension of the concept of point processes is that of a *marked* point process, which is a point process Φ with each point $x \in \Phi$ associated with a random vector (mark) $m \in \mathbb{R}^d$, where $d \geq 1$ is the dimension of the space of marks. A marked point process is a useful tool that allows for a compact representation of not only the (random) positions of a set of elements but also of their individual attributes. For example, when Φ represents the positions of APs of a cellular network, a mark may represent parameters of interest such as transmit power and the number of associated (served) UEs, whose values are typically random and (jointly) depend on the locations of all the points of Φ as well as other external parameters.

Note that a marked point process is equivalent to a non-marked point process by trivially setting all the marks equal to some arbitrary fixed value. Of particular interest for analytical purposes is the *independently marked* PPP, where the marks are independent and identically distributed, with the mark m_x of a point $x \in \Phi$ depending only on x, i.e.,

$$p(m_x|\Phi) = p(m_x|x), \text{for all } x \in \Phi,$$

where $p(m_x|\cdot)$ denotes the conditional probability density function of m_x (assumed to exist).

5.2.2 Computation of functionals

Typically, one wants to investigate the statistical properties of functionals whose values depend on the points and marks of a point process Φ. This is in general a difficult task. However, for the case of an independently marked PPP, the expected value of certain types of "separable" functionals can be obtained in integral form, which, in many cases of practical interest, lead to simple closed form expressions.

Consider any (measurable) function f:$\mathbb{R}^2 \times \mathbb{R}^d \mapsto [0, \infty)$ and a functional of the form $f_s(\{(x, m_x)\}_{x \in \Phi \cap \mathscr{A}}) \triangleq \sum_{x \in \Phi \cap \mathscr{A}} f(x, m_x)$, where Φ is an independently marked PPP and $\mathscr{A} \subseteq \mathbb{R}^2$. The Campbell's theorem [17] states that:

$$\mathbb{E}(f_s) = \lambda \int_{\mathbb{R}^2 \cap \mathscr{A}} \mathbb{E}(f(x, m_x)|x) dx. \tag{5.1}$$

As an application of (5.1), consider the functional $I = \sum_{x \in \Phi} P_x l(|x|)$, where $\{P_x\}_{x \in \Phi}$ is a bounded sequence of positive numbers, and l :$[0, \infty) \mapsto [0, \infty)$ is a monotonically decreasing function. This functional may represent the total power received by a receiver located at the origin of the plane from transmissions of APs whose positions are distributed as an PPP $\mathbf{\Phi}$ of density λ, each AP transmitting with a random i.i.d. power P_x and $l(\cdot)$ is a deterministic path-loss model which depends only on the Euclidean distance between source and destination. Direct application of (5.1) shows that the average received power equals:

$$\begin{aligned}\mathbb{E}(I) &= \lambda\int_{\mathbb{R}^2}\mathbb{E}(P_x|x)l(|x|)dx \\ &= 2\pi\lambda\mathbb{E}(P_x)\int_0^{\infty} rl(r)dr,\end{aligned} \tag{5.2}$$

where the second equality holds in the case where P_x is independent of x and by switching to polar coordinates for the integral. Note that $\mathbb{E}(I)$ requires evaluation of an integral, which, in the worst case, may be evaluated numerically (if it exists). Even so, the corresponding computational complexity is much smaller than a straightforward, simulation-based (Monte Carlo) estimate of the functional. Interestingly, for the common path-loss model $l(|x|) = |x|^{-\alpha}$, $\alpha > 0$, [20] the above integral does not exist, i.e., the average received power is unbounded.[2] This is due to the fact that the PPP model allows for realizations where there exist points arbitrarily close to the origin (or any other point for that matter).

The second type of functionals for which integral-form expressions for their expectation can be found are of the form $f_p(\{(x, m_x)\}_{x\in\mathbf{\Phi}\cap\mathscr{A}}) \triangleq \prod_{x\in\mathbf{\Phi}\cap\mathscr{A}} f(x, m_x)$, with $f(\cdot,\cdot)$ any positive-valued measurable function as the ones considered in Campbell's theorem, $\mathbf{\Phi}$ an independently marked PPP, and $\mathscr{A} \subseteq \mathbb{R}^2$. It can be shown that [17]:

$$\mathbb{E}(f_p) = \exp\left(-\lambda\int_{\mathbb{R}^2\cap\mathscr{A}}[1 - \mathbb{E}(f(x, m_x)|x)]dx\right). \tag{5.3}$$

Equation (5.3) is often referred to as the probability generating functional (PGFL) formula for a PPP of density λ [17]. The PGFL formula is probably the most powerful analytical tool for PPPs and is routinely employed in almost every wireless network study based on PPP models. Many examples of its application will be provided in the following.

5.2.3 Conditioning, stationarity, and ergodicity

Another important property of a PPP $\mathbf{\Phi}$ concerns its statistical characterization given that there exists a point $x_0 \in \mathbf{\Phi}$ for some arbitrary $x_0 \in \mathbb{R}^2$. For example, one may be interested in examining the properties of a wireless network of APs distributed as a PPP, given the existence of an AP at a certain position of interest.

[2]Note that an infinite mean value of a random variable does not necessarily imply that the realizations of the variable are unbounded. Actually, for the case of I with $l(|x|) = x^{-\alpha}$, it can be shown that I is finite with probability 1, when $\alpha > 2$, even though $\mathbb{E}(I)$ is infinite [15, Remark 2.3.5].

Note that this configuration extends the well-known notion of conditioned, vector random variables [21] to the space of point processes.

In principle, one expects that this conditioning will result in a change of the statistical properties of $\mathbf{\Phi}$, similar to the case of "conventional" random variables. Remarkably, this is not the case when $\mathbf{\Phi}$ is a PPP. In particular, one can rigorously show under the framework of Palm theory [15] that the statistical properties of the points of a conditioned PPP *excluding* the one at x_0 are the same as those of the points of an unconditioned PPP. In other words, in order to generate a sample of a PPP that contains a point at x_0, one should draw a sample from an unconditioned PPP and add an extra point at x_0 (the sample of the unconditioned PPP will not contain x_0 with probability 1).

An important application of this property of the PPP is the extension of the Campbell's theorem with the positive-valued function f depending not only on the position of a certain point and its mark, but also on the positions of all the other points of the PPP. In particular, it holds:

$$\mathbb{E} \sum_{x \in \mathbf{\Phi} \cap \mathscr{A}} f(x, m_x, \mathbf{\Phi} \backslash \{x\}) = \lambda \int_{\mathscr{A}} \mathbb{E}(f(x, m_x, \mathbf{\Phi}))dx, \tag{5.4}$$

where the expectation inside the integral is w.r.t. the statistics of m_x (given x) and $\mathbf{\Phi}$.

The last two important properties of the PPP possesses are stationarity and ergodicity. Stationarity means that the statistical properties of $\mathbf{\Phi}$ remain unchanged when its points are shifted by a vector $v \in \mathbb{R}^2$. Due to stationarity, it follows that the properties of a PPP conditioned on the existence of a point at x_0 are independent of x_0. This is the reason that x_0 is referred to as the position of the "*typical point*," usually considered to be the origin, i.e., $x_0 = (0, 0)$, for mathematical convenience. It can be shown that a PPP is also isotropic, i.e., its statistical properties remain unchanged by a common rotation of its points in $\mathbb{R}^2$.

Ergodicity means that *for any* realization ϕ of a PPP $\mathbf{\Phi}$, the spatial average of functionals of the form $f(\phi \backslash \{x\})$, $x \in \phi$, equals the ensemble average of this functional evaluated at the typical point, i.e., [15]:

$$\lim_{n \to \infty} \frac{1}{N_\phi(\mathscr{A}_n)} \sum_{x \in \phi \cap \mathscr{A}_n} f(\phi \backslash \{x\}) = \mathbb{E}(f(\mathbf{\Phi} \backslash \{x_0\})),$$

where $\{\mathscr{A}_n\}$ is any well-behaved sequence of convex subsets of $\mathbb{R}^2$ with $\mathscr{A}_n \subset \mathscr{A}_{n+1}$, $N_\phi(\mathscr{A})$ is the number of points of ϕ contained in $\mathscr{A}$, and x_0 is the (arbitrarily positioned) typical point. As an example, consider again the case where the positions of transmitting APs are distributed as a PPP. The (spatial) average interference experienced at the AP positions for any realization of the PPP equals the (ensemble) average interference experienced by an AP located at the typical point, which is equal to (5.2). The importance of the ergodicity property cannot be emphasized enough. For simplicity, the same notation, $\mathbf{\Phi}$, will be used in the following to denote both a PPP and its realizations.

5.3 System model and performance metrics

In order to allow for a tractable analysis of cellular networks, it is necessary to consider certain modeling assumptions. These assumptions, which may be regarded as oversimplifying in some cases, should take into account critical aspects of (future) networks, e.g., system topology and signal propagation conditions. Analysis of these models will hopefully provide system design insights for realistic system deployments, even though quantitative performance predictions may not be very accurate.

Clearly, the common practice of system simulation is the only method for obtaining accurate estimates of performance taking into account multiple parameters and/or impairments that are difficult to be incorporated in an analytical study. However, as noted in [16], the increasing complexity of the future network architecture in terms of, e.g., irregularity of AP positions and multiple co-existing tiers, makes the simulation approach extremely difficult as the range of scenarios that need to be investigated becomes extremely large. The analytical modeling approach allows to identify trends of certain parameter of critical interest, thus limiting the range of scenarios for which computationally intensive simulations under a more detailed system model should be performed.

This section provides a description of the most common assumptions employed in cellular network studies under the stochastic geometry framework. This system model will form the basis for modeling and analyzing various scenarios of interest in the remainder of this chapter. In addition, this section discusses appropriate performance metrics that measure the efficiency of the network as well as AP–UE association criteria used in the analysis.

5.3.1 System model

5.3.1.1 Distribution of access points and users

A heterogeneous cellular network (HetNet) will be considered, composed of multiple tiers, each consisting of a countable set of access points (APs) distributed over $\mathbb{R}^2$. Note that an infinite number of APs is considered, which, although unrealistic, is appropriate as a model for *large scale and/or ultra dense system deployments*. The APs belonging to the same tier have the same properties such as radio access technology and transmit power capabilities. Traditional studies of (single-tier) cellular networks consider a hexagonal pattern for the position of APs (base stations) as illustrated in Figure 5.1(a), which, apart from requiring computationally intensive simulations as the main tool for performance evaluation, has become less relevant for modeling of the future cellular network. The latter observation is based on the introduction of dense small-cell tiers, i.e., tiers composed of many low-cost, easily deployed APs, underlaid together over the same geographical areas and incorporated into the system in an opportunistic manner, which makes a regular-pattern model for the AP positions unrealistic [4]. This irregularity was shown in [13] to also hold to a certain extent for actual macro-cell deployments.

It is therefore natural to model the AP positions of each tier as a random point process which, for analytical tractability, is considered to be an independent (from everything else) PPP. In particular, the random AP positions of the ith tier of the network will be represented by a PPP $\Phi_{a,i} \subset \mathbb{R}^2$ of density $\lambda_{a,i} > 0$. For the case of a single-tier network the index i will be dropped from notation. A realization of a single-tier cellular network is shown in Figure 5.1(b). Note that realizations of the PPP model will typically result in areas where the density of APs is *locally* greater than other areas, resembling the existence of hot spots in practical deployment.

Modeling the positions of UEs as another point process in $\mathbb{R}^2$ can be supported by similar arguments as for the case of APs, even more so considering the mobility of some types of UEs. In the following, the positions of UEs will be considered as another independent (from everything else) PPP $\Phi_u \subset \mathbb{R}^2$ of density $\lambda_u > 0$. It is noted that the independence of APs and UEs positions is not really valid in practical deployments, where it is expected that areas with locally large density of UEs will be complemented by a locally larger density of APs (hot-spot areas). However, this independence assumption has been exclusively employed in the literature for analytical convenience with the understanding that the model may not be sufficiently accurate under certain scenarios.

5.3.1.2 Path-loss and small-scale fading

All transmitters and receivers in the network are assumed, for simplicity, to be equipped with a single omni-directional antenna. The path-loss model governing the reduction of signal power from transmitter to receiver of a wireless link is considered to be a decreasing function $l : [0, \infty) \mapsto [0, \infty)$ of the distance between the two nodes. The simplest and most common model employed in the literature is [20]:

$$l(r) = r^{-\alpha}, \tag{5.5}$$

where $r \geq 0$ is the link distance and $\alpha > 0$ is the *path-loss exponent*. Note that this is a normalized version of the model with $l = 1$ for $r = 1$, i.e., the actual path-loss model is of the form $(\beta r)^{-\alpha}$, with $\beta \neq 1$ [15]. The normalized model is typically employed as it is mathematically convenient and does not affect the analysis qualitatively. The major caveat of this model is that the path *loss* actually becomes path *gain* for $r < 1$ and grows unbounded as $r \to 0$, which, of course, is unrealistic. For this reason, path-loss models such as $l(r) = (r + \varepsilon)^{-\alpha}, \varepsilon > 0$, or $l(r) = \max(r^{-\alpha}, 1)$ which eliminate this issue have been employed in the literature. However, as will be demonstrated in the following, the simple model of (5.5) yields reasonable results when single-user decoding schemes are considered (where interference is treated as noise) and the path-loss exponent is greater than 2 (the latter is a technical requirement that guarantees that the interference power is finite with probability 1 [17]).

In addition to path loss, link quality is also affected by small-scale (fast) fading due to the (non-)coherent superposition of electromagnetic waves arriving at the receiver from multiple paths. The well-known Rayleigh model employed in communication/information theory [20,22], where the small-scale fading gain is assumed an exponentially distributed random variable of unit mean, is also widely adopted in cellular system studies. Small-scale fading is assumed to be independent among links. Extensions of the Rayleigh model such as Nakagami fading and lognormal fading can also be found in the literature, however, with the cost of unduly complicating analysis without offering any significant insights.

Remark: The literature on cellular studies based on the stochastic geometry framework exclusively considers *flat* fading, i.e., the channel multipath dispersion is much smaller than the baud period, or, equivalently, narrowband transmissions are only considered, which is not in accordance with current broadband networks [23]. The corresponding results can therefore be viewed as an approximation of the performance under frequency selective channels, or, more accurately, as the performance of transmissions restricted to a single resource block experiencing flat fading [23]. In any case, extension of analysis to the frequency selective case remains an open topic that is critical in order to investigate issues such as exploitation of frequency diversity gain and jointly scheduling UEs over frequencies.

5.3.1.3 Multiple access scheme and signal decoding

Information theory suggests that the ultimate limits of the downlink cellular network performance in terms of transmission rates (capacity) are achieved by a scheme where the data of each UE are available to every AP and transmissions are performed jointly, over the same time and frequency resources [24]. For the uplink, joint decoding of all UEs should be performed by all APs of the system. Unfortunately, the performance of these schemes are very difficult to characterize in general, and, most importantly, the associated coding/de-coding complexity and requirements (e.g., infinite back-haul capacity, multi-user detection techniques, etc.) are unrealistic. It is therefore necessary to employ in the analysis simpler yet more practical transmission (multiple access) schemes.

The majority of works in the stochastic geometry framework assume an orthogonal multiple access scheme where each UE is served by a single AP (the latter determined under some AP–UE association criterion to be discussed later). This is also the approach employed by current cellular networks [23]. A simple time-division-multiple-access (TDMA) scheme will be considered in the following for serving the UEs associated with a certain AP (for either uplink or downlink), although any other orthogonal scheme, e.g., FDMA, CDMA, could be used with no effect on the analysis and results. No priorities are considered among UEs (round-robin scheduling) and each AP takes independent scheduling decisions, i.e., no coordination among APs is assumed.

Single-user decoding is performed at the receivers, i.e., interference is treated as noise. This is again a sub-optimal but nonetheless practical detection method. Considering universal frequency reuse, i.e., all transmissions are performed in the same bandwidth,[3] and taking into account the densification of cellular network infrastructure [25] as well as the multitude of active UEs, it is clear that (inter-cell) interference becomes the limiting factor for successful transmissions. It is therefore reasonable to ignore the effect of thermal noise in system performance, i.e., to consider interference-limited operational conditions, which also simplifies the analysis.

[3]Downlink and uplink transmissions are assumed orthogonal by employing frequency or time division duplex, as is the case in practical systems [23].

As demonstrated in many studies, neglecting thermal noise has a marginal effect on the accuracy of results. Finally, in line with an information theoretic framework [26], full buffers are considered, i.e., data is always available for transmission to a certain receiver.

5.3.1.4 Summary of system model

In summary, the common system model assumptions for all the cases considered in this chapter (and indeed in most of the related literature) are the following, unless stated otherwise:

1. UEs and tiers of APs are distributed according to independent PPPs.
2. Path-loss model as per (5.5), same path-loss exponent $\alpha > 2$ for all tiers.
3. Rayleigh flat fading channels, independent over AP–UE links.
4. Single, omni-directional antenna terminals.
5. Fixed transmit power for the APs of the same tier (different transmit powers among tiers, in general).
6. TDMA per AP, no coordination/cooperation among APs.
7. Single-user decoding.
8. Universal frequency reuse, interference-limited operational conditions.
9. Full-buffer transmissions.

5.3.2 System performance metrics

In order to evaluate the performance of the cellular network, appropriate performance metrics should be defined, which, traditionally, are related to the concepts of coverage and capacity, reflecting the ability of the network to provide (minimum quality) services to a large geographical area and large transmission rates to a large number of users, respectively. Even though there is a lot of discussion on incorporation of additional metrics reflecting other attributes of interest for the future network, such as latency [27], the underlying theory is not mature, making analysis and design tailored explicitly to these concepts cumbersome. Therefore, the conventional, rate-based metrics of coverage and capacity remain as popular as ever in the cellular literature and are reviewed in the following.

5.3.2.1 Signal-to-interference ratio

The natural metric to associate with the concept of coverage is the signal-to-interference ratio (SIR),[4] which, considering a typical receiver (UE for downlink, AP for uplink transmissions) located w.l.o.g. at the origin, is equal to:

$$\mathrm{SIR} = \frac{P_{x^*} g_{x^*} |x^*|^{-\alpha}}{\sum_{x \in \Phi \setminus \{x^*\}} P_x g_x |x|^{-\alpha}}, \tag{5.6}$$

where $\Phi \subset \mathbb{R}^2$ is the set of positions of transmitters ($\Phi = \cup_i \Phi_{a,i}$ or $\Phi = \Phi_u$, for downlink and uplink, respectively), $P_x > 0$ and $g_x > 0$ are the transmit power and

[4] As stated above, interference-limited conditions are considered, therefore SIR is equivalent to the more general signal-to-interference-plus-noise-ratio (SINR).

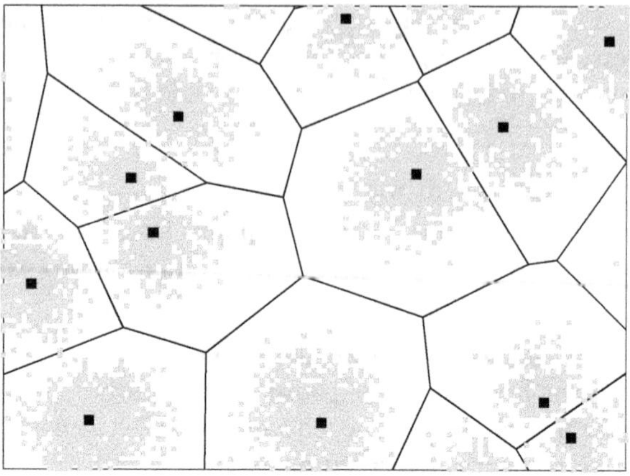

Figure 5.2 Network coverage for a random realization of channel fading gains. For visualization purposes, the plane is partitioned into small-area squares with constant channel gain within each square area (channel fading gains are independent among squares). Lightly shaded squares represent the positions where a UE experiences an SIR $\geq$ 10 dB, under nearest AP association (see Section 5.3.3). Dark squares indicate the AP positions and lines the cell boundaries. Note that positions near cell edges are not covered in general

Rayleigh fading gain of transmitter located at $x \in \mathbf{\Phi}$, respectively, and $x^* \in \mathbf{\Phi}$ is the location of the serving transmitter. Clearly, under the system model assumptions described above, the SIR is a random variable due to the randomness of transmitter locations and channel fading gains.

In order to characterize the coverage capabilities of the network, it is of interest to find the *coverage probability*:

$$\mathbb{P}(\mathrm{SIR} \geq \theta) = \mathbb{E}(\mathbb{I}(\mathrm{SIR} \geq \theta)), \quad \theta > 0.$$

The coverage probability completely characterizes the statistical properties of the SIR as it is equal to the complementary cumulative distribution function (CCDF) of SIR. The SIR CDF, usually referred to as the *SIR outage probability,* is of course equal to $\mathbb{P}(\mathrm{SIR} \leq \theta) = 1 - \mathbb{P}(\mathrm{SIR} \geq \theta)$. Note that, by the ergodicity property of the PPP, evaluating the coverage probability $\mathbb{P}(\mathrm{SIR} \geq \theta_0)$ for some $\theta_0 > 0$ is equivalent to computing the portion (coverage percentage) of $\mathbb{R}^2$ for which SIR $\geq \theta_0$ for any realization of $\mathbf{\Phi}$. Figure 5.2 shows an example of a network and its coverage area for which SIR $\geq$ 10 dB.

5.3.2.2 Link spectral efficiency

The notion of cellular network capacity is directly related to the per-link spectral efficiency of the system, which provides an upper bound on the rates in units of

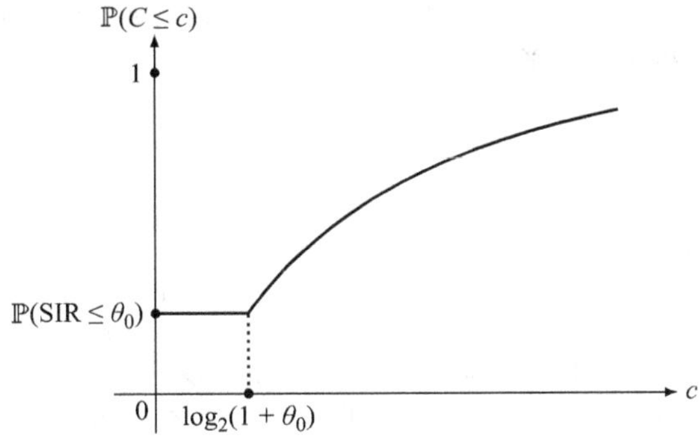

Figure 5.3 Typical CDF of link spectral efficiency

bits/s/Hz that an AP may reliably transmit (downlink case) or receive (uplink case). Under the assumption of Gaussian signaling for all transmitters, the spectral efficiency of a typical link between a receiver positioned at the origin and its associated transmitter is directly related to the receiver SIR according to the well-known capacity formula

$$C(\mathrm{SIR}) = \log_2(1 + \mathrm{SIR})\mathbb{I}(\mathrm{SIR} \geq \theta_{\mathrm{th}}),$$

where $\theta_{\mathrm{th}} \geq 0$ is the *SIR threshold*, dictated by application, system, and synchronization requirements, below which reliable communications are considered unsuccessful.[5] Since the SIR of the cellular link is random, so is the spectral efficiency with a CDF equal to:

$$\mathbb{P}(C \leq c) = \begin{cases} \mathbb{P}(\mathrm{SIR} \leq 2^c - 1) & c \geq \log_2(1 + \theta_{\mathrm{th}}) \\ \mathbb{P}(\mathrm{SIR} \leq \theta_0) & c < \log_2(1 + \theta_{\mathrm{th}}). \end{cases} \tag{5.7}$$

A qualitative example of the spectral efficiency CDF is shown in Figure 5.3.

5.3.2.3 UE rate

The link spectral efficiency is a widely adopted metric in the cellular literature as it translates the SIR capabilities of the network to AP rate capabilities by a simple transformation of variables. However, although link spectral efficiency is a valid and useful measure/indicator of the network performance [23], it provides only partial information about the actual rate experienced by the served UEs. This is because the UE rate depends not only on the link spectral efficiency but also on the resource sharing among multiple UEs served by the same AP.

[5]For example, LTE requires a minimum operational SIR of about −6 dB for data transmissions [23]. Note that LTE also specifies a maximum spectral efficiency (corresponding to an operational SIR of about 20 dB). Imposing a maximum spectral efficiency limits performance when a (very) large SIR is experienced and therefore is not considered in this analysis.

Considering a typical UE located at an arbitrary position on the plane, the actual rate it experiences is equal to:

$$R \triangleq \frac{1}{1+K_0} C,$$

in bits/s/Hz, where $K_0 \geq 0$ is the number of additional UEs served by the same AP as the typical UE and C is the AP–UE link spectral efficiency, as described above. The term $(1/(1+K_0)) \leq 1$ will be referred to in the following as the *degrees of freedom* of UE transmissions, borrowing the terminology from the communication theory literature [22]. It reflects the rate loss due to TDMA and its value depends on the positions of UEs and APs, as well as the AP–UE association scheme.

5.3.3 AP–UE association schemes

Various AP–UE association schemes have been proposed/investigated in the literature. Considering a typical receiver (UE or AP for downlink or uplink transmissions, respectively) and denoting the position of the associated (serving) transmitter as x^*, all schemes can be viewed as special cases of the *weighted-path-loss association principle*, for which:

$$x^* = \arg\max_{x \in \Phi} w_x |x|^{-\alpha}, \tag{5.8}$$

where Φ represents the locations of transmitters and $w_x \in \mathbb{R}$ is the *association weight* for the transmitter positioned at $x \in \Phi$. In general, the association weights are random variables distributed according to some joint pdf, with different receivers "observing" different realizations of these weights.

The most practical choices for the weights are the following.

- $w_x = 1$: geometrical (nearest transmitter) association
- $w_x = P_x$: maximum average received power association
- $w_x = P_x g_x$: maximum instantaneous received power association

For the first two cases, the same w_x is observed by all APs, whereas the last case results in different values of w_x due to the independent fading gains among AP–UE links. An important characteristic of the weighted-path-loss association procedure is that it can be exclusively implemented at the UE side by measuring the power of dedicated pilot (reference) signals, as is also the case in practice [23]. Note that geometrical and maximum average received power association will result in different associations only in the case where the downlink of a multi-tier network with different transmit powers among tiers is considered. For single-tier downlink and uplink connections, these two association rules are essentially the same. An interesting interpretation of the maximum instantaneous received power association can be obtained by examining the resulting SIR, which is equal to:

$$\frac{\max_{x\in\Phi} P_x g_x |x|^{-\alpha}}{\sum_{y \in \Phi \setminus \{\arg\max_{x\in\Phi} P_x g_x |x|^{-\alpha}\}} P_y g_y |y|^{-\alpha}} = \max_{x \in \Phi} \frac{P_x g_x}{\sum_{y \in \Phi \setminus \{x\}} P_y g_y |y|^{-\alpha}},$$

showing that this association rule is equivalent to a *maximum instantaneous SIR association*.

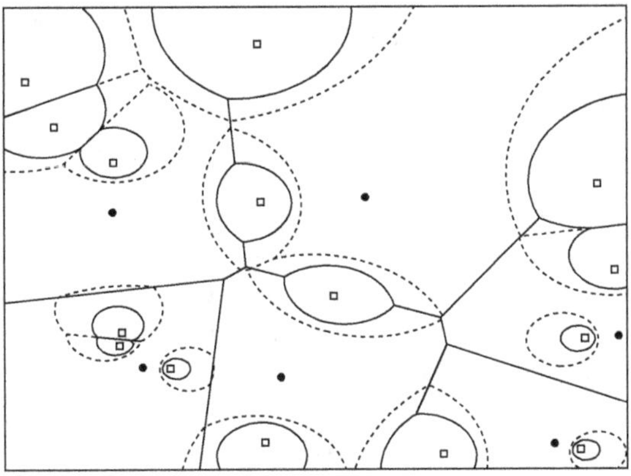

Figure 5.4 Association regions without (solid lines) and with (dotted lines) pilot power bias of a two-tier HetNet. Points and squares denote positions of macro- and micro-APs, respectively

The above association rules only consider the received power for making association decisions. However, this approach may be highly suboptimal when the metric of interest is the UE rate, which, in addition to received power, depends heavily on the traffic load experienced by an AP. This issue is particularly evident in the downlink of a two-tier HetNet composed of macro- and micro-APs with significantly different transmit power capabilities. Due to the large transmit power difference, the maximum SIR association rule will result in most UEs connecting to the macro-tier with micro-tier APs only serving the few UEs that happen to be in close proximity. An example of the resulting association regions is shown in Figure 5.4 (solid lines).

Clearly, it would be beneficial to more evenly distribute the load among tiers, even though this will result in some UEs experiencing a smaller SIR than possible. However, AP–UE association decisions that also take into account traffic balance can only be taken by the network side since traffic load information is unknown to the UEs (and is not typically provided by the network). On the other hand, it is generally desirable to implement association decisions at the UE side (at least for the initial network attachment procedure) for complexity and scaling reasons. Toward this end, an extension of the maximum instantaneous received power association has been proposed in the literature where the weights corresponding to each AP belonging to a certain tier are multiplied by a tier-dependent scaling factor (bias), i.e., $w_x = P_x = P_i$ is replaced by $B_i P_i$ for an AP located in $x \in \Phi_{a,i}$, where $B_i \geq 1$ is a fixed common bias for the weights of APs belonging to tier i [28,29]. By increasing the bias corresponding to tiers with smaller power the association regions among tiers become more balanced, as shown in the example of Figure 5.4 (dashed lines), thus achieving a more even UE distribution among tiers.

5.4 SIR performance analysis

This section investigates system performance w.r.t. to the SIR and link spectral efficiency metrics. Various operational scenarios are considered that differ in terms of UE load, transmission direction (uplink or downlink) and AP association criterion. Derivations are presented in a fairly large amount of detail in order to clearly demonstrate how the analytical tools presented in Section 5.2 can be applied for cellular systems analysis. Special emphasis is given on how system performance depends on the densities of APs and UEs and if (and how) densification of infrastructure can be beneficial.

5.4.1 Fully loaded network: downlink

5.4.1.1 Single tier—nearest AP association

A single-tier network is considered under the same system model assumptions as those described in Section 5.3. Nearest AP association is assumed, with all APs transmitting at every instant with constant power P (independent of AP position). The latter assumption suggests that each AP is associated with at least one UE, implying an infrastructure deployment scenario where $\lambda_a \ll \lambda_u$, referred to in the following as a *fully loaded network.*

It is of interest to obtain the SIR coverage probability of a typical UE positioned at the origin and associated with its closest AP located at $x^* \in \Phi_a$. This problem was first considered in [13]. Conditioned on the position x^* and denoting by $I \triangleq \sum_{x\in\Phi_a\backslash\{x^*\}} Pg_x|x|^{-\alpha}$ the interference power, the SIR distribution experienced by the typical UE is equal to:

$$
\begin{aligned}
\mathbb{P}(\mathrm{SIR} \geq \theta | x^*) &= \mathbb{P}\left(\frac{Pg_{x^*}|x^*|^{-\alpha}}{I} \geq \theta \middle| x^*\right) \\
&= \mathbb{E}(\mathbb{P}(g_{x^*} \geq \theta |x^*|^{\alpha}\, I/P | x^*, I)) \\
&\overset{(a)}{=} \mathbb{E}\left(e^{-\theta|x^*|^{\alpha}\, I/P} \middle| x^*\right) \\
&= \mathbb{E}\left(\prod_{x\in\Phi_a\backslash\{x^*\}} e^{-\theta|x^*|^{\alpha} g_x |x|^{-\alpha}} \middle| x^*\right) \\
&\overset{(b)}{=} \mathbb{E}\left(\prod_{x\in\Phi_a\backslash B(0,|x^*|)} e^{-\theta|x^*|^{\alpha} g_x |x|^{-\alpha}} \middle| x^*\right) \\
&\overset{(c)}{=} \exp\left(-\lambda_a \int_{\mathbb{R}^2\backslash B(0,|x^*|)} \left[1 - \frac{1}{1+\theta|x^*|^{\alpha}|x|^{-\alpha}}\right] dx\right),
\end{aligned}
\tag{5.9}
$$

for any $\theta > 0$, where (a) is due to g_{x^*} being exponentially distributed with unit mean, (b) follows by noting that $\Phi_a\backslash\{x^*\} = \Phi_a\backslash B(0, |x^*|)$ (see Figure 5.5), where $B(x_0, R) \subset \mathbb{R}^2$ denotes the ball centered at $x_0 \in \mathbb{R}^2$ of radius $R > 0$ (guard zone for

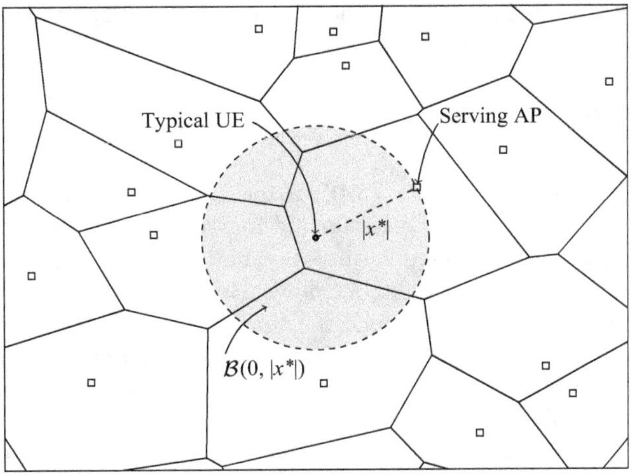

Figure 5.5 Association with the nearest AP implies a circular guard zone centered at the UE for downlink transmissions (shaded area) where no interfering APs can be present

downlink transmissions), and (c) is a direct application of the PGFL formula (5.3) where the marks are equal to $m_x = \theta|x^*|^\alpha g_x, x \in \mathbf{\Phi}_a$.[6]

Transforming the last integral into polar coordinates and after some algebra, it can be shown that [13,29]:

$$\mathbb{P}(\text{SIR} \geq \theta|x^*) = \exp(-\pi\lambda_a|x^*|^2 \rho(\theta, \alpha)), \tag{5.10}$$

where $\rho(\theta, \alpha) \triangleq \theta^{2/\alpha} \int_{\theta^{-2/\alpha}}^{\infty}(du/(1 + u^{\alpha/2})) = (2\theta/(\alpha - 2))\ {}_2F_1(1; 1 - (2/\alpha); 2-(2/\alpha); -\theta)$, and ${}_2F_1(\cdot)$ is the Gauss hypergeometric function [30].

This relatively simple expression verifies the intuition that the SIR given x^* (stochastically) degrades with increasing AP density and/or increasing distance from x^*. However, the conditional SIR distribution is of little practical value for characterization of system performance. In order to obtain the (unconditional) coverage probability of the typical UE, expression (5.10) must be averaged over the distribution of $|x^*|$. The latter can be obtained by noting that:

$$\begin{aligned}\mathbb{P}(|x^*| > r) &= \mathbb{P}(N(\mathscr{B}(0, r)) = 0)\\ &= e^{-\pi\lambda_a r^2}, 0 \leq r < \infty,\end{aligned} \tag{5.11}$$

where $N(\mathscr{A})$ is the number of points of $\mathbf{\Phi}_a$ included in $\mathscr{A} \subset \mathbb{R}^2$ and the second equality follows by Property 1 of the PPP (see Section 5.2). This expression verifies the intuition that denser AP deployments result in a smaller distance from the

[6]A subtle point in this argument is that the integration region of (5.9) is determined by the PPP itself, whereas the PGFL formula discussed in Section 5.2 considers a region that is independent of the PPP. It can be shown that application of the PGFL formula is valid in this case by the strong Markov property of the PPP and the region $B(0,|x^*|)$ being a random stopping set of $\mathbf{\Phi}_a$ [15, p. 17].

associated AP (in a statistical sense). The coverage probability can now be obtained from (5.10) and (5.11) as the simple formula:

$$\mathbb{P}(\text{SIR} \geq \theta) = \frac{1}{1+\rho(\theta,\alpha)}. \tag{5.12}$$

The most important aspects of (5.12) is that the coverage probability is parameterized only by the path-loss exponent α and *does not depend on the density of APs*. The latter is a rather unexpected result since both the received power from the associated AP as well as the interference power depend on λ_a. Informally, it can be stated that a change in λ_a equally affects the numerator and denominator of the SIR, resulting in a constant coverage probability. This independence from λ_a is a very important result as it suggests that, under the system model considered, arbitrarily large spatial reuse levels can be achieved by infrastructure densification without any effect on the perceived SIR, thus constituting a strong argument in favor of (ultra) dense networks.

Figure 5.6(a) shows the coverage probability for various values of α. It can be seen that operating under a large path-loss exponent is beneficial since, in that case, the interference is essentially generated from only a limited number of nearby APs due to the increased path loss. The average link spectral efficiency $\mathbb{E}(C)$ under various SIR thresholds θ_{th}, obtained numerically as $\mathbb{E}(C) = \int_0^\infty \mathbb{P}(C > c)dc$ using the distribution expressions of (5.7) and (5.12), is shown in Figure 5.6(b). In line with the coverage probability trend, $\mathbb{E}(C)$ increases with α. As expected, larger values of θ_{th} translate into smaller average spectral efficiencies.

In order to understand the consequences of a random infrastructure deployment compared to the traditional periodic (hexagonal) deployment pattern, the SIR coverage probability under the PPP model and the hexagonal pattern for Φ_a are compared in Figure 5.7. The performance of the hexagonal pattern is obtained via Monte Carlo simulation, with the typical UE uniformly distributed within a random cell and all other system model assumptions the same as in the PPP model. A density of 1 was used for the hexagonal pattern of the APs, although simulations suggest that, as in the PPP case, coverage probability is independent of AP density. It can be seen that the hexagonal pattern provides better coverage performance by a 4–5 dB margin. This is expected as the periodic pattern of APs limits the number of interferers that are close to the typical UE, whereas under the PPP model many interferers in close proximity may exist. Since regular/periodic deployment of dense infrastructure is rather impossible, this performance loss is a cost that system designers have to live with.

5.4.1.2 Multiple tiers—maximum average received power association

The analysis of the previous section can be straightforwardly extended to the case of a HetNet consisting of $K \geq 2$ tiers where now the term "fully loaded network" corresponds to the condition $\lambda_u \gg \sum_i \lambda_{a,i}$. All APs of tier k transmit with a fixed

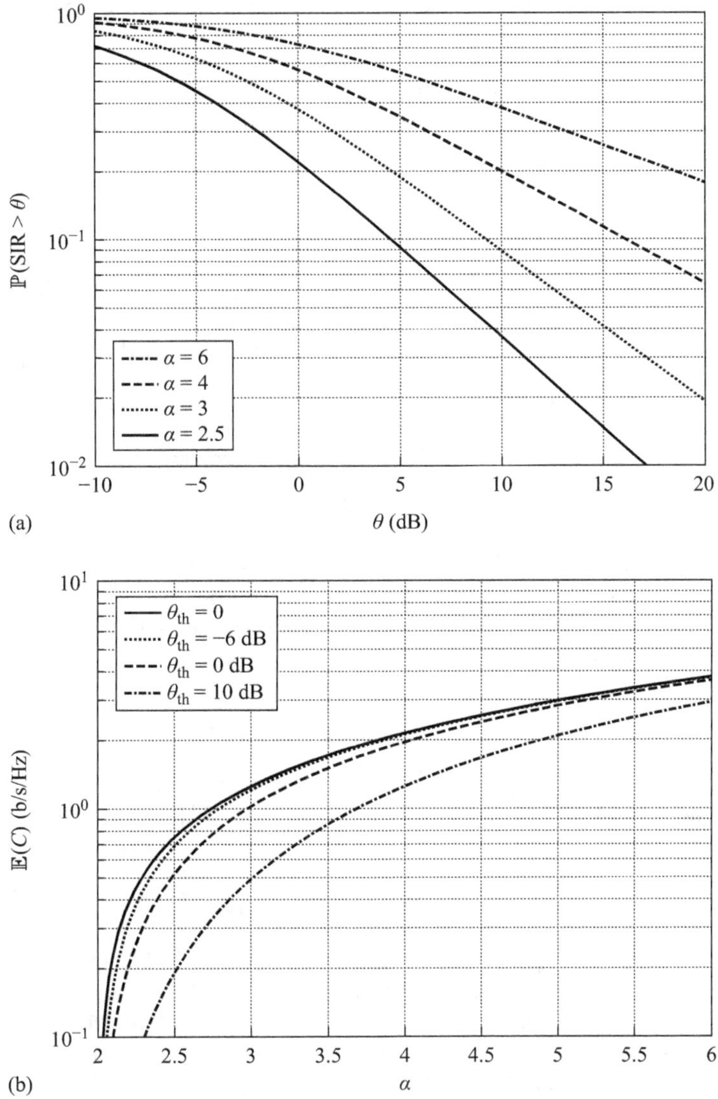

Figure 5.6 Downlink coverage probability (a) and average spectral efficiency (b) of fully loaded single-tier network

power $P_k > 0$ whereas the pilot signal power equals $B_k P_k$, where $B_k \geq 1$ is the pilot power bias of tier k, chosen appropriately for offloading purposes as discussed in Section 5.3.3. A UE is associated to the AP with the maximum biased average received pilot power. Under the exact same system model assumptions and

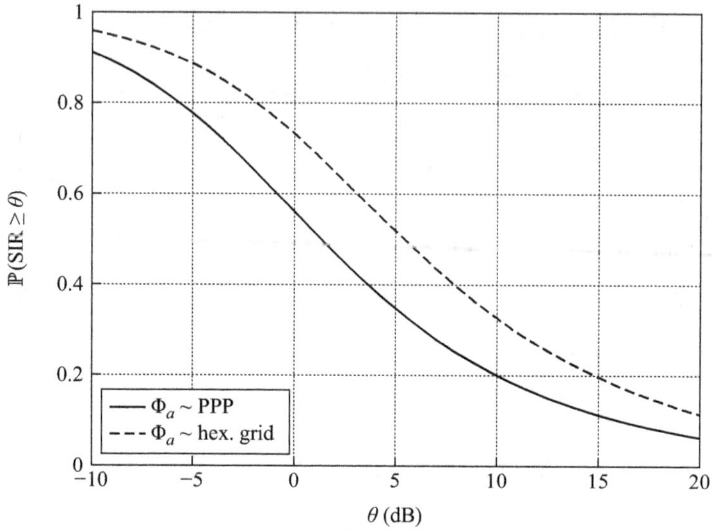

Figure 5.7 Comparison of downlink SIR coverage probability of single-tier network with $\mathbf{\Phi}_a$ *distributed as a PPP and a hexagonal grid* ($\alpha = 4$)*. Hexagonal grid performance is obtained via Monte Carlo simulation with* $\lambda_a = 1$

employing a similar methodology as in the single-tier case, the coverage probability of the typical UE can be shown to be equal to [29]:

$$\mathbb{P}(\text{SIR} \geq \theta) = \sum_{k=1}^{K} \left(\sum_{l=1}^{K} \frac{\lambda_{a,l}}{\lambda_{a,k}} \left(\frac{P_l B_l}{P_k B_k} \right)^{2/\alpha} (1 + \rho(\theta B_k / B_l, \alpha)) \right)^{-1}, \tag{5.13}$$

with $\rho(\cdot, \cdot)$ as defined in (5.10). Of course, (5.13) coincides with (5.12) for $K = 1$.

Interestingly, setting $B_k = B_l$ for all k, l in (5.13), results in the same expression for the coverage probability as in (5.12), i.e., when no bias is introduced in the AP–UE association procedure, the SIR performance of the HetNet is the same as of a single-tier network. In particular, the HetNet performance becomes independent of number, densities, and transmit power of tiers, suggesting that spatial reuse can be arbitrarily increased by introducing to the network APs belonging to existing or new tiers, essentially in a random manner.

Note that with non-equal biases, the SIR coverage probability degrades from the case of equal biases as some UEs will associate with APs that do not provide the maximum received power for data transmission. However, this observation should not be considered as a strong argument against introducing biases since the SIR degradation may be smaller compared to the gains offered by a more even association of UEs among tiers.

5.4.1.3 Maximum SIR association

The SIR performance of the downlink network under maximum SIR UE–AP association is considered in this subsection. Exact analytical expressions for the coverage probability have been obtained in the literature [16]; however, they are unduly complicated and will not be presented here. Instead, it is more instructive, both from a technical perspective as well as in order to get some insights, to consider an upper bound on the coverage probability first derived in [31].

Considering a single-tier network first, and denoting as $\mathrm{SIR}(x)$ the SIR of the typical UE located at the origin when associated with the AP at $x \in \Phi_a$, the coverage probability of the maximum SIR association scheme can be upper bounded as

$$\begin{aligned}
\mathbb{P}\left(\max_{x\in\Phi} \mathrm{SIR}(x) \geq \theta\right) &= \mathbb{P}\left(\bigcup_{x\in\Phi} \{\mathrm{SIR}(x) \geq \theta\}\right) \\
&\leq \sum_{x\in\Phi} \mathbb{P}(\mathrm{SIR}(x) \geq \theta) \\
&= \mathbb{E}\left(\sum_{x\in\Phi} \mathbb{I}(\mathrm{SIR}(x) \geq \theta)\right) \qquad (5.14) \\
&= \lambda_a \int_{\mathbb{R}^2} \mathbb{P}(\mathrm{SIR}(x) \geq \theta)dx, \qquad (5.15)
\end{aligned}$$

where the last equality was obtained by application of Campbell's theorem. Note from the right-hand side of (5.14) that this upper bound has a physical meaning, namely, it is equal to the average number of APs which are capable of providing the typical UE an SIR greater than θ. Using a similar procedure as that for obtaining (5.10), the coverage probability by an AP positioned at $x \in \mathbb{R}^2$ equals [31]:

$$\mathbb{P}(\mathrm{SIR}(x) \geq \theta) = \exp\left(-\pi\lambda_a|x|^2\, \kappa(\alpha)\theta^{2/\alpha}\right),$$

where $\kappa(\alpha) \triangleq (2\pi/\alpha)/\sin(2\pi/\alpha)$. Substituting in (5.15) results in

$$\mathbb{P}\left(\max_{x\in\Phi} \mathrm{SIR}(x) \geq \theta\right) \leq \frac{1}{\kappa(\alpha)\theta^{2/\alpha}}. \qquad (5.16)$$

Note that this simple bound is independent of the density of APs. It can be shown that this density independence holds true also for the exact coverage probability of the maximum SIR association scheme [16], similar to the case of the coverage probability with nearest AP association. The exact coverage probability with maximum SIR association is plotted in Figure 5.8 (obtained by Monte Carlo simulation) along with the upper bound of (5.16) and the coverage probability of the nearest AP association rule given by (5.12). As expected, the maximum SIR association scheme outperforms nearest AP association, especially for low SIRs, where the AP that provides the largest SIR is not always the nearest one. For large SIRs (greater than 5 dB), both schemes show the same performance as

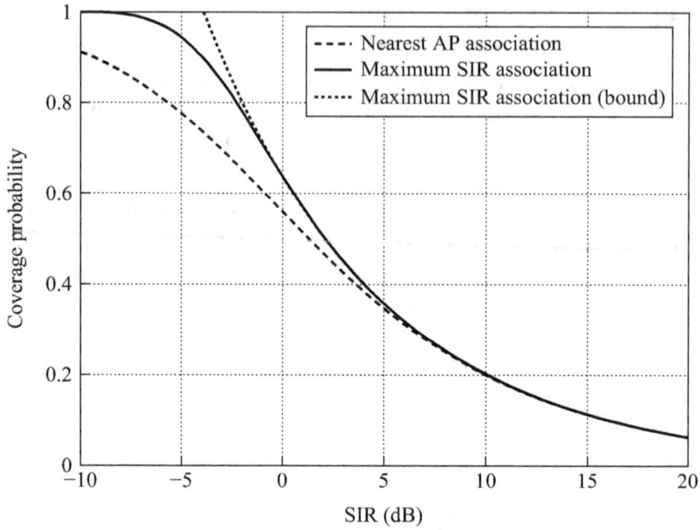

Figure 5.8 Comparison of coverage probabilities with nearest AP and maximum SIR association schemes ($\alpha = 4$)

the SIR-maximizing AP coincides in most cases with the nearest one. Note that the upper bound of (5.16) provides an excellent indication of the exact coverage probability for SIR values greater than about −2 dB. Actually, the upper bound expression is indeed equal to the coverage probability for SIR values greater or equal to 0 dB. This follows from the observation that if there exists an AP that can provide an SIR greater than or equal to 0 dB, it is unique [15,31], rendering the events $\{\mathrm{SIR}(x) \geq \theta\}_{x\in\Phi}$ mutually exclusive for any $\theta \geq 0$ dB.

The analysis for the single-tier network can be straightforwardly extended to the multi-tier case. It turns out that the upper bound on the coverage probability in this case is equal to the expression for the single-tier case, i.e., under the maximum SIR association scheme, network performance in terms of SIR is invariant to number of tiers and tier transmit powers, similar to the observation under maximum average received power association with equal biases.

5.4.2 Fully loaded network: uplink

In contrast to the multitude of works under the stochastic geometry framework that investigate the downlink performance of the cellular network, the uplink performance has received much less attention, even though it may appear on first sight as a "dual" study to the downlink investigation. This is not only due to the downlink being considered as more critical for performance,[7] but also due to analytical difficulties associated with the uplink configuration.

[7]For example, the LTE performance requirements set by IMT are more stringent for the downlink than the uplink.

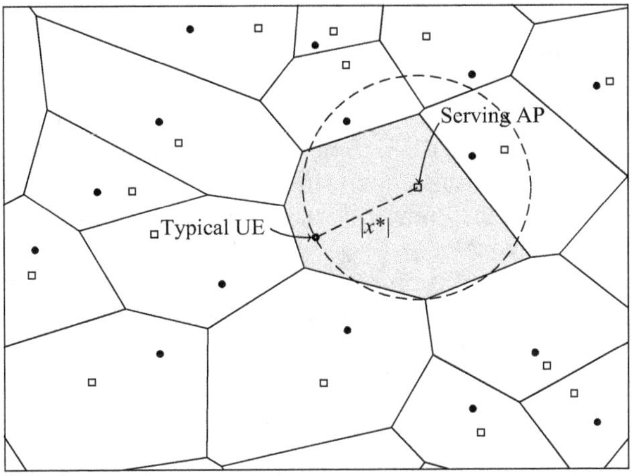

Figure 5.9 Demonstration of guard zone (shaded area) for uplink transmissions with fixed power. Points and squares represent UEs and APs, respectively

These difficulties can be understood by considering a parallel setup to the one considered in Section 5.4.1.1, where uplink UEs associate to their nearest AP and transmit with the same fixed power. Attempting to follow the same steps as in the downlink analysis faces critical challenges due to the following facts.

1. The positions of interfering UEs are not distributed as a PPP.
2. The "guard zone" within which the associated AP is free of interfering UEs is its (Voronoi) cell, which does not have a simple mathematical representation.

The first fact can be understood by noting that the interfering UEs location is a point process generated from Φ_u by dropping all points belonging to the same AP cell except a randomly selected one. However, the AP cells are not independent and, therefore, the events of two points of Φ_u being located within two different (and disjoint) cells are not independent. This is, of course, in contrast with the fundamental properties of the PPP (see Section 5.2). The second fact is obvious by simple inspection of a realization of the considered setup (as the one shown in Figure 5.9). Note that, in contrast to the downlink case, the geometric guard zone of the uplink does not guarantee that all interferers are in greater distance from a specific AP than its associated UE.

Although the point process of the interfering UEs can in principle be approximated as a PPP of density equal to λ_a, the irregularity of the guard zone does not allow further analytical progress. Toward obtaining an analytical characterization of uplink performance, a subtle but critical modification of the system model was proposed in [32]. In particular, the fixed transmit power assumption is dropped in favor of a per-link power control scheme, where the transmit power of each uplink UE is such that the average received power at its associated AP is equal to some predefined value $P^* > 0$, i.e., for a link distance $r > 0$, the uplink UE transmits with a power $P^* r^\alpha$.

The rationale of this approach is that it guarantees that the average received power at an AP generated from its associated UE is larger than the interference power generated from any other UE, even in the case when the latter happens to be geometrically closer (verify this claim). In other words, the uplink power control generates a guard zone in the "power domain," which is different from the natural geometric uplink guard zone.

Considering the uplink of the typical UE, with the origin of the plane shifted to its serving AP position,[8] the uplink SIR is equal to:

$$\mathrm{SIR} = \frac{P^* g_{x^*}}{\sum_{x \in \tilde{\Phi}_u} \mathbb{I}(P_x |x|^{-\alpha} < P^*) P_x g_x |x|^{-\alpha}},$$

where $\tilde{\Phi}_u \subset \Phi_u$ is the point process of interfering UEs, g_{x^*} is the fading gain of the typical link, $\{g_x\}_{x \in \tilde{\Phi}_u}$ are the interfering channel gains, and $\{P_x\}_{x \in \tilde{\Phi}_u}$ are the transmit powers of interfering UEs. The use of the indicator function in the denominator of SIR formula, although redundant as it trivially evaluates to unity for all $x \in \tilde{\Phi}_u$, is convenient for the following computations as it explicitly represents the effective guard zone of the serving AP.

Under the assumptions of (a) $\tilde{\Phi}_u$ distributed as a PPP of density λ_a and (b) $\{P_x\}_{x \in \tilde{\Phi}_u}$ are i.i.d. random variables, the derivation of the uplink coverage probability proceeds similar to the downlink case leading to:

$$\begin{aligned}
\mathbb{P}(\mathrm{SIR} \geq \theta) &= \mathbb{E}\left(\prod_{x \in \tilde{\Phi}_u} e^{-\frac{\theta}{P^*} \mathbb{I}(P_x |x|^{-\alpha} < P^*) P_x g_x |x|^{-\alpha}} \right) \\
&\stackrel{(a)}{=} \exp\left(-2\pi\lambda_a \mathbb{E}\left(\int_{(P_x/P^*)^{-1/\alpha}}^{\infty} \frac{\theta r^{-\alpha} P_x / P^*}{1 + \theta r^{-\alpha} P_x / P^*} r dr \right) \right) \\
&\stackrel{(b)}{=} \exp\left(\pi\lambda_a \mathbb{E}\left(\left(\frac{P_x}{P^*} \right)^{2/\alpha} \right) \rho(\theta, \alpha) \right),
\end{aligned}$$

where (a) is a direct application of the PGFL formula, followed by taking the expectation over g_x, moving the expectation over P_x out of the integral, switching to polar coordinates, and taking into account the guard zone specification and (b) follows by changing the integration variable to $u = r^2 (\theta P / P^*)^{-2/\alpha}$. Quantity $\rho(\cdot, \cdot)$ was defined in (5.10). It is straightforward to show that $\mathbb{E}\left((P_x/P^*)^{2/\alpha} \right) = (1/\pi\lambda_a)$ [32], finally leading to:

$$\mathbb{P}(\mathrm{SIR} \geq \theta) = e^{-\rho(\theta, \alpha)}. \tag{5.17}$$

[8]This is allowed by the stationarity property of the PPP.

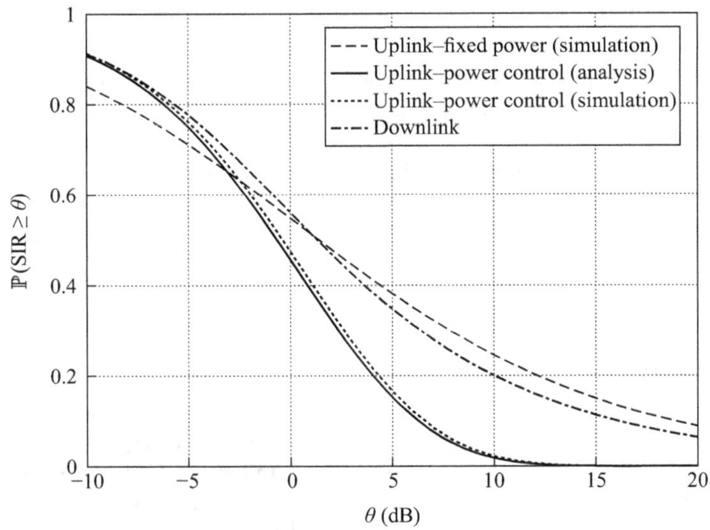

Figure 5.10 Coverage probabilities for downlink and uplink under the nearest AP association scheme ($\alpha = 4$)

This simple expression shows that the uplink SIR performance with closest AP association is independent of P^* and infrastructure density λ_a, exactly as was the case for the downlink performance. Comparison of (5.17) with the downlink SIR expression (5.12) shows that the downlink SIR stochastically dominates the power-controlled uplink SIR. These expressions are compared in Figure 5.10 along with Monte Carlo simulation for both the fixed power and power-controlled uplink schemes.[9] It can be seen that the simulation for the power controlled uplink matches the analytical expression of (5.17) suggesting that the approximations employed in its derivation are accurate. Comparison with the downlink performance shows that uplink and downlink have similar trends although the power-controlled uplink scheme is significantly inferior at high SIRs. In contrast, the fixed power uplink scheme resembles closer the downlink performance showing degradation at low SIRs due to the presence of interferers being closer to an AP than its associated UE.

5.4.3 Lightly loaded network

The previous sections considered an operational environment where each AP is associated with at least one UE, effectively assuming $\lambda_u \gg \sum_i \lambda_{a,i}$ with the actual

[9]Simulation considers $\lambda_u/\lambda_a = 10$, with a random UE selected for transmission within each Voronoi cell (if there are more than one UEs within the cell).

value of λ_u being irrelevant. However, under practical operational conditions, some of the APs will be associated with no UEs. This implies that the number of interferers is actually smaller than the one considered in the previous sections. Equivalently, the density of the point process representing the interferers positions is smaller than λ_a in the case of single-tier downlink and similarly for the other cases. It is clear that the value of λ_u plays an important role here and it is of interest to investigate how it affects the properties of the positions of the interferers as well the SIR coverage performance of the network. This section considers this question for the case of a single-tier downlink network with the analysis of multi-tier and/or uplink networks following the same principles.

5.4.3.1 Activity probability of APs

The natural starting point for this investigation is the determination of the statistics of $K \geq 0$, representing the number of UEs associated with a random AP.[10] Note that $K = K(\Phi_u, \Phi_a, \{w_x\}_{x\in\Phi_a})$, i.e., the number of UEs associated with a random AP depends on the AP and UE positions, as well as the (random) weights employed by the association scheme. Unfortunately, an expression for $\mathbb{P}(K = k), k = 0, 1, \ldots$, is not available, even for the simplest, nearest AP association scheme ($\{w_x = 1\}_{x\in\Phi_a}$). However, description of the first moments of the distribution of K can be obtained [9], which provide some insights on how K depends on AP and UE densities. In particular, under the mild condition that $\mathbb{E}(w_x^{2/\alpha})$ is finite, it is shown in Appendix A that:

$$\mathbb{E}(K) = \frac{\lambda_u}{\lambda_a}, \tag{5.18}$$

which shows that the average K *depends only on the ratio of UE and AP densities* and not on their actual values. Note that this result holds for *any* weighted-path-loss association scheme, including the ones explicated in Section 5.3.3. It is interesting that $\mathbb{E}(K)$ is the same with either $w_x = 1$ (nearest AP association), $w_x = Pg_x$ (maximum instantaneous received power association), or, for that matter, any other random association weight w_x.

Expression (5.18) suggests that increasing λ_u over λ_a increases the probability of an AP associated with at least one UE, which is, of course, intuitive. This was rigorously demonstrated in [33], where the *AP activity probability*, $\mathbb{P}(K > 0) = 1 - \mathbb{P}(K = 0)$, was shown to be bounded as

$$\frac{\lambda_u}{\lambda_u + \lambda_a} < \mathbb{P}(K > 0) \leq 1 - e^{-\lambda_u/\lambda_a}. \tag{5.19}$$

[10]Quantity K, representing the number of UEs associated with *a random AP,* is different from quantity K_0 considered in Section 5.3.2.3, representing the number of additional UEs associated with *the same AP as the typical UE*. This difference will be highlighted in Section 5.5.1.

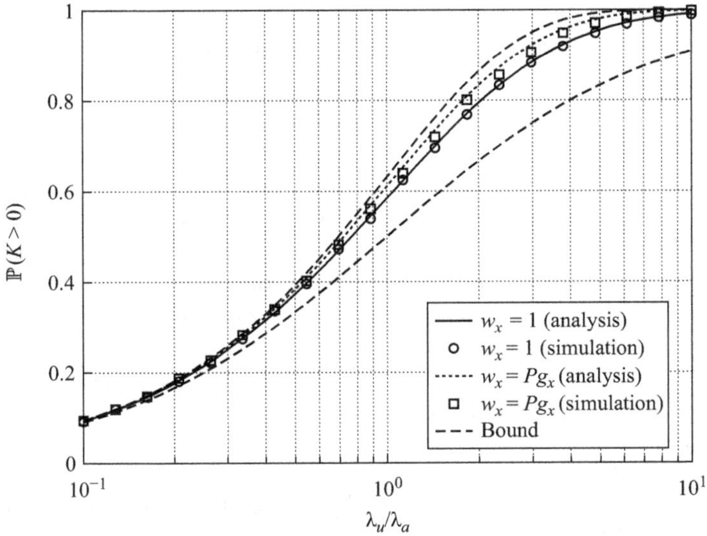

Figure 5.11 AP activity probability as a function of UE load conditions ($\alpha = 3$ for the maximum received power association)

The lower bound is not achievable, whereas the upper bound can be achieved by appropriately selecting the association weights distribution. An approximate expression for the AP activity probability was also proposed in [33] as

$$\mathbb{P}(K > 0) \approx 1 - \left(1 - \frac{\lambda_u}{\zeta\lambda_a}\right)^{-\zeta}, \tag{5.20}$$

where $\zeta \triangleq (7/2)\mathbb{E}(w_x^{2/\alpha})\mathbb{E}(w_x^{-2/\alpha})$. Note that both (5.19) and (5.20) imply that $\mathbb{P}(K > 0) \approx \mathbb{E}(K)$ when $\lambda_u \ll \lambda_a$.

Figure 5.11 shows the AP activity probability for the nearest AP association ($w_x = 1$) and maximum instantaneous received power association ($w_x = Pg_x$) obtained by Monte Carlo simulation, along with their approximate expressions (5.20),[11] as a function of the average UE load of a random cell, λ_u/λ_a. The bounds of (5.19) are also depicted. It can be seen that both association schemes result in similar, although different, activity probabilities, with the approximate expression providing a very good match in both cases, especially for the nearest AP association scheme. Note that for an operational environment with $\lambda_a = \lambda_u$, i.e., an ultra dense network, the activity probability is only about 0.6, i.e., about 40% of the APs are not transmitting, which is a significant deviation from the fully loaded network

[11] $\zeta = \frac{7}{2}$ and $\zeta = \frac{7}{2}\Gamma(1 + \frac{2}{\alpha})\Gamma(1 - \frac{2}{\alpha})$ for $w_x = 1$ and $w_x = Pg_x$, respectively, where $\Gamma(\cdot)$ is the Gamma function [30].

assumption employed in Section 5.4.1.1. This observation strongly suggests that system performance under large infrastructure density should be re-evaluated accordingly.

5.4.3.2 Coverage probability

Having obtained an (approximate) expression for the AP activity probability, the problem of system performance evaluation can now be attacked. Let $\tilde{\Phi}_a \subseteq \Phi_a$ denote the positions of transmitting APs, i.e., APs associated with at least one UE. The SIR experienced by a typical UE located at the origin is the same as in (5.6), with the summation in the denominator performed over $\tilde{\Phi}_a \backslash \{x^*\}$ instead of $\Phi_a \backslash \{x^*\}$. In order to determine the SIR distribution, the statistics of $\tilde{\Phi}_a$ must be determined, which, however, is an intractable task due to correlation between the events of two APs associated with (at least) one UE. To overcome this difficulty, a natural approach is to assume that association events are independent and consider $\tilde{\Phi}_a$ as the point process resulting from a thinning of Φ_a, namely, by independently removing each point $x \in \Phi_a$ with a probability $\mathbb{P}(K=0)$ [34,35]. It follows from Property 5 of PPPs (see Section 5.2), that $\tilde{\Phi}_a$ is a PPP of density $\mathbb{P}(K>0)\lambda_a$.

Under this approximation for the statistics of $\tilde{\Phi}_a$, the exact same procedures of Section 5.4.1.1 can be employed here, using $\mathbb{P}(K>0)\lambda_a$ instead of λ_a for the density of interfering APs. For example, following the same steps as in the derivation of (5.12), the coverage probability of the downlink cellular network with finite λ_u and nearest AP association can be shown to be (approximately) equal to (compare with (5.12)):

$$\mathbb{P}(\text{SIR} \geq \theta) = \frac{1}{1 + \mathbb{P}(K>0)\rho(\theta, \alpha)}. \tag{5.21}$$

Replacing $\mathbb{P}(K>0)$ with the approximate expression of (5.20) shows that the coverage probability is a decreasing function of λ_u/λ_a as expected, due to the increased interference levels generated by the larger number of active APs. This provides another argument in favor of network densification, i.e., increasing λ_a for a fixed value of λ_u not only allows for arbitrarily large spatial reuse levels (as argued in Section 5.4.1.1) but also inherently improves coverage probability without resorting to other methods such as bandwidth expansion and advanced transmission techniques.

Figure 5.12(a) depicts the downlink coverage probability with nearest AP association as a function of λ_u/λ_a. Both the exact probability (obtained via Monte Carlo simulation) and the analytical approximation of (5.21) are depicted. It can be seen that the approximate expression provides an excellent match of the actual probability for all range of values of λ_u/λ_a considered, suggesting that the approximation of $\tilde{\Phi}_a$ as a PPP is accurate. As mentioned above, system performance is benefited by smaller values of λ_u/λ_a and the benefit is quantified in this figure, e.g., reducing λ_u/λ_a from 10 to 1 provides an SIR gain of about 3 dB.

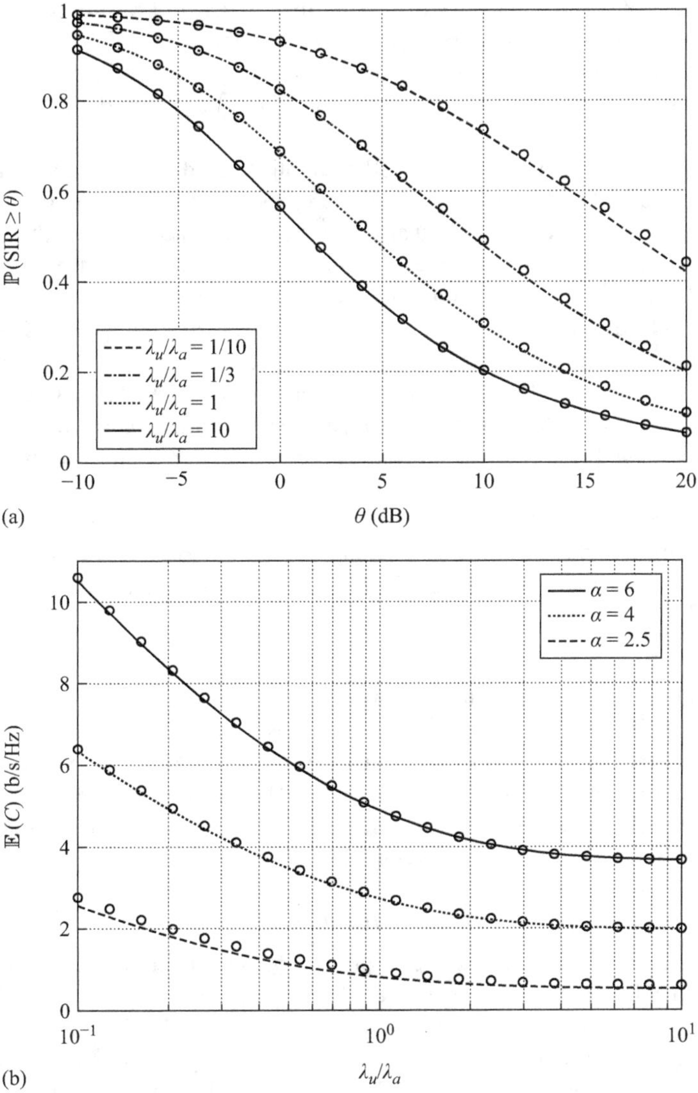

Figure 5.12 Downlink coverage probability for $\alpha = 4$ (a) and average AP spectral efficiency with $\theta_0 = 0$ dB (b) with nearest AP association under various load conditions. Lines denote analytic approximations whereas markers are Monte Carlo evaluations

Figure 5.12(b) shows the average link spectral efficiency $\mathbb{E}(C)$ obtained numerically based on the approximate analytical expression for the coverage probability along with Monte Carlo simulations. It can be seen that for values of λ_u/λ_a of about 5, $\mathbb{E}(C)$ reaches a plateau, suggesting that for these values of UE load very few of the APs are inactive. This means that the results obtained under the fully loaded network assumption are actually accurate for $\lambda_u \approx 5\lambda_a$ or greater. Smaller UE load values correspond to increased spectral efficiency, with most significant improvements shown in the regime of $\lambda_u \leq \lambda_a$ (ultra dense network) and for larger values of α.

Note that the simulations appear to be slightly biased toward larger values than the ones predicted by the analysis. This discrepancy is not only due to the approximations employed in the analysis but also due to practical limitations of the simulation setup which necessarily must consider a finite subset of $\mathbb{R}^2$ (whereas the model extents to the whole plane—see the related discussion in [17, p. 150]).

Remark: In the case of inactive APs the notion of maximum SIR association is not clear. This is because the interference level experienced by a certain UE depends on the active APs, which, in turn, depend on the association decisions of the other UEs. This implies that the performance of lightly loaded networks under maximum received power association must be attacked in a different manner than the one considered in Section 5.4.1.3 where the equivalence with the maximum SIR association was exploited. Although analysis can in principle be done using the techniques discussed so far, it appears that no simple, closed form expressions can be obtained that would provide additional insights to a numerical evaluation of performance.

5.5 User rate performance analysis

This section considers system performance with respect to UE rate, *R*, as defined in Section 5.3. Both performance analysis and system design aspects are discussed. Only nearest AP association will be discussed. For simplicity, analysis and results for a single-tier network will be presented.[12]

5.5.1 Degrees of freedom distribution

The new element introduced by the definition of *R* compared to the other performance metrics is quantity K_0, i.e., the number of additional UEs associated with the AP serving the typical UE, which determines the UE degrees of freedom, $1/(1 + K_0)$. Considering a typical UE located at the origin, K_0 can be equivalently viewed as the number of elements of Φ_u included within the cell generated by the points of Φ_a that *includes the origin*. It might appear at first sight that (a partial) characterization of K_0 has already been performed in Section 5.4.3, where some of the statistical properties of *K*, the number of UEs associated with a random AP, were presented. However, K_0 and *K* are *not* identically distributed, hence the difference in notation. This is because the statistical properties of the cell containing the origin (or any other point for that matter) are different from the properties of a random cell (with no additional requirements imposed, e.g., on including certain points).

[12]See [28] for extensions in the multi-tier case.

As an illustration of this difference, let $|\mathscr{C}_0| > 0$ and $|\mathscr{C}| > 0$ denote the areas of the cell containing the origin and any other random cell.[13] It can be shown that $\mathbb{E}(|\mathscr{C}_0|) > \mathbb{E}(|\mathscr{C}|)$ [15], from which it follows that:

$$\begin{aligned}\mathbb{E}(K_0) &= \mathbb{E}(\mathbb{E}(K_0||\mathscr{C}_0|)) \\ &\stackrel{(a)}{=} \mathbb{E}(\lambda_a|\mathscr{C}_0|) \\ &> \lambda_a\mathbb{E}(|\mathscr{C}|) \\ &= \mathbb{E}(K),\end{aligned} \tag{5.22}$$

where (a) follows by Property 1 of $\mathbf{\Phi}_a$ (Section 5.2). Therefore, the number of UEs associated with the cell including the origin is greater on average than the number of UEs served by any other random cell. This apparent "paradox" can be viewed as a manifestation to the two-dimensional case of the well-known "waiting-time paradox" occurring in the study of single-dimensional Poisson point processes [36, p. 454].

As in the case of K, exact closed form expressions for the distribution of K_0 are not available, rendering approximations necessary. A good approximation for the distributions of both K_0 and K, depending only on the ratio λ_u/λ_a, is provided in [37] as

$$\mathbb{P}(\mathscr{K} = k) = \frac{(7/2)^\eta\Gamma(k+\eta)(\lambda_u/\lambda_a)^k}{\Gamma(\eta)k!(\lambda_u/\lambda_a + 7/2)^{k+\eta}}, \quad k = 0, 10, \ldots, \tag{5.23}$$

where $\eta = 7/2$ for $\mathscr{K} = K$ and $\eta = 9/2$ for $\mathscr{K} = K_0$ and $\Gamma(\cdot)$ denotes the Gamma function [30]. Note that evaluation of $\mathbb{E}(K)$ and $\mathbb{P}(K = 0)$ using (5.23) leads to the same expressions as (5.18) and (5.20), respectively, the latter obtained by a direct approach without explicit consideration of the complete distribution of K. Evaluation of the mean of K_0 using (5.23) gives $\mathbb{E}(K_0) = (9/7)(\lambda_u/\lambda_a)$, i.e., the average K_0 is about 1.28 times the average K.

The accuracy of (5.23) is demonstrated in Figure 5.13, where an excellent match is observed with the empirical distribution obtained by Monte Carlo simulation for two test values of λ_u/λ_a. Experimental evidence implies that this level of accuracy holds for any value of λ_u/λ_a.

With the distribution of K_0 available, the distribution of the degrees of freedom $(1/1 + K_0)$ is also completely characterized. A natural metric to obtain insights on how the degrees of freedom depend on system parameters is their average $\mathbb{E}(1/(1 + K_0))$, which turns out to be available in closed form (no need to perform a direct numerical evaluation using the distribution of K_0). It is shown in Appendix A that:

$$\mathbb{E}\left(\frac{1}{1+K_0}\right) = \frac{\lambda_a}{\lambda_u}\mathbb{P}(K > 0), \tag{5.24}$$

[13]Non-zero cell areas are guaranteed since no two points of a PPP coincide with probability 1.

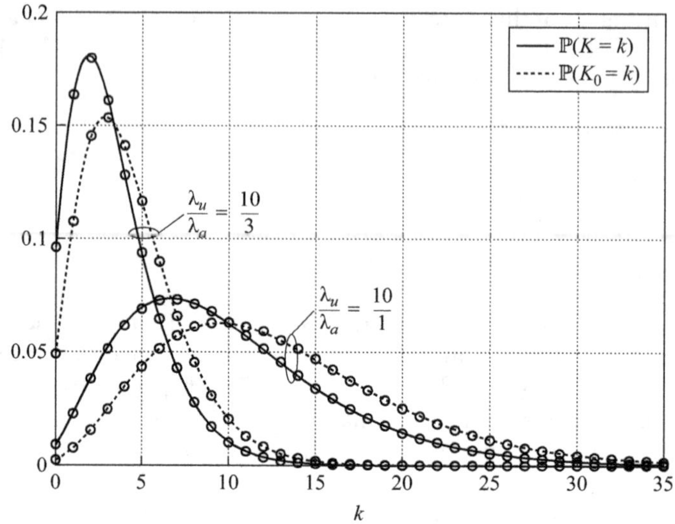

Figure 5.13 Distribution of K and K_0 for two test cases of λ_u/λ_a. Marks denote probabilities obtained by Monte Carlo simulation, whereas the approximate analytical expression is shown as continuous line for clarity

where $\mathbb{P}(K > 0)$ is (approximately) given by (5.20) with $\zeta = 7/2$. It is of interest to compare this value with the corresponding one for the case where $\mathbf{\Phi}_a$ is distributed as a periodic hexagonal grid of the same density λ_a. Denoting as $\tilde{K}_0$ the number of UEs served by the same AP as the typical UE under the regular AP deployment, it is easy to show that $\tilde{K}_0$ is a Poisson random variable of mean λ_u/λ_a and it follows that:

$$\begin{aligned}\mathbb{E}\left(\frac{1}{1+\tilde{K}_0}\right) &= \sum_{k=0}^{\infty} \frac{1}{1+k}\frac{(\lambda_u/\lambda_a)^k}{k!}e^{-\lambda_u/\lambda_a}\\ &= \frac{\lambda_a}{\lambda_u}\left(1 - e^{-\lambda_a/\lambda_u}\right)\\ &\overset{(a)}{\geq} \mathbb{E}\left(\frac{1}{1+K_0}\right),\end{aligned}$$

where (a) follows from (5.19) and (5.24). This shows that the randomness of the infrastructure deployment results in a degradation of UE performance in terms of average degrees of freedom in addition to the loss in terms of SIR performance, which was demonstrated in Section. 5.4.1.1. This degradation is quantified in Figure 5.14 where it can be seen that it is rather limited and mostly notable in the regime of $\lambda_u < 2\lambda_a$.

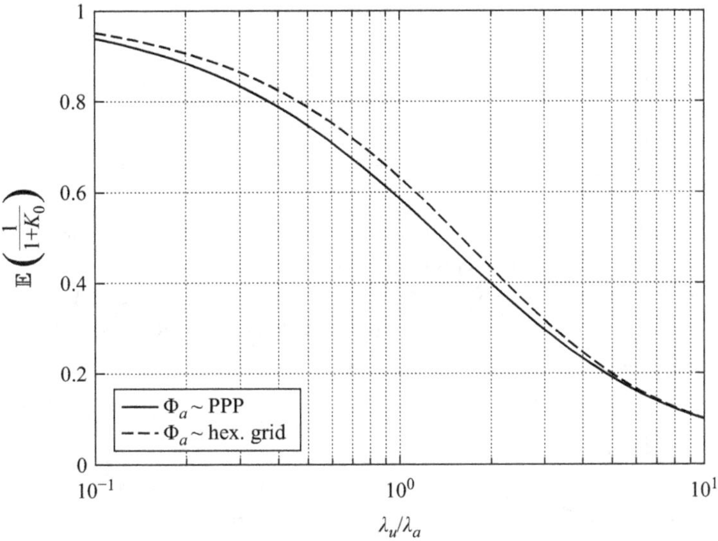

Figure 5.14 Variation of $\mathbb{E}(1/(1+K_0))$ *as a function of system load, for random and regular AP deployments*

5.5.2 User rate CDF and system design issues

With the statistics of K_0 determined, the problem of statistical characterization of the (typical) UE rate R can now be attacked. It directly follows from the definition of R that its CDF equals:

$$\mathbb{P}(R \leq r) = \mathbb{E}\left(\mathbb{P}\left(\frac{1}{1+K_0}\log_2(1+\mathrm{SIR})\mathbb{I}(\mathrm{SIR} \geq \theta_{\mathrm{th}}) \leq r \middle| K_0\right)\right)$$

for $r > 0$. In order to proceed, $\mathbb{P}(\mathrm{SIR} \leq \theta|K_0)$, i.e., the CDF of the SIR *conditioned on* K_0, must be determined. Note that in the analysis of the traditional hexagonal AP deployment pattern this poses no problem as SIR and K_0 are independent, i.e., $\mathbb{P}(\mathrm{SIR} \leq \theta|K_0) = \mathbb{P}(\mathrm{SIR} \leq \theta)$. However, this is not the case when $\mathbf{\Phi}_a$ is a PPP (or any other random point process for that matter). This can be understood intuitively by noting that larger values of K_0 suggest a larger area of the serving AP, which in turn implies that interfering APs are located further away and a correspondingly larger SIR, i.e., SIR and K_0 are positively correlated.

Unfortunately, determination of $\mathbb{P}(\mathrm{SIR} \leq \theta|K_0)$ appears to be an intractable problem, and ignoring the dependence between SIR and K_0 seems to be the only way to proceed [28,35]. This allows the use of the closed form expression of the (marginal) distributions of SIR and K_0 given in (5.21) and (5.23), respectively, which results in:

$$\mathbb{P}(R \leq r) = \sum_{k=0}^{\infty} \mathbb{P}(K_0 = k)(1 - F(k,r))$$

where

$$F(k,r) \triangleq \begin{cases} \dfrac{1}{1+\mathbb{P}(K>0)\rho(2^{r(1+k)}-1,\alpha)}, & r > \dfrac{1}{1+k}\log_2(1+\theta_{\text{th}}) \\ \dfrac{1}{1+\mathbb{P}(K>0)\rho(\theta_{\text{th}},\alpha)}, & r \le \dfrac{1}{1+k}\log_2(1+\theta_{\text{th}}), \end{cases} \tag{5.25}$$

with the distributions of K_0 and K as in (5.23). This expression for the UE rate CDF can be easily evaluated numerically (only a moderate number of terms of the sum are required for high accuracy). The assumed independence of SIR and K_0 also allows for an integral expression for $\mathbb{E}(R)$, without explicit usage of the full distribution (5.25), as

$$\begin{aligned} \mathbb{E}(R) &= \mathbb{E}\left(\frac{1}{1+K_0}\right)\mathbb{E}(\log_2(1+\text{SIR})\mathbb{I}(\text{SIR} \ge \theta_{\text{th}})) \\ &= \frac{\lambda_a \mathbb{P}(K>0)}{\lambda_u}\left(F(0,0) + \int_{\log_2(1+\theta_{\text{th}})}^{\infty} F(0,x)dx\right), \end{aligned}$$

where (5.24) has been employed.

The rate CDF is depicted in Figure 5.15, evaluated using (5.25) as well as via Monte Carlo simulation for various values of λ_u/λ_a, ranging from 10 (heavy loaded network) to $1/10$ (ultra dense network). The approximate expression of (5.25) shows very good accuracy, suggesting that the dependence of K_0 and SIR is indeed

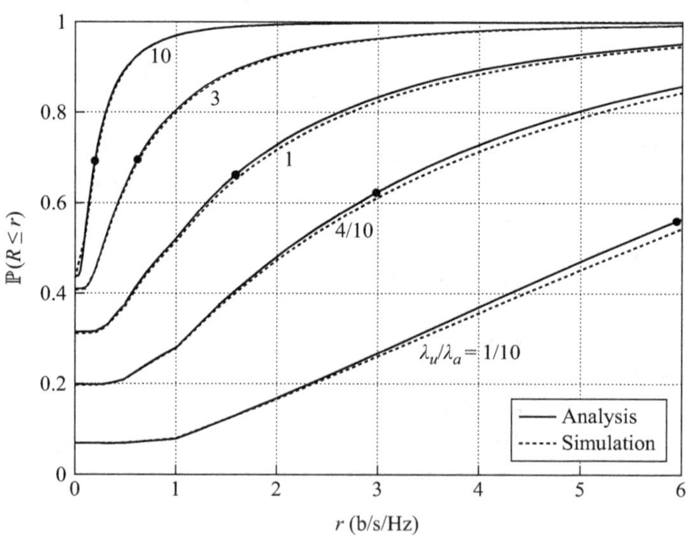

Figure 5.15 CDF of R for various λ_u/λ_a ($\alpha = 4, \theta_{th} = 0$ dB). For each CDF, a point indicates the coordinate ($\mathbb{E}(R)$,$\mathbb{P}(R \le \mathbb{E}(R))$)

very small. Heavier traffic load results in reduced rates as expected, whereas, an ultra dense network provides significant performance gains leveraging the increased degrees of freedom available due to the smaller number of UEs associated with the same AP. Note that network performance is highly unbalanced with most of the UEs experiencing rates smaller than the average (about 70% and 55% for heavy loaded and ultra dense networks, respectively) and a significant portion of UEs actually achieving zero rate due to SIR outage (i.e., SIR $< \theta_{\text{th}}$), especially for highly loaded networks.

5.5.3 System design

The theoretical CDF expression can of course be employed not only for efficiently predicting system performance but, more importantly, for obtaining system design insights. For example, a system designer targeting a certain quality of service (QoS) for the UEs (expressed as some function of the CDF of R) and with λ_u fixed and known, can employ the above formulas to efficiently determine the (minimum) infrastructure density λ_a required to provide this QoS without resorting to computationally intensive computer simulations (an example will be discussed later). This is an important network design problem given the current trend of network densification.

A question that arises at this point is if (and how much) the minimum λ_a requirements can be relaxed under a more complicated system model than the one considered so far, taking into account, e.g., advanced transmission techniques achieving higher spectral efficiency. Answering this question, of course, requires system performance characterization under these techniques. To this end, various works have appeared recently, extending the system model framework considered so far to include, e.g., beamforming [34,38], maximum ratio combining [39], and cooperative transmission from multiple APs [40–42]. As expected, these methods provide an (statistical) increase of SIR, which is reflected as an increase of link spectral efficiency and UE rate. The benefits are quantified by analytical expressions which, inevitably, are more complicated than the ones presented in this chapter.

However, another critical parameter determining UE rate, which is not addressed at all by these approaches, is the degrees of freedom $1/(1 + K_0)$, which depends exclusively on the multiple access scheme. Even though the simple TDMA assumed so far is entirely plausible and allows for simple analysis, one may conjecture that (jointly) scheduling over time *and* frequency dimensions can increase performance, at least under certain operational conditions. This will be investigated in the following by appropriately modifying the system model and introducing scheduling schemes that exploit both time and frequency dimensions.

5.5.3.1 Downlink scheduling in time and frequency

The downlink bandwidth will be considered to be partitioned into $N \geq 1$ subchannels (SCs) of equal bandwidth, each of which can be used for transmission to a single UE per AP within a certain time slot (all time slots are of equal length).

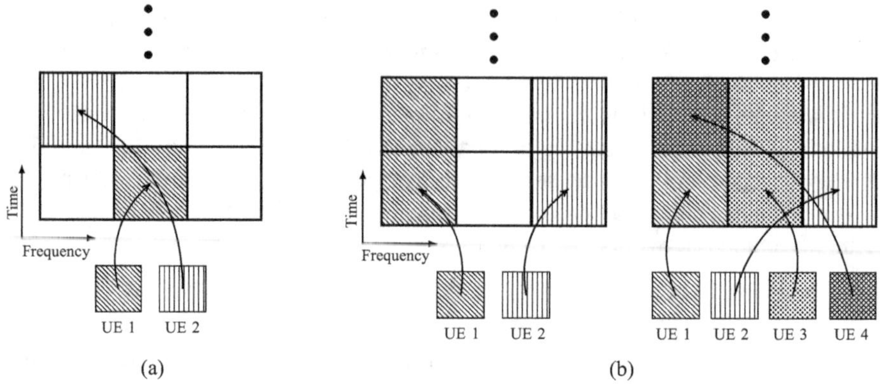

Figure 5.16 Resource allocation examples with $N = 3$ for (a) TDMA-FH and (b) FDMA with $K \leq N$ (left), where each UE is assigned an exclusive SC, and $K > N$ (right), where there exist SCs shared by multiple UEs. Blank time–frequency resource elements are not used for transmission. Only two time slots are depicted, as, for the considered cases of K, the scheduling pattern is periodically repeated in the subsequent time slots. Note that, in general, this scheduling period equals K (in time slots) for TDMA-FH and the smaller integer greater than N/K for FDMA, respectively

The number N is a free variable that remains fixed once its (optimal) value has been determined at the system design stage. To exploit the available frequency dimensions, a simple, analytically tractable extension of the baseline TDMA scheme was proposed in [13], which can be informally described as introducing a "frequency hopping (FH) element" to TDMA. In particular, each UE is served sequentially by its associated AP as in the TDMA case, however, transmission to a given UE is performed on a single, randomly selected SC out of the total N available. An example of this scheduling procedure, referred to in the following as "TDMA-FH," is depicted in Figure 5.16(a) for $N = 3$ and an AP serving $K = 2$ UEs.

Note that TDMA-FH introduces two effects on the achieved UE rate, compared to the TDMA scheme:

- The degrees of freedom of each UE are reduced by a factor of $1/N$, i.e., the UE rate expression becomes $R = \log_2(1 + \text{SIR})/(N(1 + K_0))$.
- The SIR is increased since part of the (active) APs will not transmit at the same frequency as the one used by the AP serving the typical UE.

It is easy to see that the density of interfering APs employing TDMA-FH becomes equal to $(1/N)\mathbb{P}(K > 0)\lambda_b$. Approximating their positions as a PPP, performance analysis of TDMA-FH is only a trivial modification of the corresponding one for the TDMA case [13,35].

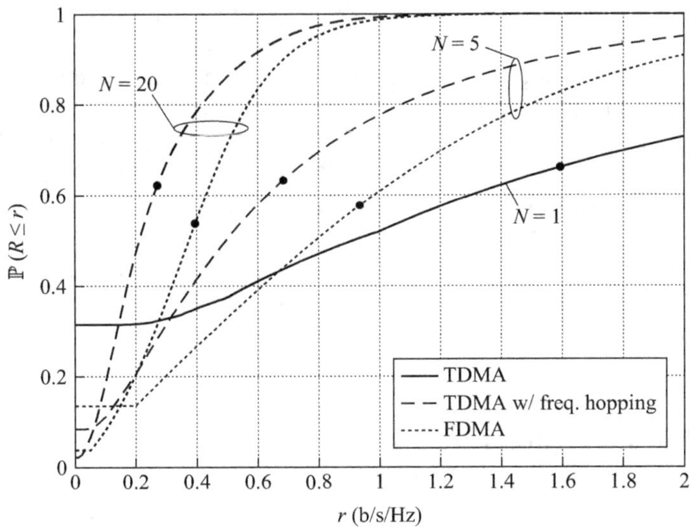

Figure 5.17 Rate CDF under TDMA, TDMA-FH, and FDMA ($\lambda_u/\lambda_a = 1, \alpha = 4, \theta_{\text{th}} = 0$ dB)

A different approach was proposed in [35] attempting to more efficiently utilize bandwidth by allowing for multiple concurrent transmissions within a single time slot. The basic principle of this scheduling scheme is that UEs are evenly and randomly distributed over SCs (only a single SC for each UE is allowed), and in the case where multiple UEs are assigned to the same SC, time sharing is performed. An example of this scheduling scheme, referred to in the following as "FDMA" is shown in Figure 5.16(b). In comparison with TDMA-FH, FDMA achieves an improved utilization of the system bandwidth, reflected by the UE rate expression, which becomes:

$$R = \frac{1}{N(1 + L_0)} \log_2(1 + \text{SIR}),$$

where $0 \leq L_0 \leq \lfloor K_0/N \rfloor \leq K_0$ is the number of additional UEs sharing the same SC as the typical UE. Effectively, FDMA increases the degrees of freedom compared to TDMA-FH. On the other hand, the more efficient bandwidth utilization of FDMA results in an increased density of interferers compared to TDMA-FH. Performance analysis for FDMA is more involved compared to TDMA-FH, however, it follows the same principles as in all the cases examined previously (see [35] for details).

The CDF of R provided by TDMA, TDMA-FH, and FDMA is shown in Figure 5.17 for $N = 5$ and $N = 20$ (of course, TDMA can be viewed as a special case of the other two schemes with $N = 1$). A UE load corresponding to $\lambda_u/\lambda_a = 1$

is considered with performance trends similar to other values of λ_u/λ_a. It can be seen that TDMA-FH significantly improves performance compared to TDMA in the low rate regime (cell edge UEs), as its inherent interference reduction results in reduced probability of a UE experiencing an SIR outage. As expected, increasing N further enhances this effect. At the same time, performance with TDMA-FH degrades in the higher rates regime (as well as in terms of average UE rate) compared to TDMA due to the reduction of available transmission degrees of freedom. Performance of FDMA follows the same trends. As it might be expected, FDMA provides (slightly) worse performance in the low rate regime compared to TDMA-FH due to the increased interference produced by concurrent transmissions, whereas it is able to provide improved performance at higher rates due to the increased degrees of freedom.

5.5.3.2 Infrastructure density requirements

In order to view the potential of scheduling in frequency, consider the problem of determining the minimum value of λ_a/λ_u, i.e., AP density normalized by the UE density, necessary to provide a UE rate $R \geq r_0 > 0$ with a probability $0 < \varepsilon < 1$ (equivalently, $\varepsilon 100\%$ of UEs experiencing rates equal or greater than r_0). This minimum λ_a/λ_u can be determined (numerically) for the TDMA case by directly using the analytical performance formulas presented previously. For the cases of TDMA-FH and FDMA, the minimum λ_a/λ_u depends on parameter N, which can be also be optimized numerically.[14] For the FDMA case, a closed form upper bound on the minimum requirements (under the optimal selection of N) can be obtained [35], which asymptotically scales as

$$\lambda_a/\lambda_u \leq \begin{cases} \dfrac{(1-\varepsilon)\rho(\theta_{\text{th}}, \alpha)}{\varepsilon \log_2(1+\theta_{\text{th}})} r_0, & r_0 \ll \log_2(1+\theta_{\text{th}}) \\[2ex] \dfrac{(1-\varepsilon)}{\varepsilon} \kappa(\alpha) 2^{2r_0/\alpha}, & r_0 \gg \log_2(1+\theta_{\text{th}}), \end{cases} \tag{5.26}$$

i.e., linearly and exponentially with r_0 for small and large r_0, respectively.

Figure 5.18 shows the minimum λ_a/λ_u with TDMA, TDMA-FH, and FDMA, for various r_0 with $\varepsilon = 0.1$ and $\theta_{\text{th}} = -6$ dB (SIR threshold of LTE [23]). The asymptotic upper bound for the case of FDMA is also depicted. It can be seen that the rate value $\log_2(1+\theta_{\text{th}})$ is a "transition point" on how minimum system density requirements scale, as also suggested by (5.26). In particular, for values of $r_0 > \log_2(1+\theta_{\text{th}})$, all three schemes provide the same performance as it turns out that the optimal N equals 1 for both TDMA-FH and FDMA. On the other hand, for values of $r_0 < \log_2(1+\theta_{\text{th}})$, bandwidth partitioning is beneficial and actually allows for a dramatic (orders of magnitude) decrease of infrastructure requirements

[14] In principle, the optimal value of N is different for TDMA-FH and FDMA.

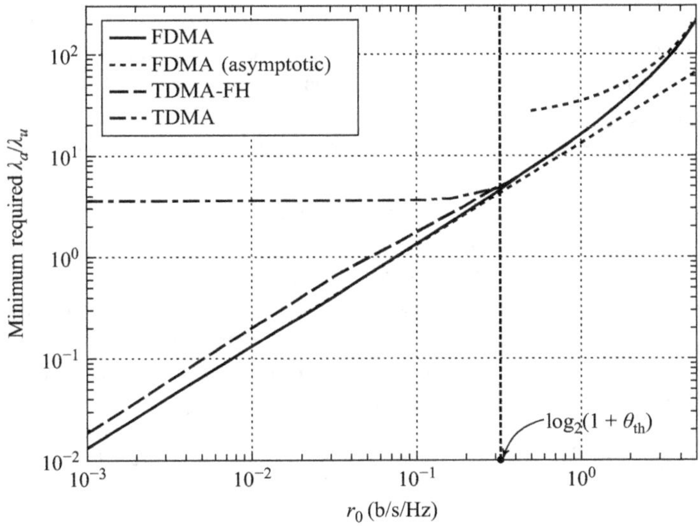

Figure 5.18 Minimum required λ_a/λ_u for providing a UE rate outage probability $\varepsilon = 0.1$ ($\theta_{th} = -6$ dB, $\alpha = 3$)

compared to TDMA. FDMA has the minimum density requirements, followed closely by TDMA-FH. These results indicate the following:

1. For small rate requirements, the simple TDMA scheme is highly inefficient and special attention is required when interpreting performance analyses derived under a TDMA assumption.
2. For large rate requirements, infrastructure density must increase exponentially, i.e., an ultra-dense network deployment is necessary. However, these requirements may be unfeasible in practice, rendering employment of advanced transmission techniques and/or increase of available bandwidth essential for future cellular networks.

5.6 Conclusion

This chapter examined the effects of densification and randomness of infrastructure in future cellular networks by employing analytical tools from stochastic geometry. Various operational scenarios were investigated, including uplink and downlink transmissions, multi-tier networks, and variable traffic load, as well as parameters and metrics of interest such as SIR, user rate, degrees of freedom, and multiple access schemes. It was shown that randomness of AP positions introduces an unavoidable but moderate performance degradation compared to a regular placement pattern. Surprisingly enough, densification of randomly placed infrastructure does not (statistically) degrade performance in terms of SIR or user rate, even under the worst case scenario where all nodes are transmitting. Taking into account inactive transmitters (due to having no associated receivers), densification results in performance improvements. Performance gains were explicitly quantified with simple closed-form expressions in most cases.

These results, even though corresponding to a simplified, yet plausible and close to practical, system model, provide a rigorous validation of the merits of densification in cellular networks. In addition, they can be applied for (initial) system design, with an example shown in the last section with respect to the minimum infrastructure density required to support a given downlink rate performance per UE. Even though the presented cases constitute only a small part of the results on this topic, it is believed that they provide a good basis for the newcomer to build upon and enter this very active research field, which has still many open topics to address, both of theoretical and of practical importance. These include:

- Consideration of more elaborate point processes than the PPP for more accurately describing real-world distribution of network nodes.
- Use of more complex channel models, taking into account issues such as multipath propagation and location-dependent path losses.
- Opportunistically exploiting UEs as relays or even content providers themselves (device-to-device communications, caching).
- Mobility in ultra dense networks.
- Cooperative transmissions.

Appendix A

A.1 Derivation of (5.18)

Derivation of (5.18) is somewhat technical, however, it is a good example of application of both Campbell's theorem and the PGFL formula. Consider a random (typical) AP of a single-tier cellular network, and shift the origin of the plane to its position. The average number of UEs connected to the typical AP is

$$\begin{aligned}\mathbb{E}(K) &= \mathbb{E}\left(\sum_{x_u\in\Phi_u}\mathbb{I}\{\text{UE positioned at } x_u \text{ associates with typical AP}\}\right)\\ &= \mathbb{E}\left(\sum_{x_u\in\Phi_u}\prod_{x_a\in\Phi_a\backslash\{0\}}\mathbb{I}\{w_0(x_u)|x_u|^{-\alpha} > w_{x_a}(x_u)|x_u - x_a|^{-\alpha}\}\right)\\ &= \mathbb{E}\left(\sum_{x_u\in\Phi_u} p(x_u, w_0(x_u))\right)\end{aligned}$$

where the second equality takes into account the general weighted-path-loss association rule (5.18), $\{w_x(x_u)\}_{x\in\Phi_a}$ are the AP weights observed by the UE positioned at $x_u \in \mathbf{\Phi}_u$, and

$$p(x_u, w_0(x_u)) \triangleq \mathbb{E}\left(\prod_{x_a \in \Phi_a \setminus \{0\}} \mathbb{I}\{w_0(x_u)|x_u|^{-\alpha} > w_{x_a}(x_u)|x_u - x_a|^{-\alpha}\} \middle| x_u, w_0(x_u) \right) \tag{A.1}$$

is the probability of a UE located at $x_u \in \mathbf{\Phi}_u$ to associate with the typical AP given the weight $w_0(x_u)$. Since the weights are i.i.d., index x_u will be dropped from the weight notation in the following for simplicity. Application of the PGFL formula (5.3) gives:

$$\begin{aligned} p(x_u|w_0) &= \exp\left(-\lambda_a \mathbb{E}\left(\int_{\mathbb{R}^2} \left[1 - \mathbb{I}\{|x - x_u| > (w_x/w_0)^{1/\alpha}|x_u|\}\right] dx \middle| x_u, w_0\right)\right) \\ &= \exp\left(-\lambda_a \mathbb{E}\left(\int_{B(x_u,(w_x/w_0)^{1/\alpha}|x_u|)} dx \middle| x_u, w_0\right)\right) \\ &= \exp\left(-\lambda_a \mathbb{E}\left(\pi (w_x/w_0)^{2/\alpha}|x_u|^2 \middle| x_u, w_0\right)\right) \\ &= \exp\left(-\lambda_a \pi \mathbb{E}\left(w_x^{2/\alpha}\right) w_0^{-2/\alpha}|x_u|^2\right). \end{aligned}$$

where $\mathrm{B}(c, r) \in \mathbb{R}^2$ denotes the ball centered at $c \in \mathbb{R}^2$ with radius $r \geq 0$. Substituting (A.2) into (A.1) gives

$$\begin{aligned} \mathbb{E}(K) &= \mathbb{E}\left(\mathbb{E}\left(\sum_{x_u \in \Phi_u} \exp\left(-\lambda_a \pi \mathbb{E}(w_x^{2/\alpha}) w_0^{-2/\alpha}|x_u|^2\right) \middle| w_0\right)\right) \\ &\overset{(a)}{=} \mathbb{E}\left(\lambda_u \int_{\mathbb{R}^2} \exp\left(-\lambda_a \pi \mathbb{E}(w_x^{2/\alpha}) w_0^{-2/\alpha}|x|^2\right) dx\right) \\ &\overset{(b)}{=} \mathbb{E}\left(\frac{\lambda_u w_0^{2/\alpha}}{\lambda_a \mathbb{E}\left(w_x^{2/\alpha}\right)}\right) \\ &= \frac{\lambda_u}{\lambda_a}, \end{aligned} \tag{A.2}$$

where (a) is an application of Campbell's formula (5.1) to the inner expectation and (b) follows by evaluating the integral over polar coordinates and noting that the statistics of w_x are independent of x.

Note that alternative derivations considering only the nearest AP association rule can be found in [9,15].

A.2 Derivation of (5.24)

The proof first appeared in [43]. Normalize without loss of generality the unit area such that $\lambda_a = 1$. Then the density of Φ_u becomes equal to λ_u/λ_a. Let $p_{|\mathscr{C}_0|}(\cdot)$ and $p_{|\mathscr{C}|}(\cdot)$ denote the probability density functions of the area of the cell containing the origin, $|\mathscr{C}_0|$, and the area of any other random cell, $|\mathscr{C}|$, respectively, under a normalized AP density $\lambda_a = 1$. It can be shown that $p_{|\mathscr{C}_0|}(x) = xp_{|\mathscr{C}|}(x), x > 0$ [37].

$$
\begin{aligned}
\mathbb{E}\left(\frac{1}{1+K_0}\right) &= \sum_{k=0}^{\infty} \frac{1}{1+k}\mathbb{P}(K_0 = k) \\
&\overset{(a)}{=} \sum_{k=0}^{\infty} \frac{1}{1+k} \int_0^{\infty} p_{|C_0|}(x) \frac{\left(\frac{\lambda_u}{\lambda_a}x\right)^k}{k!} e^{-\frac{\lambda_u}{\lambda_a}x}\, dx \\
&\overset{(b)}{=} \frac{\lambda_a}{\lambda_u} \int_0^{\infty} p_{|C|}(x) e^{-\frac{\lambda_u}{\lambda_a}x} \left(\sum_{k=0}^{\infty} \frac{\left(\frac{\lambda_u}{\lambda_a}x\right)^{k+1}}{(k+1)!} \right) dx \\
&= \frac{\lambda_a}{\lambda_u} \int_0^{\infty} p_{|C|}(x) e^{-\frac{\lambda_u}{\lambda_a}x} \left(e^{\frac{\lambda_u}{\lambda_a}x} - 1 \right) dx \\
&= \frac{\lambda_a}{\lambda_u} \left(1 - \int_0^{\infty} p_{|C|}(x) e^{-\frac{\lambda_u}{\lambda_a}x}\, dx \right) dx,
\end{aligned}
$$

holds, where (a) is due to K_0 being Poisson distributed with mean $(\lambda_u/\lambda_a)x$, given $|\mathscr{C}_0| = x$ and (b) follows by interchanging integration and summation (Fubini's theorem). Noting that the last integral equals $\mathbb{P}(K = 0)$ completes the proof.

References

[1] J. G. Andrews, S. Buzzi, W. Choi, *et al.*, "What will 5G be?", *IEEE Journal on Selected Areas in Communications*, vol. 32, no. 6, pp. 1065–1082, Jun. 2014.

[2] F. Boccardi, R. W. Heath, A. Lozano, T. L. Marzetta, and P. Popovski, "Five disruptive technology directions for 5G," *IEEE Communications Magazine*, vol. 52, pp. 74–80, Feb. 2014.

[3] W. H. Chin, Z. Fan, and R. J. Haines, "Emerging technologies and research challenges for 5G wireless networks," *IEEE Wireless Communications*, vol. 21, no. 2, pp. 106–112, Apr. 2014.

[4] J. G. Andrews, "Seven ways that HetNets are a cellular paradigm shift," *IEEE Communications Magazine*, vol. 51, no. 3, pp. 136–144, Mar. 2013.

[5] N. Brushan, J. Li, D. Malladi, *et al.*, "Network densification: the dominant theme for wireless evolution into 5G," *IEEE Communications Magazine*, vol. 52, no. 2, pp. 82–89, Feb. 2014.

[6] I. Hwang, B. Song, and S. S. Soliman, "A holistic view on hyper-dense heterogeneous and small cell networks," *IEEE Communications Magazine*, vol. 51, no. 6, pp. 20–27, Jun. 2013.

[7] S. F. Yunas, M. Valkama, and J. Niemeläm, "Spectral and energy efficiency of ultra-dense networks under different deployment strategies," *IEEE Communications Magazine*, vol. 53, no. 1, pp. 90–100, Jan. 2015.

[8] S. N. Chiu, D. Stoyan, W. S. Kendall, and J. Mecke, *Stochastic Geometry and Its Applications*, 3rd ed., Wiley, Hoboken, NJ, 2013.

[9] S. G. Foss, and S. A. Zuyev, "On a Voronoi aggregative process related to a bivariate poisson process," *Advances in Applied Probability*, vol. 28, no. 4, pp. 965–981, Dec. 1996.

[10] F. Baccelli, M. Klein, M. Lebourges, and S. Zuyev, "Stochastic geometry and architecture of communication networks," *Telecommunication Systems*, vol. 7, no. 1, pp. 209–227, 1997.

[11] J. G. Andrews, R. K. Ganti, N. Jindal, M. Haenggi, and S. Weber, "A primer on spatial modeling and analysis in wireless networks," *IEEE Communications Magazine*, vol. 48, no. 11, pp. 156–163, Nov. 2010

[12] S. Weber, J. G. Andrews, and N. Jindal, "An overview of the transmission capacity of wireless networks," *IEEE Transactions on Communications*, vol. 58, no. 12, pp. 3593–3604, Dec. 2010.

[13] J. G. Andrews, F. Baccelli, and R. K. Ganti, "A tractable approach to coverage and rate in cellular networks," *IEEE Transactions on Communications*, vol. 59, no. 11, pp. 3122–3134, Nov. 2011.

[14] H. ElSawy, E. Hossain, and M. Haenggi, "Stochastic geometry for modeling, analysis, and design of multi-tier and cognitive cellular wireless networks: a survey," *IEEE Communications Surveys & Tutorials.*, vol. 15, no. 3, pp. 996–1019, Jul. 2013.

[15] F. Baccelli, and B. Blaszczyszyn, *Stochastic Geometry and Wireless Networks*. Delft, The Netherlands: Now Publishers Inc., 2009.

[16] S. Mukherjee, *Analytical Modeling of Heterogeneous Cellular Networks*, Cambridge University Press, Cambridge, UK, 2014.

[17] M. Haenggi, and R. K. Ganti, *Interference in Large Wireless Networks*, NOW (Foundations and Trends in Networking, vol. 3, pp. 127–248), Delft, The Netherlands, 2008.

[18] S. Weber, and J. G. Andrews, *Transmission Capacity of Wireless Networks*, NOW (Foundations and Trends in Networking, vol. 5, pp. 109–281), Delft, The Netherlands, 2012.

[19] M. Haenggi, *Stochastic Geometry for Wireless Networks*, Cambridge University Press, Cambridge, UK, 2012.

[20] T. S. Rappaport, *Wireless Communications: Principles and Practice*, 2nd ed., Prentice Hall, Upper Saddle River, NJ, 2002.

[21] A. Papoulis, and S. U. Pillai, *Probability, Random Variables and Stochastic Processes*, 4th ed., McGraw-Hill, New York, NY, 2001.

[22] D. Tse, and P. Viswanath, *Fundamental of Wireless Communications*, Cambridge University Press, Cambridge, UK, 2005.

[23] S. Ahmadi, *LTE-Advanced: A Practical Systems Approach to Understanding 3GPP LTE Releases 10 and 11 Radio Access Technologies*, Elsevier Science, Amsterdam, The Netherlands, 2013.

[24] D. Gesbert, S. Hanly, H. Huang, S. Shamai, O. Simeone, and W. Yu, "Multi-cell mimo cooperative networks: a new look at interference," *IEEE Journal on Selected Areas in Communications*, vol. 28, no. 9, pp. 1380–1408, Dec. 2010.

[25] I. Hwang, B. Song, and S. S. Soliman, "A holistic view on hyper-dense heterogeneous and small cell networks," *IEEE Communications Magazine*, vol. 51, no. 6, pp. 20–27, Jun. 2013.

[26] A. El-Gamal, and Y.-H. Kim, *Network Information Theory*, Cambridge University Press, Cambridge, UK, 2012.

[27] G. P. Fettweis, "The tactile internet: applications and challenges," *IEEE Vehicular Technology Magazine*, vol. 9, no. 1, pp. 64–70, Mar. 2014.

[28] S. Singh, H. S. Dhillon, and J. G. Andrews, "Offloading in heterogeneous networks: modeling, analysis and design insights," *IEEE Transactions on Wireless Communications*, vol. 12, no. 5, pp. 2484–2497, May 2013.

[29] H.-S. Jo, Y. J. Sang, P. Xia, and J. G. Andrews, "Heterogeneous cellular networks with flexible cell association: a comprehensive downlink SINR analysis," *IEEE Transactions on Wireless Communications*, vol. 11, no. 10, pp. 3484–3495, Oct. 2012.

[30] F. W. J. Olver, D. W. Lozier, R. F. Boisvert, and C. W. Clark, Eds., *NIST Handbook of Mathematical Functions*, Cambridge University Press, New York, NY, 2010.

[31] H. S. Dhillon, R. K. Ganti, F. Baccelli, and J. G. Andrews, "Modeling and analysis of K-tier downlink heterogeneous cellular networks," *IEEE Journal on Selected Areas in Communications*, vol. 30, no. 3, pp. 550–560, Apr. 2012.

[32] H. ElSawy, and E. Hossain, "On stochastic geometry modeling of cellular uplink transmission with truncated channel inversion power control," *IEEE Transactions on Wireless Communications*, vol. 13, no. 9, pp. 1536–1276, Aug. 2014.

[33] C.-H. Liu, and L.-C. Wang, "Optimal cell load and throughput in green small cell networks with generalized cell association," *IEEE Journal on Selected Areas in Communications*, vol. 34, no. 5, pp. 1058–1072, May 2016.

[34] C. Li, J. Zhang, and K. B. Letaief, "Throughput and energy efficiency analysis of small cell networks with multi-antenna base stations," *IEEE Transactions on Wireless Communications*, vol. 13, no. 5, pp. 2505–2517, May 2014.

[35] S. Stefanatos, and A. Alexiou, "Access point density and bandwidth partitioning in ultra dense wireless networks," *IEEE Transactions on Communications*, vol. 62, no. 9, pp. 3376–3384, Sep. 2014.

[36] D. J. Daley, and D. Vere-Jones, *An Introduction to the Theory of Point Processes, Volume I*, 2nd ed., Springer, New York, NY, 2003.

[37] S. M. Yu, and S.-L. Kim, "Downlink capacity and base station density in cellular networks," in *Proceedings of the IEEE WiOpt Workshop SpaSWiN*, Tsukuba Science City, Japan, pp. 119–124, May 2013.

[38] H. S. Dhillon, M. Kountouris, and J. G. Andrews, "Downlink MIMO HetNets: modeling, ordering results and performance analysis," *IEEE Transactions on Wireless Communications*, vol. 12, no. 10, pp. 5208–5222, Oct. 2013.

[39] R. Tanbourgi, H. S. Dhillon, J. G. Andrews, and F. K. Jondral, "Effect of spatial interference correlation on the performance of maximum ratio combining," *IEEE Transactions on Wireless Communications*, vol. 13, no. 6, pp. 3307–3316, Jun. 2014.

[40] G. Nigam, P. Minero, and M. Haenggi, "Coordinated multipoint joint transmission in heterogeneous networks," *IEEE Transactions on Communications*, vol. 62, pp. 4134–4146, Nov. 2014.

[41] Y. Lin, and W. Yu, "Downlink spectral efficiency of distributed antenna systems under a stochastic model," *IEEE Transactions on Wireless Communications*, vol. 13, no. 12, pp. 6891–6902, Dec. 2014.

[42] R. Tanbourgi, S. Singh, J. G. Andrews, and F. K. Jondral, "A tractable model for non-coherent joint-transmission base station cooperation," *IEEE Transactions on Wireless Communications*, vol. 13, no. 9, pp. 4959–4973, Sep. 2013.

[43] S. Stefanatos, A. G. Gotsis, and A. Alexiou, "Operational region of D2D communications for enhancing cellular network performance," *IEEE Transactions on Wireless Communications*, vol. 14, no. 11, pp. 5984–5997, Nov. 2015.

Chapter 6

Wireless device-to-device (D2D) links for machine-to-machine (M2M) communication

Nuno K. Pratas[1] *and Petar Popovski*[1]

Abstract

Device-to-Device (D2D) communications will play an important role in the fifth generation (5G) cellular networks, by increasing the spatial reuse of spectrum resources and by enabling communication links with low latency. D2D is composed of two fundamental building blocks: proximity discovery and direct communication between nearby users. Another emerging trend in wireless cellular systems is Machine-to-Machine (M2M) communications, often characterized by fixed, low transmission rates. In this chapter we motivate the synergy between D2D and M2M, and present technologies that enable M2M-via-D2D communication to operate as an underlay to the ordinary cellular transmissions.

6.1 Introduction

Among the large number of wireless technologies being developed towards 5G, wireless cellular networks embrace two technological trends: *Device-to-Device (D2D)* communications and *Machine-to-Machine (M2M)* communications. These trends are different from the mainstream technology trend, inherited from the previous wireless generations, which targets to achieve higher data rates and improved coverage. D2D communications enable direct links between the wireless devices that use the same spectrum and radio interface used for cellular communications. The second trend of M2M communication, also known by the name *Machine-Type Communications (MTC)*, [1,2], has the objective to attach a large number of low-rate low-power devices, termed *Machine-Type Devices (MTDs)* to the cellular network. There are multiple factors that demand increased number of connected MTDs: the smart-grid, large-scale environment and structure monitoring, asset and health monitoring, etc.

[1]Fredrik Bajers Vej 7A, A3-204, 9220 Aalborg Øst, Denmark

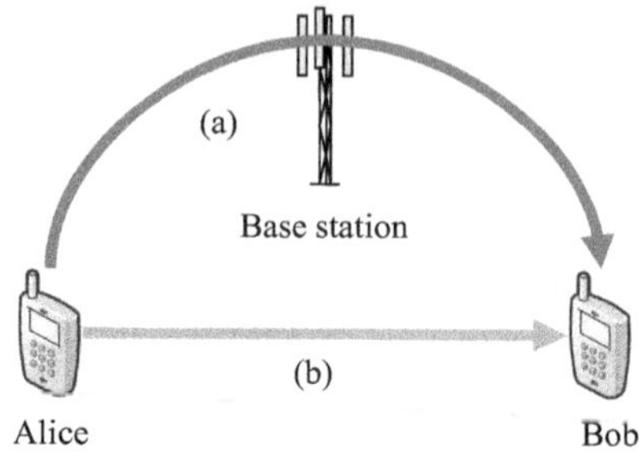

Figure 6.1 Example of communication between devices (a) via the cellular infrastructure and (b) direct device-to-device

6.1.1 Key concepts in D2D communication

The key concept behind Device-to-Device communications can be summarized as follows. Consider two users Alice and Bob, each with an associated cellular device, that want to communicate with each other using the cellular network resources,[1] as depicted in Figure 6.1. In a traditional cellular network, such as GSM, UTMS, and LTE-A (prior to Release 12), this communication would take place over the cellular network infrastructure. In contrast, with D2D communications the devices communicate directly, which is only feasible if the two devices are in proximity with each other. To ease the explanation of the key concepts, we assume to be under a LTE-A cellular network. Within a cellular network, the D2D communications establishment between proximal cellular devices is assisted by the network [3–6]; and is expected to play an important role in 5G systems [7].

The motivation to introduce D2D communications within a cellular context, comes from (i) the need to improve the use of network resources, i.e., the *area spectral efficiency* and (ii) to enable new cellular services that require a direct communication between devices. The potential increase of the spectrum usage efficiency comes from the insight [8] that overall network capacity can be increased, if direct connectivity between nearby devices is allowed within a cell. This is also the motivation for the introduction of small cells, sometimes termed as *cell densification*; while D2D takes this concept further via *link densification.* Figure 6.2 illustrates the potential gains of allowing D2D connections within a cellular network, in comparison with the capacity increase by just introducing smaller cells. Furthermore, the introduction of D2D communications opens the possibility to offload the cellular infrastructure whenever identical content needs to be served in the same localized area, as well as the possibility to have reliable peer-to-peer links in a licensed spectrum for sharing e.g., multimedia content. In these cases the data volumes are relatively large and therefore is worthwhile to take advantage of the high-rate D2D links.

[1] Although a direct communication between the two users' devices can be established using other access technologies such as Bluetooth and Wi-Fi direct, these use the ISM band spectrum resources that due to the lack of interference coordination do not offer quality of service guarantees.

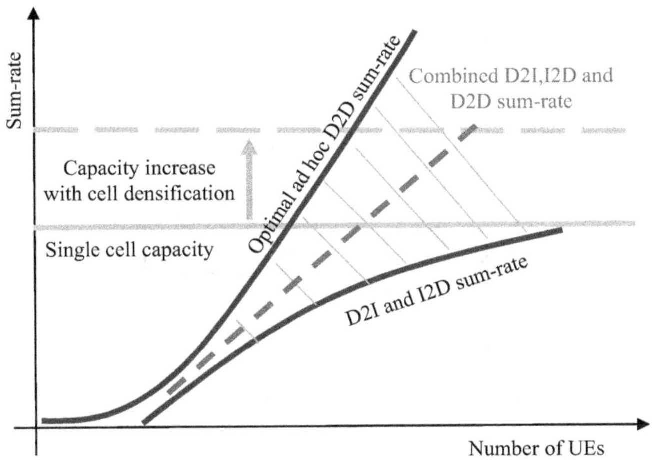

Figure 6.2 The introduction of D2D links within a cellular network can potentially increase the overall network capacity, if proper medium access mechanisms are in place, based on [8]

For connections that traverse the cellular infrastructure, communication can take place only after both communicating devices are connected to the network and are put in an RRC_Connected state. This connection establishment is accomplished using the cellular Access Reservation protocol and the additional signalling exchanges required to switch the device from RRC_Idle state to RRC_Connected. We illustrate this protocol in Figure 6.3. After this procedure has been completed by both devices, the communication link can then be established between the two cellular devices via the cellular infrastructure.

In the D2D case, prior to the communication, it is necessary for the devices to discover if they are in the proximity of each other. Proximity discovery can be supported by the network, in which case the devices first need to be connected to the network (using the described access procedure in Figure 6.3). On the other hand, if there is a lack of network support or the devices are outside the cellular network coverage, proximity discovery is done in an ad hoc manner. Whatever the approach used for proximity discovery, the devices only establish direct communication if they are proximal to each other. In case they are not proximal and are within cellular network coverage, then the communication takes place via the cellular network infrastructure. In case there is no cellular network coverage, then the only option to achieve connectivity is via multi-hop relaying supported by other cellular devices.

6.1.2 How can D2D help MTC?

The essential feature of using D2D for MTC is the requirement for a low-rate D2D link. For example, consider on-body sensors as MTC devices: each device requires low rate and a very low transmission power, insufficient to connect directly to

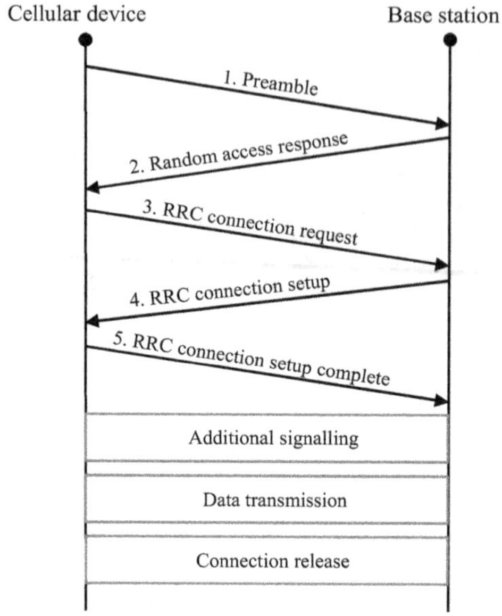

Figure 6.3 Illustration of the cellular access reservation procedure

the Base Station (BS). In that case, the MTC device uses a D2D link to a smart phone, which further relays the MTC traffic to the BS. In general, this type of D2D-relayed MTC can be used for other MTC applications that do not have stringent latency requirements.

We observe that it is desirable to carry the machine-type traffic not through direct links to a BS, but through a nearby relay. Therefore, in this chapter we explore the opportunity opened by D2D links for supporting MTDs, for which the motivation is two-fold: (1) When the MTD has the opportunity to connect to the infrastructure through a nearby device, then it can lower the transmission power; (2) there could be potentially thousands of MTDs connected to the same cell [9,10], such that the management of many individual connections can incur a large overhead and inefficiency in the cellular system. If multiple MTDs are connected to a relaying node and the relaying node accesses the BS on behalf of those devices, then the connection management load on the BS can drastically decrease. Critics may point out two arguments against the use of relaying for machine-type traffic. The first is that relaying increases the connection latency; however, there is a large variety of machine-type applications and not all of them require delay in milliseconds [1]. Examples include heat meters, health sensors, etc. and relaying can be restricted to the traffic that has a higher delay tolerance. The second argument is that it is too costly to introduce infrastructure relays, and this sets the motivation to relay the MTC traffic through another, more capable device, such as smart phone, using the D2D link. Another important observation, as we will describe later in the chapter, is that the communications generated during the proximity discovery procedure can also be classified as an MTC-like traffic. The setup considered in

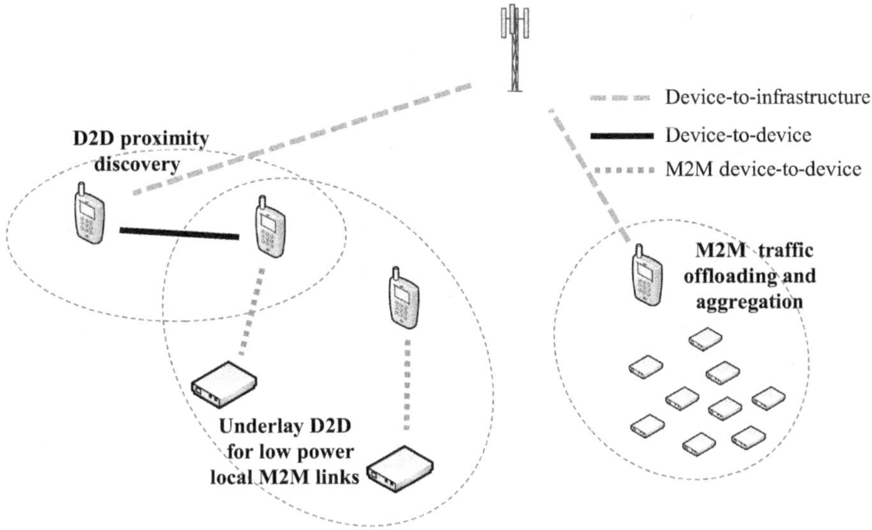

Figure 6.4 Scenario illustration

this chapter is shown in Figure 6.4. The difference with the common usage of the D2D transmission is that here the D2D link is used for sending at a low, fixed rate and the objective is to attain certain outage probability guarantees.

Recently, several potential synergies between D2D and M2M have been investigated [11–13], aiming to increase the spectral reuse, range, and coverage. In [14] these synergies were explored in the context of group-based operations, i.e., of a D2D-enabled cellular device that acts as a clusterhead for a group of MTDs. In the uplink the clusterhead aggregates and forwards the gathered requests, data packets and status information from MTC devices to the connected 3GPP BS. In the downlink, the clusterhead forwards management messages and data packets from the BS to the MTDs in the group. The use of D2D communications reduces the signalling congestion on the air interface and the network management load, which goes in line with the spirit of the group-based management defined by 3GPP [15]. In [16] it was considered how direct communication within a cluster of devices can improve the performance of a conventional cellular system, by leveraging the selection between direct D2D and infrastructure relaying. In [17], the impact of the MTDs on the cellular network is mitigated by randomly deploying data collectors to gather the traffic from the MTDs. Finally, in [18] a trunking and aggregation protocol was proposed along the same lines as the one depicted in Figure 6.4.

Standardization considers D2D under the name Proximity Services (ProSe) [19], where among several other use cases, it highlights the use case in which a cellular device improves the coverage by acting as a relay on behalf of one or more other cellular devices that are outside the network range. From architecture viewpoint, D2D relaying in ProSe is referred to as *range extension* and is specific to public safety use cases [20]. The report [21], besides providing design aspects for the physical and the upper layers, it also states that one of the requirements is the support of a large number of concurrently participating ProSe-enabled users.

6.1.3 Chapter organization

In Section 6.2 we expose recent results that enable underlay operation, while offering zero-outage to the link being underlayed. In Section 6.3 we present the key general concepts behind proximity discovery and we show how to apply underlay technique in order to implement discovery. In Section 6.4 we present and discuss two underlay communication topologies and derive the transmission power penalty required to meet the same performance as in the case without underlay operation. The chapter is concluded with a summary of the presented results.

6.2 Underlay D2D links

The D2D links, from a cellular network perspective, can use in-band and/or out-band spectrum resources (see Figure 6.5). The in-band resources typically belong to the spectrum licensed for cellular access; while out-band resources correspond to unlicensed spectrum, such as the ISM bands. In the out-band resources there is no interference with the cellular network operation and in the case where there is no central entity in charge of controlling the spectrum usage, then there is also no interference coordination. In in-band resources the channel access can be classified either as *overlay* or *underlay*. In overlay access, the cellular and D2D links are given dedicated and orthogonal resources; while in the underlay access, the cellular and D2D links share the same radio resources. In this chapter we focus on exploiting underlay D2D access.

For underlay operation, the D2D links introduce interference and are affected by the interference of the neighbouring network nodes. By assuming the cellular links and the D2D links to be in two separate tiers, the interference can be classified as: (1) cross-tier, which refers to situations in which the aggressor (e.g., a D2D user) and the victim (e.g., a cellular user) of interference belong to different network tier; (2) co-tier, where the aggressor (e.g., a D2D user) and the victim (e.g., a neighbouring D2D user) belong to the same network tier.

The underlay communication can take place in both the cellular uplink or in the downlink, as depicted in Figure 6.6. When the underlay is done in the uplink resources, then the D2D interference occurs at the BS; while when the underlay uses the downlink resources, the interference is at the cellular user. Traditionally, the BS is more able to cope with the interference than a cellular user, which explains why most of the work in the literature is based on underlay of the uplink [4,5]. In contrast, in this chapter we explore the underlay of D2D links in the downlink resources, motivated by the introduction of cellular devices with advanced multi-user decoding capabilities, such as successive interference cancellation and joint decoding.

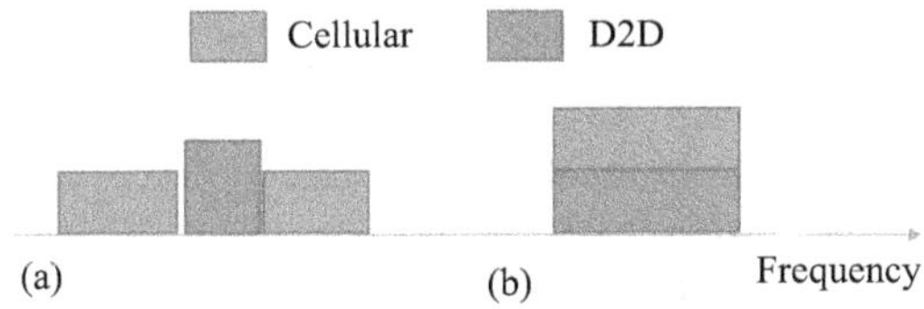

Figure 6.5 Overlay and underlay spectrum sharing example

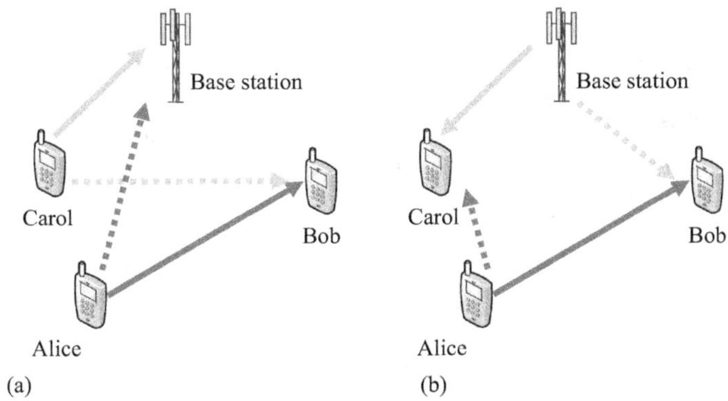

Figure 6.6 Cellular underlay in the (a) uplink and (b) in the downlink

6.2.1 Underlay signal model

In the considered downlink underlay setting in Figure 6.6, Carol and Bob receive simultaneously the transmissions from the BS and Alice. The composite signal y received at the receiver (Bob or Carol) is,

$$y = h_A x_A + h_B x_B + Z \tag{6.1}$$

where h_B and h_A denote respectively the complex channel gains between the BS, Alice and the receiver, while Z is the complex-valued Gaussian noise with variance $E[|Z|^2] = \sigma^2$. x_B and x_A are given by the circular zero-mean Gaussian complex signal transmitted by the BS and Alice, where the respective variances are $E[|x_A|^2] = P_A$ and $E[|x_B|^2] = P_B$. For simplicity, Alice's transmission power is assumed to be constant and dimensioned to meet the system outage target.

All communication links are assumed to be characterized by block Rayleigh fading, i.e., the channel fading gains will not vary within a scheduling epoch. $g_i = |h_i|^2$ is the channel envelope fading gain, which in the presence of Rayleigh fading follows an exponential distribution. The associated Probability Density Function (PDF), $f_g(x)$, is given by,

$$f_g(x) = \frac{1}{\overline{g}} \exp\left(-\frac{x}{\overline{g}}\right), \quad \text{with } \overline{g} = 1 \tag{6.2}$$

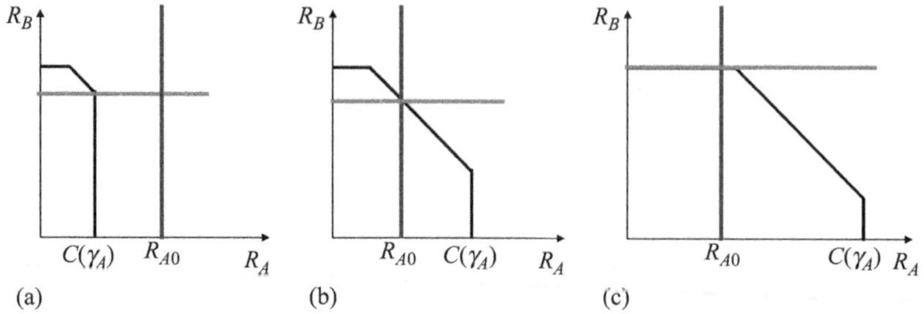

Figure 6.7 Illustration of the working regimes of a Joint decoder, where one of the links has a fixed rate. The transmission from the BS: (a) is decoded while treating Alice's transmission as noise; (b) is jointly decoded with Alice's transmission; (c) is decoded in the presence of noise, after Alice's transmission has been decoded and subtracted from the composite signal

All communications are performed using single-user point-to-point capacity-achieving Gaussian codebooks and the instantaneous achievable rate R_i is given by the asymptotic Shannon capacity in AWGN, $R_i = C(x) = W\log_2(1+x)$, where W is the channel bandwidth. The target Signal-to-Noise Ratio (SNR) Γ_x, for a given target rate R_x, is defined as $\Gamma_x = C^{-1}(R_x) = 2^{R_x/W} - 1$.

6.2.2 Optimal underlay decoding

As shown in (6.1), a two-user Gaussian Multiple Access Channel (MAC) [22] is created at both Bob's and Carol's receivers. In the standard MAC the assumption is that the receiver should decode successfully the signals of all the transmitters. Denoting the rates of the signals present as R_B, for the BS, and R_A, for Alice, then the MAC defining inequalities are

$$\begin{aligned} &R_A \leq C(\gamma_A) \\ &R_B \leq C(\gamma_B) \\ &R_A + R_B \leq C(\gamma_A + \gamma_B) \end{aligned} \tag{6.3}$$

where γ_A and γ_B denote signal-to-noise ratio corresponding to the signals from Alice and the BS at the receiver. In our MTC scenario, we assume that Alice transmits with a fixed rate $R_{A,0}$. Therefore, the analysis of our model is different from the classical MAC treatment, since $R_{A,0}$ is fixed and we need to analyse the decodability of R_B, regardless of whether $R_{A,0}$ is decodable.

In Figure 6.7 we depict the three working regimes of the Joint Decoder. We now take as a point of view Carol's receiver. If $\gamma_A < \Gamma_{A,0}$, then the first inequality in (6.3) is violated, such that the signal from Alice is treated as noise and the maximal achievable downlink rate is

$$R_B \leq C\left(\frac{\gamma_B}{1+\gamma_A}\right) \tag{6.4}$$

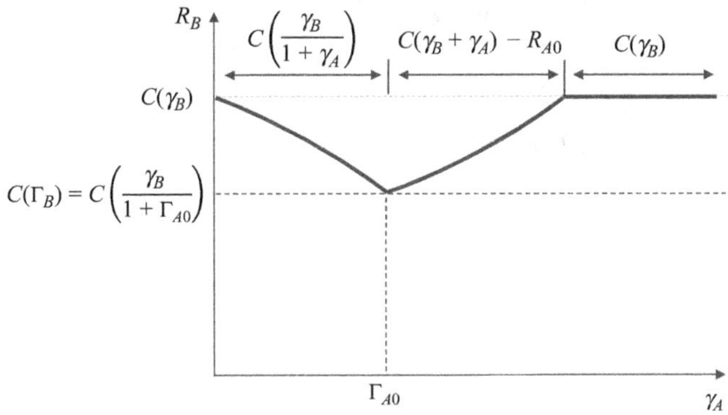

Figure 6.8 Achievable R_B when using Joint User Decoding (JD) given known γ_B that does not put the link between the BS and the receiver in outage, according to the channel realization γ_A of the link between Alice and the receiver

On the other hand, when $\gamma_A \geq \Gamma_{A,0}$ then there are two options: (a) either continue to treat the signal from Alice as noise and use (6.4) to determine the maximal R_B or (b) employ the MAC inequalities (6.3), which leads to the following:

$$R_B \leq \min\{C(\gamma_B), C(\gamma_A+\gamma_B) - R_{A,0}\} = \min\{C(\gamma_B), C(\gamma_A+\gamma_B) - C(\Gamma_{A,0})\} \tag{6.5}$$

However, note that if for given $\gamma_A \geq \Gamma_{A,0}$ the expression (6.4) leads to a higher bound on R_B compared to (6.5), then the receiver should treat the signal from Alice as noise (but we will see that this is not the case). With a slight abuse of notation, we can define $R_B(\gamma_A)$ as the maximal achievable R_B for given γ_A, which can then be written compactly as

$$R_B(\gamma_A) = \begin{cases} C\left(\dfrac{\gamma_B}{1+\gamma_A}\right) & \gamma_A < \Gamma_{A,0} \\ \max\left\{C\left(\dfrac{\gamma_B}{1+\gamma_A}\right),\ \min\{C(\gamma_B), C(\gamma_A+\gamma_B) - C(\Gamma_{A,0})\}\right\} & \gamma_A \geq \Gamma_{A,0} \end{cases} \tag{6.6}$$

Figure 6.8 depicts how this function looks like. When $\gamma_A < \Gamma_{A,0}$, $R_B(\gamma_A)$ decreases, since a larger γ_A means a larger noise, corresponding to the regime depicted in Figure 6.7(a). After γ_A reaches $\Gamma_{A,0}$, $R_B(\gamma_A)$ starts to increase, corresponding to the regime depicted in Figure 6.7(b). Finally, when the link between Alice and Carol becomes too strong $\gamma_A \geq \Gamma_{A,0}$, the downlink rate reaches its maximal possible value $R_B(\gamma_A) = C(\gamma_B)$ where the interference effect from Alice's transmission vanishes, corresponding to the regime depicted in Figure 6.7(c).

We can now state the following:

Lemma 6.2.1 *Let there be a transmission with fixed rate $R_{A,0}$. Let B know the rate $R_{A,0}$ and the SNR realization γ_B at the receiver, but not γ_A. Then the maximal downlink transmission rate that is always decodable by the receiver is $R_B = C(\Gamma_B)$ where*

$$\Gamma_B - \frac{\gamma_B}{1 + \Gamma_{A,0}} \tag{6.7}$$

and $\Gamma_{A,0} = C^{-1}(R_{A,0})$.

Proof. The proof, exposed in [23], is based on finding the switching point between the region where the transmission from B is decoded by treating the transmission from A as noise, to the region where both signals are jointly decoded, as depicted in Figure 6.8. The maximum downlink transmission $R_B = C(\Gamma_B)$ then corresponds to that same switching point, which occurs when $\gamma_A = \Gamma_{A,0}$. □

In other words, when adapting the rate of the signal from B to meet the defined upper bound, then B's transmission can always be decoded, regardless of the transmission from A being decoded. An interesting side effect of this rate upper bound, is how it affects the outage of the announcing device transmission. When we apply the downlink rate upper bound, all the MAC inequalities are respected, i.e., the signal from B is always decodable. Then the decodability of $R_{A,0}$ will depend solely on whether $R_{A,0} \leq C(\gamma_A)$, i.e., the probability of the transmission from A being decoded is given by $Pr\{\gamma_A \geq \Gamma_A\}$.

Corollary 6.2.2 *Let there be a transmission with rate $R_{A,0}$ and let B know the rate $R_{A,0}$ and the SNR γ_B, and select its rate such that $R_B = C(\Gamma_B)$. Then the decodability of $R_{A,0}$ will depend solely on $R_{A,0} \leq C(\gamma_A)$.*

Proof. From Lemma 6.2.1 it is stated that the transmission from B is always decodable. Therefore there are two decoding scenarios: (i) B is decoded, by treating Alice's transmission as noise, and subtracted from (6.1), allowing Alice's transmission to be decoded in the presence of noise if and only if $R_{A,0} \leq C(\Gamma_{A,0})$; (ii) Alice's transmission is jointly decoded with B, which is only true if all the inequalities in (6.3) hold. The first inequality, $R_{A,0} \leq C(\gamma_A)$, must hold for joint decoding to be possible. The second inequality holds, since $R_B \leq C(\Gamma_B) \leq C(\gamma_B)$. Finally, the third inequality also holds:

$$\begin{aligned} R_{A,0} + R_B &= C(\Gamma_{A,0}) + C(\Gamma_B) \\ &= C(\Gamma_{A,0}) + C(\gamma_B + \Gamma_{A,0}) - C(\Gamma_{A,0}) \\ &= C(\gamma_B + \Gamma_{A,0}) \leq C(\gamma_B + \gamma_A) \end{aligned}$$

where the last inequality comes from $\Gamma_{A,0} \leq \gamma_A$, one of the requirements for joint decoding to take place. □

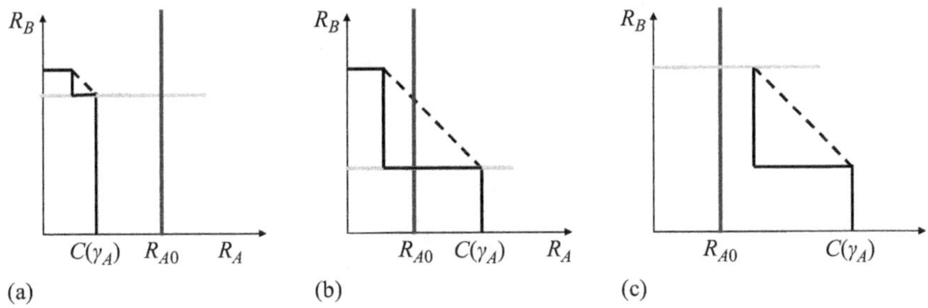

Figure 6.9 Illustration of the working regimes of a Single-User decoder with successive interference cancellation, where one of the links has fixed rate. The transmission from the BS: (a) is decoded while treating Alice's transmission as noise; (b) is decoded while treating Alice's transmission as noise, due to the Single-User decoding; (c) is decoded in the presence of noise, after the transmission from Alice has been decoded and subtracted from the composite signal

We note that the presented communication scheme is based on an information-theoretic setup and codebooks. Therefore, to bring this concept to practice one may resort to transmission techniques that are suitable for multiuser decoding. In the following sections, we will take advantage of this result to enable underlay operation.

6.2.3 Sub-optimal underlay decoding

We now consider the case in which the receiver applies Single-User Decoding (SD) in each decoding step, treating the yet-to-be-decoded user as a noise. We show now how to select the maximal downlink rate for given γ_B that guarantees that there is no outage. Recall that there are three operating regions of the JD receiver. When SD is applied, there are only two decoding options, depicted on Figure 6.9. If the link between Alice and the receiver:

- is weak, then γ_A is low and the signal from B should be decoded by treating the signal from Alice as noise. This regime is depicted in Figure 6.9(a) and (b).
- is very strong, given that Alice's transmission rate is fixed to R_{A0}, then the signal from Alice is decoded, subtracted and a "clean" signal of B is decoded at the maximal possible rate $C(\gamma_B)$. It can be shown that the signal from Alice is decodable and thus can be treated as strong when $\gamma_A = \Gamma_{A0}(1 + \gamma_B)$. This regime is depicted in Figure 6.9(c).

We can then define $R_B(\gamma_A)$, which is illustrated in Figure 6.10, as the maximal achievable R_B for given γ_A, which can be written as

$$R_B(\gamma_A) = \begin{cases} C\left(\dfrac{\gamma_B}{1+\gamma_A}\right) & \gamma_A < \Gamma_{A0}(1+\gamma_B) \\ C(\gamma_B) & \gamma_A \geq \Gamma_{A0}(1+\gamma_B) \end{cases} \tag{6.8}$$

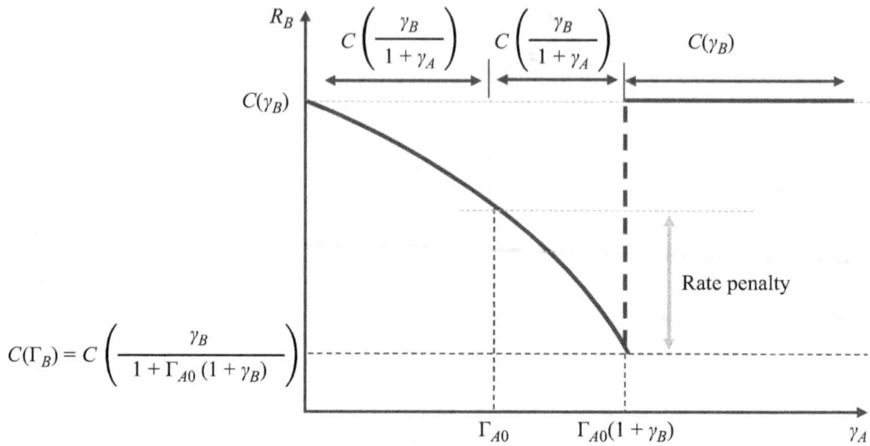

Figure 6.10 Achievable equivalent SNR Γ_B when using Single-User Decoding (SD) given known γ_B that does not put the $B-U$ link in outage

Proposition 6.2.3 *Let there be a fixed rate transmission with rate R_A and let the receiver apply successive single user decoding. If B knows the instantaneous SNR γ_B, but not γ_A, then the maximal downlink transmission rate that is always decodable by the receiver is $R_B = C(\Gamma_B^{SD})$ where*

$$\Gamma_B^{SD} = \frac{\gamma_B}{1+\Gamma_{A0}(1+\gamma_B)} \tag{6.9}$$

and $\Gamma_{A0} = C^{-1}(R_{A0})$.

Proof. The proof, exposed in [23], is based on finding the switching point between the region where the transmission from B is decoded by treating the transmission from A as noise, to the region where the signal from A can be decoded and then subtracted from the composite signal, so that the signal from B can be decoded only in the presence of noise. The maximum downlink transmission $R_B = C(\Gamma_B^{SD})$ then corresponds to that same switching point, which occurs when $\gamma_A = \Gamma_{A,0}$. □

Clearly, the zero outage rate in the SD setting is much lower than the one in the JD setting.

6.3 Underlay proximity discovery

Prior to establishing the D2D links, the network first needs to become aware if these devices want to communicate and if they are in close proximity to each other. Within a 3GPP cellular network, this information can be inferred from the complementary proximity discovery mechanisms taking place both in the Evolved Packet Core (EPC)

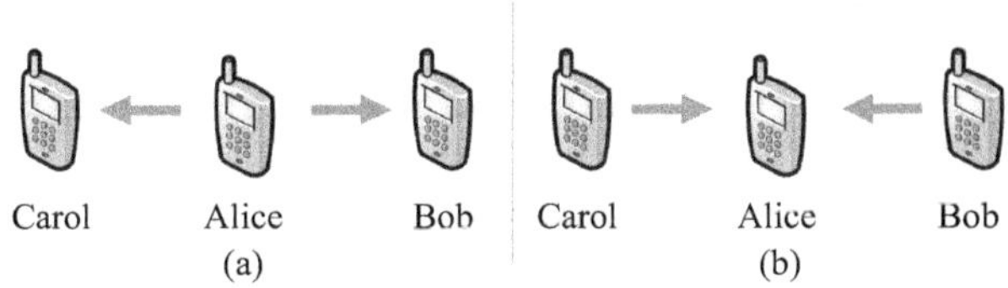

Figure 6.11 Direct proximity discovery within a device proximity region: (a) Model A ("I am here") and (b) Model B ("Who is there?")

and in the radio access network (E-UTRAN) [3]. In the EPC—denoted as "EPC-level discovery"—the proximity between devices can be extracted from the device's periodic location updates, while the interest to establish direct communication between devices can either be triggered by the devices or from the network side. The latter can be triggered from the monitoring of the ongoing communications between devices taking place over the network infrastructure [3]. On the other hand, in the E-UTRAN—denoted as "direct discovery"—the proximity is established by direct communication between the interested devices.

There are two direct proximity discovery protocols being currently specified in 3GPP [24]: Model A ("*I am here*") and Model B ("*Who is there?*").

In the Model A discovery protocol, as illustrated in Figure 6.11(a), the cellular devices can take either the role of an *announcer* or of a *monitor*. Devices in the announcer role broadcast an announcing message with information that could be of interest to the monitors within its proximity region. On the other hand, devices in the monitor role listen for certain information of interest, transmitted by the announcers within their vicinity. The announcer transmission is dimensioned with fixed power and rate such that the devices within a *proximity region* detect it with high probability.

In the Model B discovery protocol, as illustrated in Figure 6.11(b), the cellular devices can take the role of either *discoverer* or of *discoveree*. In this model, the discoverer receives the information from the nearby discoverees. In contrast with Model A, this protocol is based on the aggregation of the transmissions from the nearby cellular devices, which are not necessarily known a priori by the network to be within the proximity region.[2]

In the remaining of this section, we will focus our analysis on the Model A discovery protocol setting. The radio resource management approach in place for Model A [21] is to schedule dedicated resources to the announcer transmission, i.e., resources orthogonal to the ones assigned to the other active cellular links. In contrast, we consider the case where the announcer transmission is scheduled to underlay the ongoing downlink transmissions to the monitors. This leads to

[2] The associated radio resource management problem is the same as in random access settings without a priori knowledge, i.e., the allocation of a certain amount of resources, given an unknown number of arrivals.

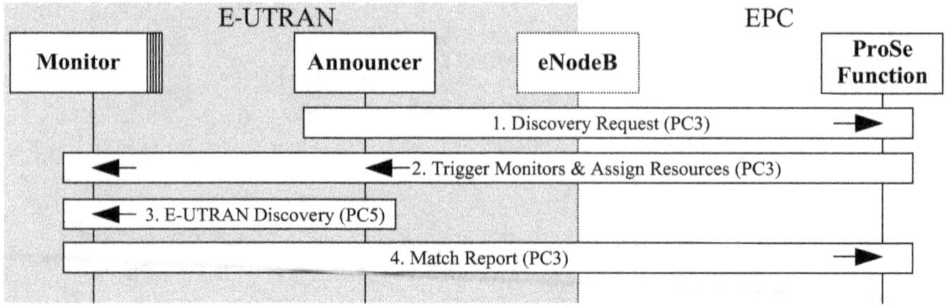

Figure 6.12 Overview of Model A network controlled ProSe direct discovery [24]. PC3 denotes the logical interface between the ProSe enabled devices and the ProSe function at the Evolved Packet Core (EPC), which takes place over the E-UTRAN and EPC. PC5 denotes the logical interface between two ProSe enabled devices, which takes place in the E-UTRAN

the creation of a multiple access channel [22], composed by the announcer and the BS transmissions.

The occurrence of D2D underlay communication links, although previously considered in the literature [11,25–27], has not been considered so far in a proximity discovery setting until our recent work in [28]. The main obstacle for the underlay proximity discovery is that the channel state between the announcer and monitor device is unknown, which makes the level of interference to the downlink transmissions unknown and therefore preventing the eNodeB to cope with it. However, as discussed in [23] and in Section 6.2, we have shown that if the interfering signal uses a fixed rate and the downlink receiver uses multi-user decoding, then it is possible to receive the downlink transmission with zero outage, without requiring the BS to have any prior Channel State Information at the Transmitter (CSIT) of the interfering link. The signals applied in the discovery process use fixed rate, which makes them perfect candidates for underlaying while guaranteeing zero outage to the downlink connection.

In this section, we reformulate the results in Section 6.2 in terms of power adaptation, such that a fixed rate zero-outage downlink transmission is achieved at the monitoring devices. The analysis and numerical results show that the underlay operation is indeed possible, but at the cost of an increased downlink power usage that increases with the rate of the underlay transmission.

6.3.1 Network-controlled proximity discovery overview

There are three network entities involved in the Model A discovery protocol, namely the announcer, the monitor and the ProSe function (located in the network infrastructure at the EPC). The ProSe function is responsible for the admission control to ProSe, coordination of the proximity discovery and the monitoring of the ongoing D2D communication links. Figure 6.12 depicts the four essential steps that an announcer takes to complete the proximity discovery protocol, when authorized to use ProSe. In the *Discovery Request* step, the announcer sends a discovery request to the ProSe function. If this request is accepted, the *Trigger Monitors & Assign Resources* step takes place where the ProSe function triggers the monitors within the announcer proximity region, assigning and informing the intervening devices (announcer and monitors) which system resources have been allocated to the transmission of the announcer message, i.e., the radio resources over which the PC5 interface will take place. In the *E-UTRAN Discovery* step, the announcer transmits its message to the monitors over the PC5 interface in the E-UTRAN. Finally, in the *Match Report* step, the monitors that were able to decode successfully the announcer message, transmit the received information to the ProSe function for confirmation of the announcer identity. After this last step, the direct D2D communication link can be established.

In the remaining of the chapter, we focus on the third step of the discovery protocol. Namely, on the transmission of the announcer message underlaying the downlink transmission targeted to one of the monitors.

6.3.2 System model

We focus our analysis on a single cell network composed by a base station and multiple cellular devices. A set of these devices takes the announcer role, while the remaining devices take the monitoring role, listening to the announcer message in the resources assigned for that purpose by the network. We assume that the announcers are sufficiently spread in space such that their proximity regions do not overlap[3] and that the network assigns different resources to each announcer. The network is assumed to have knowledge about the positions of each device via periodic device location updates, while being *unaware of the channel conditions between any two devices prior to the proximity discovery procedure*. Therefore, the network knows which monitors will potentially be in the proximity region.[4]

We assume that within the proximity region of an announcer A there are N nearby monitors. We further assume that each of these monitors has a dedicated ongoing downlink transmission from the base station B. We identify the associated downlink channel and its intended receiver by the same subscript, e.g., the ith channel intended receiver is the ith device. We assume that these downlink transmissions have a fixed rate R_B, to be met with zero outage. The fixed downlink rate assumption simplifies the introduction of our power control concept, but we note that an extension to an adaptive downlink rate setting is possible.

Finally, we assume an OFDMA-like based cellular system [21], where (i) the synchronization between network devices is ensured by the cyclic prefix in each transmission; (ii) the devices are able to receive and decode all the transmissions in the downlink subcarriers; and (iii) the devices report periodically to B the channel

[3] In other words, the set of proximal monitors to any announcer is disjoint from the set of monitors associated with any other announcers.

[4] Two devices can be in spatial proximity, but due to the local radio environment characteristics they might not have a radio channel with enough quality to warrant direct D2D communications. The verification if the radio link has enough quality is the main purpose of the proximity discovery procedure.

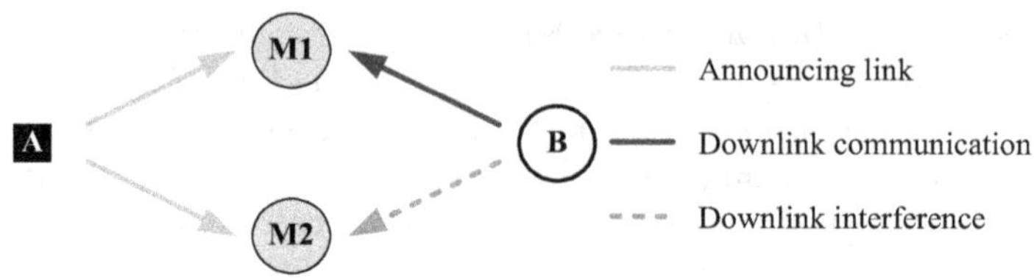

Figure 6.13 Signal and interference links model

quality of all downlink channels that are estimated based on the downlink channel pilots.

If we assume that the announcer transmits in the kth downlink channel, as illustrated in Figure 6.13, then the composite signal $y_{k,i}$ received in the kth channel at the ith device is,

$$y_{k,i} = h_{A,i}x_A + h_{B,i}x_B + Z \tag{6.10}$$

where $h_{B,i}$ and $h_{A,i}$ denote, respectively, the complex channel gains between the B and A transmitters and the ith receiver, while Z is the complex-valued Gaussian noise with variance $E[|Z|^2] = \sigma^2$. x_B and x_A are given by the circular zero-mean Gaussian complex signal transmitted by A and B, where the respective variances are $E[|x_A|^2] = P_A$ and $E[|x_B|^2] = P_B$. Here P_A is assumed to be constant and dimensioned such that the monitors within the proximity region radius are able to detect the announcers transmission with very high probability. As previously discussed, the downlink transmission is assumed to have a fixed rate R_B that is to be met with zero outage. In order to meet this requirement, P_B is adapted via the power control scheme described in Section 6.3.3, which copes with the downlink channel conditions and the interference from the underlay announcer transmission.

The baseline scheme assumes an orthogonal setting, where the downlink channels have only the downlink transmission, such that the received signal at the ith device is given by

$$y_{j,i} = h_{B,i}x_B + Z \quad \text{with } j \neq k \tag{6.11}$$

We assume that B has a fixed downlink transmission rate and is able to adapt its transmission power based on the Channel State Information available at the Transmitter (CSIT) of the $B - M_i$ links. We further assume that the absence of any channel state information about the $A - M_i$ links, since it is assumed that no communication between A and M_i has taken place prior to the announcer transmission. On the other hand, it is assumed that B knows the fixed transmission rate associated with the proximity announcement transmission, denoted as R_A.

6.3.3 Analysis

In our analysis, we focus on the downlink transmission power expenditure required to enable underlay operation of the direct proximity discovery protocol. We cast the zero-outage result provided in Lemma 6.2.1 to a power control setting, by assuming that each downlink transmission has a fixed rate R_B. We start by analysing the case where there is a single monitoring device, then expand the analysis to multiple monitoring devices. Finally, we extend the analysis to the case where the network chooses one of available downlink channels to be underlayed by the direct proximity discovery protocol, based on minimum required downlink transmission power.

6.3.3.1 Single monitoring device

We start by introducing the power control mechanism for the downlink transmission, assuming a single monitor within the proximity region. Defining the instantaneous SNR of the downlink transmission

$$\gamma_B = \frac{P_B g d_B^{-\alpha}}{\sigma^2}$$

where $d_B^{-\alpha}$ is the path loss, σ^2 is the noise power, $g = |h|^2$ is the downlink channel gain realization and P_B the downlink transmission power. The capacity of such obtained Gaussian channel is then

$$\begin{aligned} R_B &= W \log_2\left(1 + \frac{\gamma_B}{1+\Gamma_A}\right) \\ &= W \log_2\left(1 + \frac{P_B g d_B^{-\alpha}}{\sigma^2} \cdot \frac{1}{1+\Gamma_A}\right) \end{aligned}$$

where W is the bandwidth of the assigned resources. The term $1/(1+\Gamma_A)$ accounts for the rate of penalty due to the underlay announcer transmission, in order to have a guarantee for achieving zero-outage in the downlink, as stated in Lemma 6.2.1. The required power P_B to meet the fixed rate R_B is then given by

$$P_B = \frac{1}{g} \cdot K \tag{6.12}$$

where

$$K = \frac{\sigma^2(1+\Gamma_A)}{d_B^{-\alpha}} \cdot \left[2^{R_B/W} - 1\right]. \tag{6.13}$$

In order to cope with deep fades, a truncated channel inversion strategy is put in place [29]. In this setting, B will only attempt transmission when the fade depth is above a cutoff threshold μ, i.e., when $g > \mu$. This link outage event is then:

$$Pr\{g < \mu\} = \int_0^{\mu} f_g(x)dx = 1 - \exp(-\mu/\overline{g}) \tag{6.14}$$

where $\mu = -\overline{g} \log(1 - Pr\{g < \mu\})$. We note that there is always announcer transmission although the downlink transmission might not take place due to the truncated power control strategy. In other words, the announcer transmission is decodable with probability $Pr(\gamma_A \geq \Gamma_A)$. The average transmit power required to sustain R_B, is then:

$$E[P_B] = \int_{\mu}^{\infty} \frac{1}{g} \cdot Kf_g(g)dg_i = E_1\left(\frac{\mu}{\overline{g}}\right) \cdot K \tag{6.15}$$

where $E_1(.)$ is the exponential integral function.

6.3.3.2 Multiple monitoring devices

When there are multiple monitoring devices within the proximity region that receive broadcast traffic from the BS, then P_B needs to be computed for the worst link. This is the case since R_B needs to be decodable at all the monitors within the proximity region. The required power becomes:

$$P_B = \frac{1}{\min(g_1, \ldots, g_N)} \cdot K \tag{6.16}$$

where g_i is the downlink channel gain of the ith monitor and where, for analytical tractability, K is assumed to be identical for all monitors. To account for the cutoff value we introduce the nth order statistics distribution:

$$f_n(x) = Nf_g(x)\binom{N-1}{n-1}F(x)^{n-1}(1 - F(x))^{N-n}$$

from which we want $n = 1$, the distribution of the worst channel:

$$f_1(x) = \frac{N}{\overline{g}}\exp\left(-\frac{Nx}{\overline{g}}\right) \tag{6.17}$$

The outage event is then defined as

$$\begin{aligned} Pr\{\min(g_1, \ldots, g_N) < \mu\} &= \int_0^{\mu} f_1(x)dx \\ &= 1 - \exp(-N\mu/\overline{g}) \end{aligned} \tag{6.18}$$

and the associated cutoff threshold as

$$\mu = -\frac{\overline{g}}{N}\log(1 - Pr\{\min(g_1, \ldots, g_N) < \mu\}) \tag{6.19}$$

The mean required P_B in this setting is then given by

$$E[P_B|N] = E_1\left(\frac{N\mu}{\overline{g}}\right) \cdot N \cdot K \tag{6.20}$$

6.3.3.3 Choice of downlink channel to underlay

Now we consider the case where there are M downlink channels, from which the network can choose one to be underlayed. In this setting, the channel chosen for underlay operation is the one that will require the least used power.

We assume that each downlink channel has different channel gains at each monitor. Let us assume that, out of the N monitors, the one with the worst channel gain in the ith channel is given by $u_i = \min(g_{1,i}, \ldots, g_{N,i})$, where $g_{n,i}$ corresponds to the downlink channel gain of the ith channel at the $_{n\text{th}}$ device. Then, the channel that will lead to the least power expenditure is the one with the channel gain given by $u_{max} = \max(u_1, \ldots, u_M)$. In other words, assuming that the monitor with the worst conditions in each channel is the limiting factor, then the selection of the best channel in these conditions is modelled by the order statistics distribution of the maximum value.

The corresponding density function, $f_M(x)$, is given by,

$$f_M(x) = Mf_1(x)[F_1(x)]^{M-1} \tag{6.21}$$

and the cumulative function $F_M(x)$ is given by

$$F_M(x) = [F_1(x)]^M = \left[1 - \exp\left(-\frac{Nx}{\overline{g}}\right)\right]^M \tag{6.22}$$

The outage event is then defined by

$$Pr\{u_{max} < \mu\} = F_M(\mu) = \left(1 - \exp\left(-\frac{N\mu}{\overline{g}}\right)\right)^M \tag{6.23}$$

where

$$\mu = -\frac{\overline{g}}{N} \log(1 - Pr\{u_{max} < \mu\}^{1/M}) \tag{6.24}$$

In these conditions the average power used is given by

$$E[P_B|M] = K \int_{\mu}^{\infty} \frac{1}{x} f_M(x) dx \tag{6.25}$$

which can be computed by means of numerical integration.

The sum-rate per used resource, S, focused on the worst channel, is given by,

$$S = R_B \cdot Pr\{u_{max} \geq \mu\} + R_A \cdot Pr\{\gamma_A \geq \Gamma_A\}. \tag{6.26}$$

As a measure of energy efficiency we consider the amount of energy required to transmit a bit of information, ψ, given by,

$$\psi = E\left[\frac{P_B}{S}\right] = \frac{K}{S} \int_{\mu}^{\infty} \frac{1}{x} f_N(x) dx \tag{6.27}$$

which can be computed via numerical integration.

6.3.3.4 Orthogonal downlink power control

We now consider baseline as the setting where the announcer and downlink transmissions are allocated orthogonal resources by the network. In other words, the downlink and the announcer transmissions occur in two identical orthogonal channels. Starting from the asymptotic capacity on a Gaussian channel at the ith device:

$$R_B = W\log_2\left(1+\frac{P_B g_i d_B^{-\alpha}}{\sigma^2}\right) \tag{6.28}$$

the required downlink power P_B to achieve the fixed rate R_B is then given by

$$P_B = \frac{1}{g_i}\cdot\frac{\sigma^2\left[2^{R_B/W}-1\right]}{r^{-\alpha}}. \tag{6.29}$$

To provide a fair comparison, we focus on the power required to serve the worst downlink channel. In this setting, the link outage event is provided by (6.23) and the associated cutoff threshold by (6.24). The average transmit power required to sustain R_B, is then:

$$E[P_B] = \int_{\mu}^{\infty}\frac{1}{x}\cdot\frac{\sigma^2\left[2^{R_B/W}-1\right]}{d_B^{-\alpha}}f_M(x)dx \tag{6.30}$$

which can be computed numerically.

The sum-rate per used resource S, from the worst channel and the channel dedicated for the announcing device transmission, is given by

$$S = \frac{R_B\cdot Pr\{u_{max}\geq\mu\}+R_A\cdot Pr\{\gamma_A\geq\Gamma_A\}}{2}. \tag{6.31}$$

Note that the division by 2 comes from the announcer and the downlink transmissions occurring in two identical orthogonal channels.

The associated energy efficiency, ψ, is given by,

$$\psi = E\left[\frac{P_B}{S}\right] = \frac{1}{S}\int_{\mu}^{\infty}\frac{1}{x}\cdot\frac{\sigma^2\left[2^{R_B/W}-1\right]}{d_B^{-\alpha}}f_M(x)dx. \tag{6.32}$$

6.3.4 Numerical results and discussion

In this section we present the numerical results corresponding to the performance of the proximity discovery in the underlay setting, in terms of expended power $E[P_B]$ and energy efficiency ψ versus the announcer payload (computed as $R_A\cdot W\cdot T_A$, where T_A is the duration of the announcer transmission). These numerical results were obtained by evaluating the corresponding analytical expressions, derived in the previous section and through stochastic Monte Carlo simulations. Table 6.1 lists the relevant assumed system parameters. We first observe that the sum-rate achieved in the underlay setting will always be higher than in the orthogonal resources setting, as can been seen when comparing (6.26) with (6.31). This higher sum-rate comes at a cost of extra downlink transmission expenditure when compared to the orthogonal resources setting, as shown in Figure 6.14. Specifically, in

Table 6.1 Simulation scenario settings

Parameter	**Value**	**Parameter**	**Value**
W	180 kHz	T_A	5 ms
d_B	200 m	d_A	20 m
P_A	20 dBm	σ^2	−97 dBm
α	4	$Pr\{\gamma_A \geq \Gamma_A\}$	0.99
R_B	5 B/s^2	$Pr\{u_{max} \geq \mu\}$	0.99
N, M	20	A's payload	[100,1,100] (B)

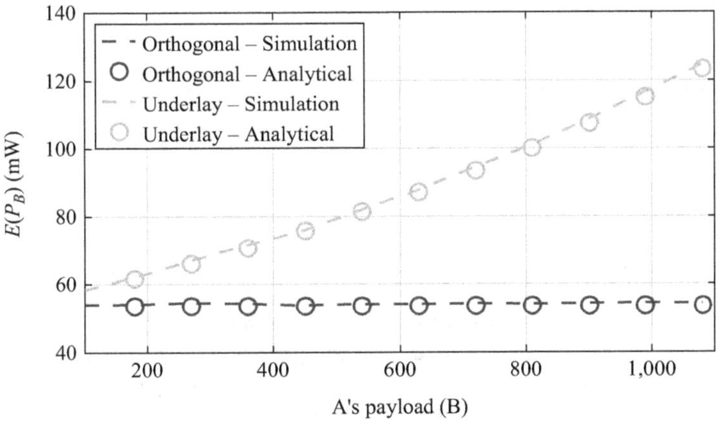

Figure 6.14 Base station power consumption comparison between underlay and orthogonal discovery, for multiple monitoring devices

the underlay setting the higher is the announcers rate (and corresponding payload), the higher will the required downlink transmission power be. On the other hand, from an energy efficiency perspective, in Figure 6.15 we observe that, for lower announcer payloads and despite the higher downlink power required, the underlay scheme is more energy efficient. This is the case, since the penalty introduced by the term $(1 + \Gamma_A)$ in (6.12) is for payloads that are low enough not to offset significantly the sum-rate gains in the underlay setting.

6.3.5 Remarks

Proximity discovery is an essential enabler for D2D communications. We have shown that reliable underlay proximity discovery is feasible, achieving higher sum-rates than in the orthogonal resources setting, at the cost of a higher amount of expended downlink power. On the other hand, we have observed that for low announcer payloads the underlay discovery is up to twice more energy efficient compared to the orthogonal setting.

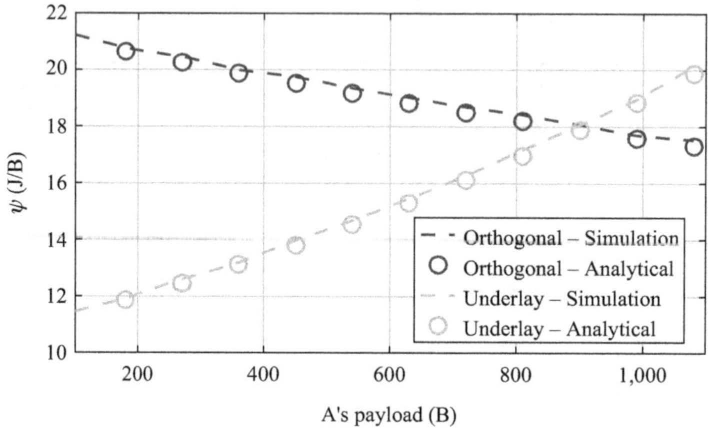

Figure 6.15 Energy efficiency comparison between underlay and orthogonal discovery, for multiple monitoring devices

6.4 Underlay direct communications

The direct communication between two devices takes place if these are proximal to each other, which can be determined with the previously described procedure. We now consider the case where these direct communications underlay the downlink transmission. We consider the two underlay topologies depicted in Figure 6.16, where the difference is in the device that acts as the intended receiver of the downlink transmission. In order to enable these two underlay communication topologies, we exploit the downlink zero-outage result for joint decoding, as presented in Section 6.2. We denote the topology in Figure 6.16(a) and (b), respectively as topologies A and B. We start by analysing topology A, in terms of the required downlink transmission power to achieve the same rate performance as in the case without underlay transmission. The goal is to ensure that the downlink transmission has zero-outage, while the MTD transmission is not affected by the downlink transmission. These two conditions are ensured by the application of the results in Lemma 6.2.1 and Corollary 6.2.2.

6.4.1 Analysis of topology A

In this network topology, as depicted in Figure 6.16(a), Alice receives downlink traffic from the BS and from the MTD.[5] To simplify the notation, we denote the BS and the MTD, respectively, as B and M. The link between the MTD and Alice is characterized by fixed rate and power, denoted respectively as R_M and P_M.

[5] We note that the MTD transmits its own data and is not acting as a relay to the BS transmission.

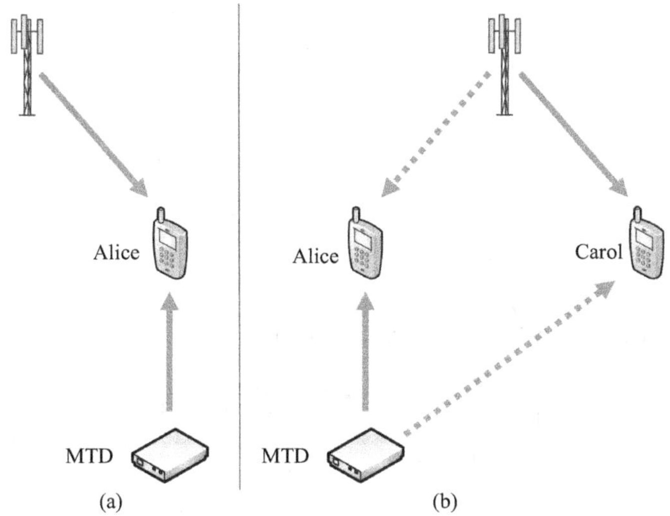

Figure 6.16 Underlay direct device communications where the destination of the downlink transmission is (a) the same device receiving the transmission from the machine and (b) another device

The MTD transmits to Alice during the downlink transmission of B, such that Alice's receiver observes a multiple access channel (MAC) with two transmitters:

$$y_A = h_{B,A}x_{B,A} + h_{M,A}x_M + Z \tag{6.33}$$

where $h_{B,A}$ and $h_{M,A}$ are the complex gains of the channels between the BS and Alice, and between the MTD and Alice, respectively. x_B and x_{M_i} are given by the circular zero-mean Gaussian complex signals transmitted by the B and M nodes, such that the respective variances are $E[|x_B|^2] = P_B$ and $E[|x_M|^2] = P_M$. Here P_B and P_M represent the constant power levels used by B and M, respectively. Finally, Z is the complex-valued Gaussian noise with variance $E[|Z|^2] = \sigma^2$.

In the following, we will define the power control mechanism for the downlink transmission that will enable the downlink transmission rate to achieve the same level as in the case without underlay transmission. Defining the instantaneous SNR of the downlink transmission:

$$\gamma_B = \frac{P_B g d_B^{-\alpha}}{\sigma^2}$$

where $d_B^{-\alpha}$ is the path loss losses, σ^2 is the noise power, $g = |h|^2$ is the downlink channel gain realization and P_B the downlink transmission power.

Taking into account the zero-outage joint decoding rate upper bound in (6.7), we can express the zero-outage downlink rate as

$$R_B = W \log_2\left(1 + \frac{P_B g d_B^{-\alpha}}{\sigma^2} \cdot \frac{1}{1 + \Gamma_A}\right) \tag{6.34}$$

which can be transformed into:

$$P_B = \frac{1}{g} \cdot \frac{\sigma^2(1+\Gamma_A)}{d_B^{-\alpha}} \cdot \left[2^{R_B/W} - 1\right]. \tag{6.35}$$

Now noting that the achievable downlink rate $\dot{R}_B$ when there is no underlay present is

$$\dot{R}_B = W \log_2\left(1 + \frac{\dot{P}_B g d_B^{-\alpha}}{\sigma^2}\right) \tag{6.36}$$

where $\dot{P}_B$ is the original downlink transmission power. Then:

$$\begin{aligned} P_B &= \frac{1}{g} \cdot \frac{\sigma^2(1+\Gamma_A)}{d_B^{-\alpha}} \cdot \left[2^{W\log_2\left(1+\frac{\dot{P}_B g d_B^{-\alpha}}{\sigma^2}\right)/W} - 1\right] \\ &= (1+\Gamma_A)\dot{P}_B. \end{aligned} \tag{6.37}$$

The extra downlink power required to ensure that the downlink transmission is able to achieve the same downlink rate is proportional to $(1+\Gamma_A) = 2^{R_A/W}$. In other words, the required power will increase exponentially with the rate of the underlay transmission. This underlay scheme is therefore suitable to low rate applications, commonly associated with MTC.

6.4.2 Analysis of topology B

We now consider the network topology depicted in Figure 6.16(b), where the communication links between the BS and Carol, and between the MTD and Alice occur simultaneously in a time slot. As in the previous topology, the link between the MTD and Alice is characterized by fixed rate and power, denoted, respectively, as R_M and P_M. While the link between the BS and Carol is characterized by variable rate and fixed power, hereafter denoted, respectively, as R_B and P_B.

The MTD transmits to Alice during the downlink transmission of B, such that Alice's and Carol's receivers observe a multiple access channel (MAC) with two transmitters:

$$y_A = h_{M,A}x_M + h_{B,A}x_B + Z, \qquad y_C = h_{M,C}x_M + h_{B,C}x_B + Z \tag{6.38}$$

where $h_{B,A}$, $h_{B,C}$, $h_{M,C}$, and $h_{M,A}$ denote the complex channel gains on the channel between the respective nodes. x_B and x_M are, respectively, the signals transmitted by the B and the MTD, such that $E[|x_B|^2] = P_B$ and $E[|x_M|^2] = P_M$. Z is the complex-valued Gaussian noise with variance σ^2.

Recall that we want to ensure that the downlink transmission does not affect the MTD underlay transmission. This is facilitated by the zero-outage result in Corollary 6.2.2, which requires the downlink transmission to be decodable at Alice's receiver. Therefore, the required downlink transmission power is then computed based on the worst link between the BS and either Alice or Carol. The required transmission power then becomes:

$$P_B = \frac{1}{\min(g_A, g_C)} \cdot \frac{\sigma^2(1+\Gamma_A)}{d^{-\alpha}} \cdot \left[2^{R_B/W} - 1\right] \tag{6.39}$$

where g_A and g_C are, respectively, the downlink channel gains towards Alice and Carol. In this scenario there is the possibility that worst link channel gain is too low, leading to an excessive large amount of power to meet the required link rate. To cope with these deep channel fades, we adopt a power control scheme based on truncated channel inversion [29]. In this setting, the BS will only attempt transmission when the fade depth is above a cutoff threshold μ, i.e., when $\min(g_A, g_C) > \mu$. To account for the cutoff value we introduce the nth order statistics distribution:

$$f_n(x) = 2f_g(x)\binom{2-1}{n-1}F(x)^{n-1}(1-F(x))^{2-n}$$

from which we want $n = 1$, the distribution of the worst channel:

$$f_1(x) = \frac{2}{\overline{g}}\exp\left(-\frac{2x}{\overline{g}}\right) \tag{6.40}$$

The outage event is then defined as

$$\begin{aligned} Pr\{\min(g_A, g_C) < \mu\} &= \int_0^{\mu} f_1(x)dx \\ &= 1 - \exp(-2\mu/\overline{g}) \end{aligned} \tag{6.41}$$

and the associated cutoff threshold as

$$\mu = -\frac{\overline{g}}{2}\log(1 - Pr\{\min(g_A, g_C) < \mu\}) \tag{6.42}$$

We assume that when in outage, the downlink transmission does not occur, to either Alice nor Carol. Now noting that the achievable downlink rate $\dot{R}_B$ when there is no underlay present is,

$$\dot{R}_B = W\log_2\left(1 + \frac{\dot{P}_B g_C d_B^{-\alpha}}{\sigma^2}\right) \tag{6.43}$$

where $\dot{P}_B$ is the original downlink transmission power. Then:

$$\begin{aligned} P_B &= \frac{1}{\min(g_A, g_C)} \cdot \frac{\sigma^2(1+\Gamma_A)}{d_B^{-\alpha}} \cdot \left[2^{W\log_2\left(1+\frac{\dot{P}_B g_C d_B^{-\alpha}}{\sigma^2}\right)/W} - 1\right] \\ &= (1+\Gamma_A) \cdot \frac{g_C}{\min(g_A, g_C)} \cdot \dot{P}_B. \end{aligned} \tag{6.44}$$

The extra power required to meet the same downlink rate, as the case without underlay, is dependent on the worst channel realization. On average, the power extra power needed is then given by

$$E\left[\frac{P_B}{\dot{P}}\right] = \int_0^{\infty}\int_{\mu}^{\infty}(1+\Gamma_A)\frac{y}{x}f_1(x)f_g(y)dxdy \\ = (1+\Gamma_A)\cdot 2\cdot E_1(2\mu) \tag{6.45}$$

where $E_1(x)$ is the exponential integral function.

As in topology A, the extra downlink power required is proportional to $2^{R_A/W}$. Therefore, this underlay scheme is also suitable for low rate applications such as the ones associated with MTC.

6.5 Conclusion

In this chapter we have discussed how the Device-to-Device communication paradigm can be used to enable low power, low rate and delay tolerant Machine-to-Machine applications. Furthermore, we have identified underlay communications as suitable for low rate underlay transmissions, which is the case of the traffic associated with Machine-to-Machine applications as well as to the two key blocks of Device-to-Device communications: the proximity discovery and the direct communication between proximal devices. To enable the underlay operation of these two communication blocks, we have applied a recent result, which exploits both the network assisted Device-to-Device operation and terminals with multiuser decoding capabilities to allow zero-outage operation of the underlayed links. We cast this zero-outage result into a power control setting that enables underlay fixed rate links under the downlink transmission, at the cost of additional downlink transmission power, which increases exponentially with the rate of the underlay transmission.

In the underlay proximity discovery setting we extended the power control scheme to support multiple monitoring devices and have shown how to select the best channel, in terms of downlink energy expenditure, for the discovery procedure. In the underlay direct communication setting, where machine-to-machine communications can take place, we have extended the zero-outage power control scheme towards two representative communication topologies and quantified the required additional downlink power to enable zero-outage underlay operation.

Some future work and current open research questions include: (i) how extra information, even if quantized, about the channel quality of the underlay link can be used and how it will alter the underlay zero-outage result; (ii) how the proposed underlay power control scheme can be adapted into general link schedulers and scalable it the number of active network nodes; and (iii) how to adapt the proposed underlay operation with other novel wireless connectivity techniques such as multi-link and decoupled uplink and downlink.

Acknowledgements

This work was performed partly in the framework of H2020 project FANTASTIC-5G (ICT-671660) and by the European Research Council (ERC Consolidator Grant No. 648382 WILLOW) within the Horizon 2020 Program. The authors acknowledge the contributions of the colleagues in FANTASTIC-5G.

References

[1] D. Boswarthick, O. Elloumi, and O. Hersent, *M2M Communications: A Systems Approach*. Hoboken, NJ: John Wiley & Sons Ltd, 2012.

[2] D. Estrin, R. Govindan, J. Heidemann, and S. Kumar, "Next century challenges: scalable coordination in sensor networks," in *Proceedings of the Fifth Annual ACM/IEEE International Conference on Mobile Computing and Networking*, ser. MobiCom'99. New York, NY: ACM, 1999, pp. 263–270. [Online]. Available: http://doi.acm.org/10.1145/313451.313556

[3] G. Fodor, E. Dahlman, G. Mildh, *et al.*, "Design aspects of network assisted device-to-device communications," *IEEE Communications Magazine*, vol. 50, no. 3, pp. 170–177, 2012.

[4] A. Asadi, Q. Wang, and V. Mancuso, "A survey on device-to-device communication in cellular networks," *IEEE Communications Surveys Tutorials*, vol. 16, no. 4, pp. 1801–1819, Fourth quarter 2014.

[5] X. Lin, J. Andrews, A. Ghosh, and R. Ratasuk, "An overview of 3GPP device-to-device proximity services," *IEEE Communications Magazine*, vol. 52, no. 4, pp. 40–48, Apr. 2014.

[6] L. Lei, Z. Zhong, C. Lin, and X. Shen, "Operator controlled device-to-device communications in LTE-advanced networks," *IEEE Wireless Communications*, vol. 19, no. 3, pp. 96–104, Jun. 2012.

[7] F. Schaich, B. Sayrac, M. Schubert, *et al.*, "FANTASTIC-5G: 5G-PPP Project on 5G air interface below 6 GHz," in *Proceedings of European Conference on Networks and Communications EUCNC*, 29 June–2 July 2015, Paris, France.

[8] P. Gupta and P. Kumar, "The capacity of wireless networks," *IEEE Transactions on Information Theory*, vol. 46, no. 2, pp. 388–404, Mar. 2000.

[9] M. Z. Shafiq, L. Ji, A. X. Liu, J. Pang, and J. Wang, "A first look at cellular machine-to-machine traffic: large scale measurement and characterization," *SIGMETRICS Performance and Evaluation Review*, vol. 40, no. 1, pp. 65–76, Jun. 2012. [Online]. Available: http://doi.acm.org/10.1145/2318857.2254767

[10] R. C. D. Paiva, R. D. Vieira, and M. Saily, "Random access capacity evaluation with synchronized MTC users over wireless networks," in *2011 IEEE 73rd Vehicular Technology Conference (VTC Spring)*, Yokohama, 2011, pp. 1–5.

[11] N. K. Pratas and P. Popovski, "Underlay of low-rate machine-type D2D links on downlink cellular links," in *2014 IEEE International Conference on Communications Workshops (ICC)*, Sydney, NSW, 2014, pp. 423–428.

[12] N. K. Pratas and P. Popovski, "Low-rate machine-type communication via wireless device-to-device (D2D) links," *ArXiv e-prints*, May 2013.

[13] S.-Y. Lien, K.-C. Chen, Y.-C. Liang, and Y. Lin, "Cognitive radio resource management for future cellular networks," *IEEE Wireless Communications*, vol. 21, no. 1, pp. 70–79, Feb. 2014.

[14] K.-C. Chen and S.-Y. Lien, "Machine-to-machine communications: Technologies and challenges," *Ad Hoc Networks*, vol. 18, pp. 3–23, 2014. [Online]. Available: http://www.sciencedirect.com/science/article/pii/S1570870513000395

[15] 3GPP, "Ts 22.368 – service requirements for machine-type communications," 3GPP, Tech. Rep., 2014.

[16] T. Koskela, S. Hakola, T. Chen, and J. Lehtomaki, "Clustering concept using device-to-device communication in cellular system," in *2010 IEEE Wireless Communication and Networking Conference*, Sydney, Australia, 2010, pp. 1–6.

[17] T. Kwon and J. Cioffi, "Random deployment of data collectors for serving randomly-located sensors," *IEEE Transactions on Wireless Communications*, vol. 12, no. 6, pp. 2556–2565, Jun. 2013.

[18] G. Rigazzi, N. Pratas, P. Popovski, and R. Fantacci, "Aggregation and trunking of M2M traffic via D2D connections," in *2015 IEEE International Conference on Communications (ICC)*, June 2015, pp. 2973–2978.

[19] 3GPP, "Tr22.803 – feasibility study for proximity services (prose)," 3GPP, Tech. Rep., 2013.

[20] 3GPP, "Tr 23.703 – study on architecture enhancements to support proximity-based services (prose)," 3GPP, Tech. Rep., 2014.

[21] 3GPP, "TR 36.843 – study on LTE device to device proximity services-radio aspects," 3GPP, Tech. Rep., 2013.

[22] A. Gamal and Y. Kim, *Network Information Theory*. New York, NY: Cambridge University Press, 2011.

[23] N. K. Pratas and P. Popovski, "Zero-outage cellular downlink with fixed-rate D2D underlay," in *IEEE Transactions on Wireless Communications*, vol. 14, no. 7, pp. 3533–3543, July 2015.

[24] 3GPP, "TS 23.303 – proximity-based services (prose); stage 2," 3GPP, Tech. Rep., 2014.

[25] P. Janis, C.-H. Yu, K. Doppler, *et al.*, "Device-to-device communication underlaying cellular communications systems," *International Journal of Communications, Network and System Sciences*, vol. 2, pp. 169–178, 2009.

[26] K. Doppler, M. Rinne, C. Wijting, C. Ribeiro, and K. Hugl, "Device-to-device communication as an underlay to LTE-advanced networks," *IEEE Communications Magazine*, vol. 47, no. 12, pp. 42–49, 2009.

[27] C. H. Yu, O. Tirkkonen, K. Doppler, and C. Ribeiro, "On the performance of device-to-device underlay communication with simple power control," in *VTC Spring 2009 – IEEE 69th Vehicular Technology Conference*, Barcelona, 2009, pp. 1–5.

[28] N. K. Pratas and P. Popovski, "Network-assisted device-to-device (D2D) direct proximity discovery with underlay communication," in *2015 IEEE Global Communications Conference (GLOBECOM)*, San Diego, CA, 2015, pp. 1–6.
[29] M.-S. Alouini and A. Goldsmith, "Capacity of Rayleigh fading channels under different adaptive transmission and diversity-combining techniques," *IEEE Transactions on Vehicular Technology*, vol. 48, no. 4, pp. 1165–1181, Jul. 1999.

Chapter 7
Caching in large wireless networks

Georgios Paschos[1] and Marios Kountouris[1]

7.1 Introduction

With more than one billion wireless subscribers today and predictions for this number being tripled over the next five years, the ICT industry is confronted with an avalanche of traffic demand. The unprecedented traffic increase of wireless network traffic is a side effect of the info-revolution we are experiencing, fuelled by the proliferation of smart mobile devices and bandwidth-greedy video applications. The quarterly Internet monitoring report of Akamai reveals that the network traffic has reached the exabyte scale and keeps increasing steadily. Reports from Cisco indicate that mobile traffic gets a larger and larger share of the pie, soon reaching 60% [1]. Therefore, a big challenge for the upcoming 5th generation (5G) wireless networks is to meet the increasing demand for network traffic.

A well-established fundamental result states that long multihop wireless communication do not scale [2], i.e., the maximum common traffic rate for all flows is inversely proportional to the average number of hops. As a result, the proposed techniques to tackle the communication challenges of future networks focus mostly on further enhancing the link capacities and densifying the wireless network with more access points (APs) or base stations (BSs), which are connected to Internet with wires. Several advanced transmission technologies have been proposed and implemented in order to boost the performance of wireless networks, including carrier aggregation, multi-antenna (MIMO) and cooperative communication, as coordinated scheduling. Despite continuous progress, all these cutting edge physical layer (PHY) technologies do not allow significant enhancements as they are reaching their theoretical limits. Ultra dense networks (UDNs) and heterogeneous networks (HetNets) are envisioned to provide a significant network performance leap by deploying heterogeneous infrastructure, e.g., remote radio heads (RRHs) and low power nodes such as picocells, small cells, femto APs, and relays. Densifying the network by installing more infrastructure has two key advantages. First, by bringing the network closer to end-users, the link quality and

[1]Mathematical and Algorithmic Sciences Lab, Paris Research Center, Huawei Technologies Co. Ltd., 17 Rue des Peupliers, 92100 Boulogne-Billancourt, France

coverage can be enhanced due to reduce distance. Second, the large number of cells allows for more efficient spectrum reuse and therefore larger data rates. Nevertheless, in large-scale dense networks, backhaul connectivity, capacity and reliability may become the performance bottleneck, without even mentioning the additional cost associated with their deployment and maintenance.

Two decades ago, the World Wide Web was experiencing a similar challenge. Under the legacy client–server model, every browser in the world needed to connect to the same remote server. The sum of traffic of these end-to-end connections was creating a catastrophic increase of traffic in the Internet backhaul. The solution to that challenge was *web caching*; a technique borrowed from computing processors and operating systems, which exploits the temporal together with the spatial vicinity of user requests to improve the user-perceived Quality of Service (QoS) and network performance metrics. The exploitation of web caching led to the proliferation of Content Delivery Networks (CDNs), responsible for replicating content across the entire Internet today, and intercepting more than 60% of traffic that is directed to remote servers, thereby eliminating a significant amount of backhaul traffic.

With motivation from the success story of CDNs, it has been recently proposed to use caching to improve the sustainability of wireless networks [3,4]. *Moving from CDNs to wireless networks does not simply produce a "wireless CDN"*. New paradigms of content replication appear, new schemes promise to improve efficiency by combining caching with properties of physical (PHY) layer, while new challenges are imposed by mobility, geometrical properties of wireless signal propagation, and interference. Many longstanding models and the associated conventional wisdom may have to be revisited in the coming years.

In this chapter we study the fundamental theoretical underpinnings of wireless networks with caching capabilities. More specifically:

1. We present an analytical tool for simplifying the calculations required to decide the optimal content replication in a wireless network. This tool is used to derive the asymptotic laws of backhaul capacity scaling for multihop wireless networks. The presented results consider (i) the number of networks nodes N, (ii) the number of contents M, and (iii) the cache size K as the key system "size" parameters that increase arbitrarily at different proportions. By studying the different cases we extract valuable intuition into the benefits of caching for the sustainability of wireless networks.
2. We investigate the performance of dense cache-enabled small cell networks and provide useful insights on how the system operating values, the network topology and the interference affect the network performance. Using tools from stochastic geometry to model the node distribution, we provide guidelines where to place the most popular content, i.e., at the mobile device or at the AP/BS. The effect of spatial correlation in content requests is also analysed.
3. The chapter is concluded with a summary of interesting research directions for the future that can play a crucial role in the proliferation of caching as a wireless network technique.

7.2 Sustainable multihop wireless networks with caching

The sustainability of wireless multihop networks can be characterized by examining the capacity scaling laws of a network that grows in size. In their seminal work [2], Gupta and Kumar studied the asymptotic behaviour of multihop wireless networks when communications take place between N pairs of randomly chosen nodes. In such a scenario the maximum data rate is $O(1/\sqrt{N})$, hence as the network grows ($N \to \infty$) the data rate vanishes to zero. This finding argues against the sustainability of multihop communications. Moreover, the $O(1/\sqrt{N})$ law was shown to arise from geometric considerations of the 2D wireless transmissions [5], hence it is impossible to breach under the classical random communicating pair model.

In this section we revisit the subject of the sustainability of wireless multihop networks with the added element of caching. Each node in the network represents a user that is interested in a particular content (instead of communicating with another node). The content can be replicated in the network caches, and thus the user can conveniently retrieve it from a nearby cache, reducing in this way the number of traversed hops. The derived asymptotic laws for capacity scaling suggest that caching has a powerful effect in the sustainability of wireless networks, and there exist interesting regimes where the $O(1/\sqrt{N})$ law is breached.

The fundamental size parameters are the number of nodes N, and the number of contents M, and thus the scaling laws will depend on how these increase to infinity. Another key parameter is the cache size K which depicts the number of contents that can be cached at each node. Taking K to infinity represents an interesting regime that reflects networks where nodes are upgraded as storage gets abundant and inexpensive [6]. Another influencing factor is the content popularity. With skewed popularity, popular files are requested multiple times and hence smaller K have a bigger impact on the system performance. As anticipated, in our analysis the popularity skewness crucially affects the sustainability of wireless networks.

The presented analysis is a summary of [7–10] and involves the solution of a hard combinatorial optimization that captures the best way of replicating contents with different popularities and retrieving them over network routes. Fortunately, by relaxing the problem to a convex optimization and rounding the solution, an order optimal performance can be obtained.

7.2.1 System model

7.2.1.1 Network model

The network is modelled by a grid topology as in Figure 7.1, which captures essential characteristics of the wireless networks (i) multihop short-range communications are preferable [2] and (ii) the network diameter scales as $\sqrt{N}$.[1] More realistic random topologies are handled in [12], but they are shown to result in the same scaling laws.

[1]Note that the grid model is not accurate for wireline networks where the diameter scales as $\log N$, as node connectivity follows power laws [11].

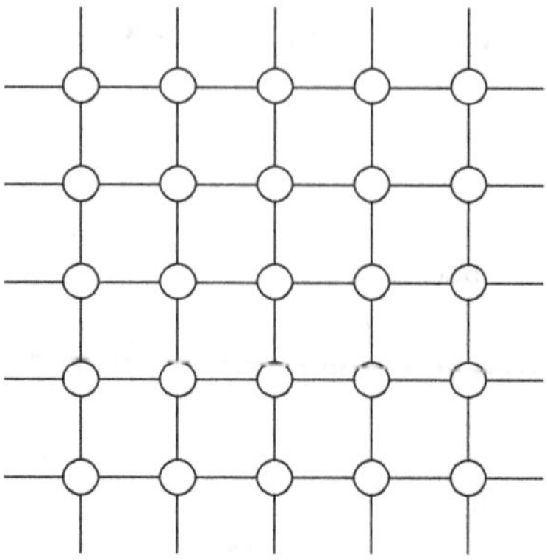

Figure 7.1 For the scaling laws analysis, the network topology is considered to be a toroidal square grid

Formally, we consider a set of N nodes indexed by $n \in \mathcal{N} \triangleq \{1, 2, \ldots, N\}$, arranged on a $\sqrt{N} \times \sqrt{N}$ square grid on the plane. Each node is connected via undirected links to its four neighbours that lie next to it on the same row or column. By keeping the node density fixed and increasing the network size N, we obtain a scaling network similar to [2], with random pair throughput scaling as $O(1/\sqrt{N})$. Moreover, to avoid boundary effects, we consider a toroidal structure as in [5].

7.2.1.2 Content requests

Nodes generate requests to a catalogue of contents, indexed by $m \in \mathcal{M} \triangleq \{1, 2, \ldots, M\}$. Normally each node requests content m with rate λ_n^m. Such a consideration would lead the analysis of sustainability to the formation of the capacity region (which is the set of all rate vectors $[\lambda_n^m]$ that are sustainable), see [13], which however is expressed through a set of inequalities and will not provide a simple expression for the scaling laws. Hence to simplify the exposition and get a quantitative result, we assume the symmetric case where all nodes have $\lambda_n^m = \lambda^m, \forall n$.

More specifically, we consider the Independent Reference Model (IRM) [14], according to which the requests for file m are due to an independent Poisson process with intensity λp_m, where p_m represents the file popularity distribution—the latter is assumed time-invariant.

Last, we fix $\lambda = 1$ and study the required link capacity to support the communications. This is the inverse of the classical approach where we fix the link capacity and study the scaling of maximum supportable throughput. For example, in a network where under fixed link capacities the throughput scales as $O(1/\sqrt{N})$, our model will yield required link capacity scaling of $\Theta(\sqrt{N})$ to support constant throughput $\lambda = 1$. More generally, the obtained scaling laws for required link capacity can be inverted to reflect the throughput scaling laws. Accordingly, for sustainability we require the link capacity to be as small as possible, ideally to be $O(1)$.

7.2.1.3 Caching model

Each node n is equipped with a cache, whose contents are denoted by the set $\mathcal{B}_n$, a subset of $\mathcal{M}$. If a request at node n regards a file m that lies in $\mathcal{B}_n$, then it is served locally. Due to the limited memory space of the cache, m will often be not available in $\mathcal{B}_n$, thus, node n will have to request m over the network from some other node w that keeps m in its cache. We denote by $\left[\mathcal{R}_{n,k}\right]$ the set of routes connecting nodes to nearby caches. The choice of the sets $[\mathcal{B}_n]$ and $\left[\mathcal{R}_{n,k}\right]$ crucially affects the network link loading.

Let, moreover, K be the storage capacity of nodes' cache measured in the number of files it can store. This means that all M files are of the same (unit) size, placing a constraint on the cardinality of cache contents $|\mathcal{B}_n| \leq K$. The important size parameters of the system are N, M, K. The interesting regime that we will study is

$$K < M \leq KN. \tag{7.1}$$

The first inequality implies that the cache size K is not enough to fit all files and hence each node needs to make a selection of files to cache. The second inequality requires that the total network cache capacity KN (summing up all individual caches) is sufficient to store all files at least once, and thus each content request can be served.

7.2.2 Multihop capacity optimization: relaxation and rounding

Due to placement constraints $|\mathcal{B}_n| \leq K$, the structure of cache optimization problems is inherently combinatorial, and thus they are hard to solve. In this section (i) we state the relevant problems for optimizing capacity scaling laws with caching, (ii) propose a relaxed problem based on file densities which is easy to solve, and in this way (iii) establish a feasible solution which is of the same order with the original combinatorial problem.

7.2.2.1 Joint replication-routing problem

To yield the correct scaling law of required link capacities, it is necessary to select the best placement and the best routing that minimizes the load at the worst link in the network. Let C_ℓ be the traffic load carried by link ℓ. The network is stable, not discarding requests, only if the capacity of link ℓ exceeds C_ℓ for all ℓ. The link loads $[C_\ell]$ can be adjusted by (i) caching popular content (by means of choosing the placement $[\mathcal{B}_n]$) and (ii) load balancing across a set of routing paths $\left[\mathcal{R}_{n,k}\right]$ from caching node n to requesting node k. Hence the correct scaling law of the required link capacities C^* is identified as the value of the following optimization:

Problem 7.1 $C^* = \text{Minimize}_{[\mathcal{B}_n],\left[\mathcal{R}_{n,m}\right]} \max_\ell C_\ell$, s.t.

1. $|\mathcal{B}_n| \leq K, \quad \forall n$ (*cache constraint*),
2. $\sum_n 1_{\{m \in \mathcal{B}_n\}} \geq 1, \quad \forall m \in \mathcal{M}$ (*all files are cached*),
3. $\left[\mathcal{R}_{n,m}\right]$ *are feasible routes*

This joint optimization turns out to be a hard combinatorial problem,[2] not amenable to an easy-to-compute solution. Therefore, we resort to simplifications and approximations that provide an order-optimal solution, i.e., whose value of the objective function is within a constant to the optimal, hard-to-compute min max C_ℓ. A first step that preserves the order-optimality of the solution is the relaxation of the objective to the average link traffic $\text{avg}_\ell C_\ell$; then routing variables $[\mathscr{R}_{n,k}]$ can be fixed to shortest paths without losing optimality; these steps are explained in detail in [8]. However, the decisive step involves breaking the coupling between the individual caches $[\mathscr{B}_n]$. To achieve this decoupling, we introduce a new notion, that of *replication density* of content m.

7.2.2.2 Relaxed density-based problem

Given a placement $[\mathscr{B}_n]$, consider the frequency of occurrence of each file m in the caches, or *replication density* d_m as the fraction of nodes that store file m in the network:

$$d_m = \frac{1}{N}\sum_{n\in\mathscr{N}} 1_{\{m\in\mathscr{B}_n\}}.$$

Based on this metric, we define a simpler problem:

Problem 7.2 $C = \text{Minimize}_{[d_m]}\sum_{m\in\mathscr{M}}\left(\frac{1}{\sqrt{d_m}} - 1\right)p_m$, s.t.

1. *For any* $m \in \mathscr{M}$, $1/N \le d_m \le 1$,
2. $\sum_{m\in\mathscr{M}} d_m \le K$.

In the above, we optimize on the densities d_m, which express the fraction of caches containing file m. In the objective, $1/\sqrt{d_m} - 1$ approximates (in-order) the average hop count from a random node to a cache containing m. Weighted by the probability p_m of requests on m, the summation expresses the average link load per request. Additionally, the constraint $\sum_{m\in\mathscr{M}} d_m \le K$ reflects another relaxation, whereby the cache size constraint is only satisfied on the average in the network. Observe that we have removed the individual cache constraints, and a feasible solution to Problem 7.2 may yield file densities that correspond to caching more than K files in one node.

It is clear that any feasible solution of Problem 7.1 yields a file density $[d_m]$ that is feasible for Problem 7.2, but not vice versa, hence the density-based formulation is a relaxed version of the original min–max problem [8] and we have

$$C = O(C^*).$$

In particular, Problem 7.2 is convex and its unique solution can be found using the Karush–Kuhn–Tucker (KKT) conditions. Regarding the constraints on d_m about its minimum and maximum values, either one of them can be an equality, or none. This partitions $\mathscr{M}$ into three subsets, the "up-truncated" $\mathscr{M}_\uparrow = \{m : d_m = 1\}$ containing files stored at all nodes, the "down-truncated" $\mathscr{M}_\downarrow = \{m : d_m = 1N\}$ containing files

[2]The cache optimization problem of interest bears resemblance to *database location problem*, see [15].

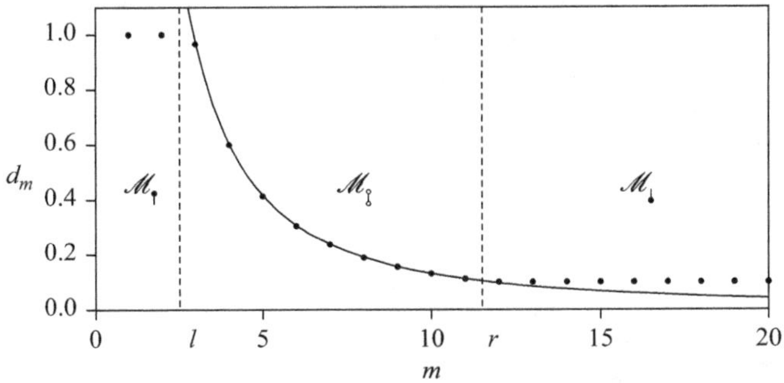

Figure 7.2 An example case of density d_m and the $\mathscr{M}_{\uparrow}$, $\mathscr{M}_{\circ}$ and $\mathscr{M}_{\downarrow}$ partitions. Solid line plots the $\sim m^{-\frac{2\tau}{3}}$ law of $m \in \mathscr{M}_{\circ}$, when p_m follows Zipf's law

stored in just one node, and the complementary "non-truncated" $\mathscr{M}_{\circ} = M \backslash (\mathscr{M}_{\uparrow} \cup \mathscr{M}_{\downarrow})$ of files with $1/N < d_m < 1$. Arranging p_m in decreasing order, the partitions become $\mathscr{M}_{\uparrow} = \{1, 2, \ldots, l-1\}$, $\mathscr{M}_{\circ} = \{l, l+1, \ldots, r-1\}$, and $\mathscr{M}_{\downarrow} = \{r, r+2, \ldots, M\}$; l and r are integers with $1 \le l \le r \le M+1$. The solution d_m is equal to:

$$d_m = \begin{cases} 1, & m \in \mathscr{M}_{\uparrow}, \quad (7.2a) \\ \dfrac{K - l + 1 - \frac{M-r+1}{N}}{\sum j \in \mathscr{M}_{\circ} p_j^{\frac{2}{3}}} p_m^{\frac{2}{3}}, & m \in \mathscr{M}_{\circ}, \quad (7.2b) \\ 1/N, & m \in \mathscr{M}_{\downarrow}. \quad (7.2c) \end{cases}$$

Figure 7.2 illustrates such an example solution, depicting the density d_m, indices l and r, as well as sets $\mathscr{M}_{\uparrow}$, $\mathscr{M}_{\circ}$, $\mathscr{M}_{\downarrow}$ when file popularities follow the Zipf law (see Section 7.2.3).

7.2.2.3 Discrete density (rounding)

The solution (7.2) is not directly mapped to a feasible solution for Problem 7.1, but we may construct one using a two-step process, (i) first rounding $[d_m]$ to $[d_m^{\circ}]$ and (ii) second placing the content symmetrically on the network according to $[d_m^{\circ}]$ so that the constraints $|\mathscr{B}_n| \le K$ are satisfied. For (i), we simply define $d_m^{\circ} \triangleq 4^{-\nu_m^{\circ}}$ rounded to the largest power less or equal to $[d_m]$:

$$d_m^{\circ} \triangleq \max\{4^{-i} : 4^{-i} \le d_m, i \in \{0, 1, \ldots, \nu\}\}. \quad (7.3)$$

Then for (ii), [8] gives an algorithm to allocate the files $\mathscr{M}$ in the caches $[\mathscr{B}_n]$ given the replication densities d_m°. The algorithm can be explained by means of

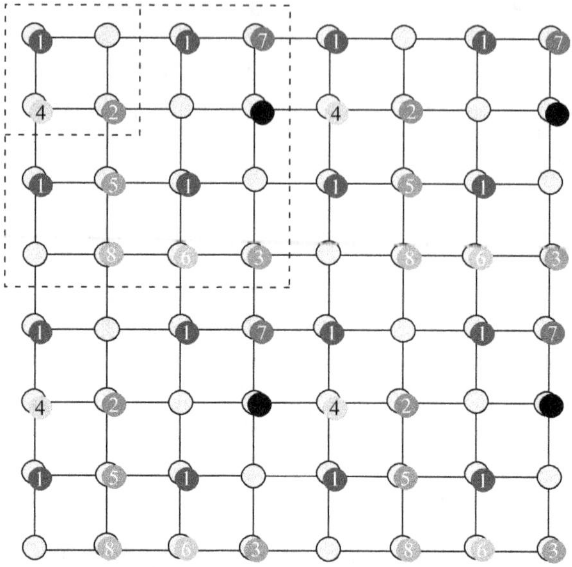

Figure 7.3 An example of the canonical rounding used to obtain a feasible cache placement solution. Here, the following items are placed $d_0^\circ = 1$, $d_1^\circ = 1/4$, $d_2^\circ = \cdots = d_8^\circ = 1/16$

Figure 7.3. We begin with the grey file which has $d_0^\circ = 1$, this file is simply cached everywhere. Then for the file with $d_0^\circ = 1/4$, we focus on a 2×2 subgrid of nodes (any such subgrid suffices but it helps to fix an origin as a reference). In this subgrid we try to fill the diagonal, which in this case is achieved by placing file 1 at the coordinate $(1, 1)$ (top left node in the grid). As a last step for this file, we place replicas by tiling the subgrid everywhere in the network. The result is that file 1 is replicated with density 1/4, as prescribed by the solution. Then for the files with density 1/16 we enlarge the subgrid to 4×4. In general the subgrid has a size $2^{\nu_m^\circ} \times 2^{\nu_m^\circ}$ and is aligned with all the considered subgrids. We then fill the subgrid with the new files starting with the diagonal, notably files 2 and 3 in the example. Then files 4, 5, 6 are filled in the second diagonal which is below the first, while 7 completes the second diagonal by wrapping up at the coordinate $(1, 4)$. During the filling, we only select nodes that have less files than the maximum. For example, when we are filling object 2 in the subgrid 4×4, we skip the node $(1, 1)$ since that node has already two files (file 0 and file 1), and we place file 2 at node $(2, 2)$ which only had one file so far (file 0). Where would we place file 10 with $d_{10}^\circ = 1/64$? We would consider the subgrid 8×8 (i.e., the entire grid), the first three diagonals are fully filled, hence we would place it in the first open spot in the fourth diagonal, that is node $(4, 1)$.

Using this placement, we ultimately obtain a feasible solution for Problem 7.1 with value $C^\circ = \Omega(C)$. Finally, the following result is established in [8].

Theorem 7.1 (Relaxation and rounding yields order optimal link loads). There exist positive constants a, b that depend on the distribution $[p_m]$, and cache capacity K, such that

$$C^* \leq C^\circ \leq aC^* + b.$$

Furthermore it is $\Theta(C^*) = \Theta(C^\circ) = \Theta(C)$.

In conclusion, we may derive the scaling laws of wireless networks with caching C^* by directly using the relaxed Problem 7.2 solution, C. We remark that since a, b above depend on K, scaling laws with respect to K need to consider this dependence carefully. This is taken into account in the analysis of [10].

7.2.3 Asymptotic laws for Zipf popularity

To study the scaling of C, we switch from the arbitrary popularity to the Zipf law, which has been observed to model well the content popularity of the traffic of www and other types of services in numerous traces in the literature [12,14,16–20]. It is defined as follows:

$$p_m = \frac{1}{H_\tau(M)} m^{-\tau}, \tag{7.4}$$

where $H_\tau(n) \triangleq \sum_{j=1}^{n} j^{-\tau}$ is the truncated zeta function evaluated at τ (also called the nth *τ-order generalized harmonic number*), and τ is the parameter of the distribution, adjusting the rate of popularity decline with m. Regarding parameter τ, values ≤ 1 are reported for web traffic [21–23]. Interestingly, Padmanabhan and Qiu [24] measure $\tau = 1$ at proxies versus 1.4–1.6 at a "busy" web server. Analysing traces from mobile browsing, Yamakami [25] reports values in the range of 1–1.5, while in User Generated Content (UGC) video, Cha *et al.* [26] fit popularity with a combination of Zipf and Exponential Cut-off, where the Zipf component has τ in the ranges of about 0.98–1.47 and 0.45–1.09 in two different scenarios. Last, content popularity in P2P systems was fitted with a $\tau = 0.95$ in [27], and in a Video on Demand system with 0.70 in [28]. As there is no conclusive answer about the value of τ, we consider all possibilities in our investigation.

We derive an approximation for $H_\tau(n)$ by bounding the sum: for $n \geq m \geq 0$,

$$\int_m^n (x+1)^{-\tau} dx \leq H_\tau(n) - H_\tau(m) \leq 1 + \int_{m+1}^n x^{-\tau} dx, \Rightarrow$$

$$\begin{cases} \dfrac{(n+1)^{1-\tau} - (m+1)^{1-\tau}}{1-\tau} \leq H_\tau(n) - H_\tau(m) \leq \dfrac{n^{1-\tau} - (m+1)^{1-\tau}}{1-\tau} + 1, & \text{if } \tau \neq 1, \\ \ln\dfrac{n+1}{m+1} \leq H_\tau(n) - H_\tau(m) \leq \ln\dfrac{n+1}{m+2}, & \text{if } \tau = 1. \end{cases} \tag{7.5}$$

Substituting the solution (7.2) and plugging in the Zipf distribution into the objective of Problem 7.2, it follows that:

$$C \triangleq \sum_{m\in\mathcal{M}} \left(d_m^{-\frac{1}{2}} - 1\right) p_m = C_{\circ} + C_{\bullet} - \sum_{j=l}^{M} p_m, \tag{7.6}$$

where $\sum_{j=l}^{M} p_m = O(1)$ (as it always lies in $[0, 1]$), and

$$C_{\circ} \triangleq \sum_{m\in\mathcal{M}_{\circ}} \frac{p_m}{\sqrt{d_m}} \overset{(1.2),(1.4)}{=} \frac{\left[H_{\frac{2\tau}{3}}(r-1) - H_{\frac{2\tau}{3}}(l-1)\right]^{\frac{3}{2}}}{\sqrt{K - l + 1 - \frac{M-r+1}{N}\, H_\tau(M)}}, \tag{7.7}$$

$$C_{\bullet} \triangleq \sum_{m\in\mathcal{M}_{\bullet}} \frac{p_m}{\sqrt{d_m}} \overset{(1.2),(1.4)}{=} \sqrt{N}\, \frac{H_\tau(M) - H_\tau(r-1)}{H_\tau(M)}. \tag{7.8}$$

To analytically compute the law of C, we need to analyse l and r. Note that since M scales to infinity, l, r may also scale to infinity, or not, depending on the actual solution (7.2). Moreover, observing that the expressions (7.7) and (7.8) depend on $H_\tau, H_{\frac{2\tau}{3}}$, we expect different cases to appear due to the form of (7.5). In fact, previous work [8] yields five cases depending on the values of τ. These in terms are mapped to scaling laws via (7.7) and (7.8). The results are presented in Table 7.1. The cases $\tau = 1$ and $\tau = \frac{3}{2}$ are omitted to avoid clutter; they are similar to the cases $\tau < 1$ and $\tau > \frac{3}{2}$ up to a logarithmic factor.

7.2.4 Scaling laws for constant cache size K

Table 7.1 shows how the solution of Problem 7.2 scales with the system size parameters N (number of nodes/users), and M (catalogue size); K (cache size) in this table is a constant. From Theorem 7.1, the scaling of C also applies to the required link capacity for sustaining a uniform request rate $\lambda = 1$. For example, a

Table 7.1 Scaling laws with constant cache size

		$N \to \infty$, then		$M \sim KN$, hence $M = \Theta(N)$	
	M	**M finite**	**$M \to \infty$**	**$KN - M = \omega(1)$**	**$KN - M = O(1)$**
C	$\tau < 1$	$\Theta(1)$	$\Theta(\sqrt{M})$	$\Theta(\sqrt{N})$	$\Theta(\sqrt{N})$
	$1 < \tau < \frac{3}{2}$	$\Theta(1)$	$\Theta\left(M^{\frac{3}{2}-\tau}\right)$	$\Theta\left(\frac{\sqrt{N}}{(KN-M)^{\tau-1}}\right)$	$\Theta(\sqrt{N})$
	$\tau > \frac{3}{2}$	$\Theta(1)$	$\Theta(1)$	$\Theta\left(\frac{\sqrt{N}}{(KN-M)^{\frac{3(\tau-1)}{2\tau}}}\right)$	$\Theta(\sqrt{N})$

scaling $\Theta(1)$ means that the network can sustain unit throughput even if the link capacities are (sufficiently large) constants; this is the desirable case. On the other hand, $\Theta(\sqrt{M})$ means that to sustain unit throughput, the link capacity needs to increase proportionally to the square root of the file catalogue. Since in practical systems M is expected to grow with the network size N, while the link capacity is expected to obey certain limits (like the Shannon limit), this means that the network will be unsustainable under such a law. Recall that we may also inverse the scaling to see how throughput will scale in a network with fixed link capacities. For example if we invert the law $\Theta(\sqrt{N})$ it gives $\lambda = \Theta(1/\sqrt{N})$, which is the Gupta–Kumar scaling for random communicating pairs in a scaling network [2]. This law appears in our analysis whenever the replication capacity $KN - M = \Theta(1)$ is low, i.e., almost all cache slots are used to store each file once—see the last column of Table 7.1.

Since K is constant, the results are derived for different cases of how N and M scale to infinity together. The second column corresponds to the case where only N scales and M is fixed, in which case the system is sustainable (required link capacity $\Theta(1)$). However, in practical systems the file catalogue grows with the user population.

The third column refers to the case where we take the limits in order, first $N \to \infty$ and then $M \to \infty$, i.e., we have $M = o(N)$ and the catalogue grows sub-linearly to the network size. The scaling laws in this case relate to the file catalogue size, which is a slight improvement over the Gupta–Kumar scaling. Different scaling is obtained for different values of τ, where larger values of τ improve the sustainability (decrease the required link capacity).

In the last two columns we set $M = \alpha KN$ for some positive $\alpha < 1$, and then take the limit $N \to \infty$. The scaling laws in this case also depend on $KN - M$, i.e., how many replication slots are left in the caches after we store each file once. There are two cases, (a) $KN - M = \omega(1)$ and (b) $KN - M = O(1)$. In the first case, despite $M = \alpha KN$, there are enough replication slots to yield scalings that represent improvements over the Gupta–Kumar scaling as long as $\tau > 1$. In the second case, we obtain the Gupta–Kumar scaling irrespective of the value of τ.

7.2.5 Scaling the cache size $K \to \infty$

As memory becomes cheaper and cheaper, we may envisage the scenario where the cache size per node K also scales to infinity. The extension of the analysis of (7.7) and (7.8) in the case $K \to \infty$ is found in [10]. Here we briefly discuss the resulting scaling laws presented in Table 7.2.

Considering how N, M, K grow, we can study the system in different regimes of operation, which complicates the exposition of scaling laws. For this reason, we focus here on two specific regimes of interest. We compare the total network memory KN to the file catalogue M, and split the analysis to two cases (i) $KN = \Omega(M)$ and (ii) $KN = O(M)$ (in fact Gitzenis *et al.* [10] provide also the asymptotic constant that separates the two regimes). The first case is called "High" K and the second "Low" K.

Table 7.2 Scaling laws with scaling cache size

		"High" K	**"Low" K**
C	$\tau < 1$	$O\left(\sqrt{\frac{M}{K}}\right)$	$\Theta(\sqrt{N})$
	$1 < \tau < \frac{3}{2}$	$O\left(\frac{M^{\frac{3}{2}-\tau}}{\sqrt{K}}\right)$	$\Theta\left(\frac{\sqrt{N}}{(KN-M)^{\tau-1}}\right)$
	$\tau > \frac{3}{2}$	$O(1)$	$\Theta\left(\sqrt{\frac{N}{KN-M}}\right)$

7.2.5.1 "High" K

The most interesting regime to explore for perfect sustainability is the one of $C = O(1)$. As the formulas show, to keep C bounded, the hardest case is on $\tau < 1$: node cache capacity K should scale as fast as content volume M. In the intermediate case of $1 < \tau < \frac{3}{2}$, node capacity K has to scale with M, but slower, at a sublinear power. The case of $\tau > \frac{3}{2}$ is quite interesting, as $C = O(1)$ always holds true.

7.2.5.2 "Low" K

This regime is characterized as unsustainable, because the increase of network node cache K and/or node count N against content volume M is not sufficient to keep C low. First, note that when the replication storage capacity beyond the storage of the primary copy is limited, i.e., $KN - M = O(1)$, C scales as fast as $\sqrt{N}$, the law of [2]. Observe, moreover, that for $\tau > 1$, $\sqrt{N}$ is scaled down by the capacity $KN - M$ available for replication beyond the primary copy at some power equal to$\sqrt{KN - M}$ or $(KN - M)^{\tau-1}$ for $\tau > \frac{3}{2}$ or $1 < \tau < \frac{3}{2}$, respectively; this quantifies the gain from adding extra capacity in the network beyond the one required to store a primary copy per file.

7.2.6 *Discussion about the sustainability of wireless caching networks*

We presented a joint delivery and replication problem that leads to the characterization of the sustainability of multihop wireless networks with caching. After formulating the precise combinatorial problem on square lattice networks, we relaxed it to a simple density-based problem, whose analysis permitted to derive the scaling laws of the required link capacity. The key factors in network sustainability are (a) the file popularity power law parameter τ and (b) the relative scaling of the numbers of files vs. network nodes.

While our treatment so far characterizes the caching benefits for multihop networks, a different line of work examines the caching scaling laws in the wireless broadcast medium using coding [29,30]. Next, we continue our exposition zooming in the cellular topology of wireless networks, considering the exploitation of caching in small cells with device–device cooperation.

7.3 Cache-enabled small cell networks

Proximity-based content caching and distribution in wireless networks has been identified as a promising traffic offloading solution for improving the capacity and the quality of experience (QoE) by exploiting content popularity and spatio-temporal request correlation.

The concept of caching at the radio access network edge, termed here as small cell (SC) caching, is driven by the bandwidth and latency gains from introducing caching capabilities to the SBSs and reducing the backhaul utilization. Cache-enabled small cell networks (SCNs) have been proposed as a potential solution for alleviating the backhaul bottleneck problem [31–33]. The main idea is to introduce cache capacity at small cell base stations (SBSs) to prefetch popular contents during off-peak hours before being requested by local users. Caching in wireless networks also exploits the high degree of asynchronous content reuse caused by information-centric applications such as video-on-demand (VoD), social networks and content sharing. When users request for some popular contents already cached in the local SBSs, the service latency is largely reduced since it does not need to pass through the backhaul to retrieve the content from remote servers. The improved energy efficiency is also an important benefit of small cell caching mainly due to the fact that repeated fetching of the same content from the core network to local SBSs are avoided.

A different approach for content reuse is to exploit the available storage space of mobile devices. Popular content is cached in devices' storage and distributed to other devices in proximity through D2D communication. Ji *et al.* [34] discuss the throughput-outage tradeoff of D2D caching networks using a protocol model for the spatial scheduling of coexisting D2D links. Golrezaei *et al.* [35] study the scaling behaviour of a D2D network and give a closed-form expression for the optimal collaboration distance as a function of the system model parameters. Kang *et al.* [36] consider a random caching strategy for stochastically distributed caching-server devices to serve nearby user devices and proposes an algorithm to find the optimal caching probabilities of contents in the request library.

7.3.1 Cache-enabled SCNs with local user interest correlation

Prior studies investigating wireless caching problems often assume similar popularity pattern for all users in the system. Considering different interest levels over requested content, a clustering algorithm is proposed in [33] to group users into clusters based on their request profile. Optimal caching decisions are then made at each SBS by using a distributed regret learning approach to minimize the total service delay.

We are interested here in the case where caching decisions are made at SBSs based on local users' request patterns rather than a global content popularity, since in cache-enabled SCNs, each SBS usually serves a small amount of users. Specifically, each SBS constructs its local regular content library by sampling regularly files requested by the users in its sampling range in its sampling range, then determining the set of files to be cached given its limited cache capacity. We use the conventional "cache the most popular files" strategy to analyse the impact of user interest correlation on the cache hit/miss probability. The results can be generalized to more advanced caching strategies [37]. We assume Zipf-like distribution for the local regular content popularity in the same region, and we derive the cache service probability as a function of the maximum sampling distance. We also consider that there is a distance-dependent sampling cost, and we study the tradeoff between caching service probability and sampling cost. The presented analysis is a summary of [38].

7.3.1.1 System model

We consider a cache-enabled SCN model with limited storage space at each SBS, serving its nearby users within a certain distance dictated by a transmit power constraint. We model the SBS distribution on the two-dimensional Euclidean plane $\mathbb{R}^2$ by a homogeneous Poisson point process (PPP) $\Phi_c = \{Y_i, i \in \mathbb{N}^+\}$ with intensity λ_c, where Y_i denotes the position of the ith SBS. Users to be served are distributed according to another independent homogeneous PPP $\Phi_u = \left\{x_j, j \in \mathbb{N}^+\right\}$ with intensity $\lambda_u \geq \lambda_c$, where x_j denotes the position of the jth user.

We introduce the notion of "local user interest pattern," which results when each SBS samples the regularly requested content of all users in its sampling region of radius R_p. The small cell cachers decide which files to cache based on the adopted caching policy. Assume there exists a maximum distance that an SBS can serve, denoted by R_v, the disk centred at the SBS with radius R_v can be seen as the service region of this SBS.

When increasing the sampling range, the SBSs may have better knowledge of local user interest pattern, but due to the limited cache storage only the most popular files can be stored. For a random user being sampled, the probability of finding its regular content cached at its covering SBSs will decrease with the sampling region size. The service range constraint also gives un upper bound on the probability of an arbitrary user being served by the small cell cachers. We denote this service probability by P_{sv}. Increasing the sampling range also leads to higher cost for learning the local request pattern at the SBSs.

Suppose that each user has a library of size J, which contains its regularly requested files. For simplicity, we assume that all the files have equal unit size. Each SBS has limited cache storage size denoted by M, which is the maximum number of files that can be stored. Denoting by N the number of users in the sampling region of an SBS, from point process theory, we have $\mathbb{E}[N] = \lambda_u \pi R_p^2$ [39].

When users in the same sampling region request similar content, different users may have overlapping files in their regular content libraries, hence the overall regular content library size will not grow linearly with the number of users in this area. Conditioning on having N users in the sampling region of a typical SBS, we denote by $\mathscr{C} = \{c_1, \ldots, c_S\}$ the local content library of the N users with S representing the library size. Denote by $g : \mathbb{N}^+ \rightarrow \mathbb{N}^+$ the mapping function from the number of users N to the local library size S. In reality $S = g(N)$ can be learned numerically at the SBSs and the approximate function can be found by data fitting. Here we assume that g is a piecewise function and follows $S = \min(JN, \lceil J(1 + \mu \log N)\rceil)$, where μ is a constant factor that characterizes the similarity level of local user interest. For a given local regular content library, we assume that the popularity distribution of files in it follows the Zipf–Mandelbrot law, that is, for the i-th most popular file we have its request probability as

$$p_i = \frac{\Omega}{(q+i)^\gamma} \tag{7.9}$$

where $\Omega = \left(\sum_{n=1}^{S} (q+n)^\gamma\right)^{-1}$ is the normalization constant, q is the shift parameter which is related to the shifting based on the Zipf distribution, and γ is the shape parameter, which defines the concentration (correlation) level of the content popularity. Similarly to the g function, the real popularity distribution can be learned by sampling and sorting the popularity of all requested files in the local region. Note that μ in the g function defines how similar the regular content libraries of users in the same area are, and γ in the popularity distribution defines the relative disparity of content popularity in the local library. The combination of these two factors gives full characterization of the correlation level of local user interest.

7.3.1.2 Performance analysis

Service Probability Analysis. We consider a "cache the most popular files" policy and assume that cache entities at the SBSs have identical and limited storage size as M files. Assume that we have an SBS at the origin with N users in its sampling region of radius R_p. When $M \geq S$, all files in the local content library can be stored at the SBS. When $M < S$, due to the limited cache storage only the M most popular files in the sampling region can be cached. The cache miss probability P_m, which is the probability that a random file in the local regular content library $\mathscr{C}$ not being cached, can be given as

$$P_m = \sum_{i=M+1}^{S} p_i = \sum_{i=M+1}^{S} \frac{1}{(q+i)^\gamma \sum_{n=1}^{S} (q+n)^\gamma}. \tag{7.10}$$

Evidently, P_m increases with S, which verifies the intuition that larger content library size leads to higher cache miss probability when the cache storage capacity is finite. Assuming that users outside the sampling range of an SBS have almost negligible probability to find their regularly requested content in this small cell cache, the service probability P_{sv} is given by the probability of the intersection of three events:

- $\mathscr{E}_1 =$ user's regular content library being sampled by at least one SBS,
- $\mathscr{E}_2 =$ the requested content being cached,
- $\mathscr{E}_3 =$ user in the service range of the sampling SBS.

Furthermore we have $\mathscr{E}_1 \subseteq \mathscr{E}_3$ if $R_p \leq R_v$ and $\mathscr{E}_3 \subset \mathscr{E}_1$ otherwise. In the case where $R_p \leq R_v$, P_{sv} can be given by

$$\begin{aligned} P_{sv} &= \mathbb{P}\left[\text{user being sampled} \cap \text{content being cached}\right] \\ &= 1 - \mathbb{P}\left[\text{not being sampled}\right] \\ &\quad - \sum_{k=1}^{\infty} \mathbb{P}\left[\text{sampled by k cachers}\right] \cdot \mathbb{P}\left[\text{k cache miss}\right] \\ &= 1 - e^{-\lambda_c \pi R_p^2} \\ &\quad - \sum_{k=1}^{\infty} \frac{\left(\lambda_c \pi R_p^2\right)^k}{k!} \exp\left(-\lambda_c \pi R_p^2\right) \prod_{j=1}^{k} P_m^j \end{aligned} \tag{7.11}$$

where P_m^j is the cache miss probability of the j-th SBS. When different small cells have different request correlation levels, which corresponds to different γ in the Zipf distribution, the cache miss probability can be given as

$$P_m^j = \sum_{i=M+1}^{S} \frac{1}{(q+i)^{\gamma_j} \sum_{n=1}^{S} (q+n)^{\gamma_j}} \tag{7.12}$$

where γ_j is the shape parameter of the content popularity distribution in the sampling region of the j-th SBS. Evidently, P_{sv} increases monotonically with R_p, meaning that larger sampling distance gives higher probability of a random user to find its regularly requested content cached in the covering SBSs.

In the case where $R_p > R_v$, the service probability is given by

$$\begin{aligned} P_{sv} &= \mathbb{P}\left[\text{user in service range} \cap \text{content being cached}\right] \\ &= 1 - e^{-\lambda_c \pi R_v^2} \\ &\quad - \sum_{k=1}^{\infty} \frac{\left(\lambda_c \pi R_v^2\right)^k}{k!} \exp\left(-\lambda_c \pi R_v^2\right) \prod_{j=1}^{k} P_m^j. \end{aligned} \tag{7.13}$$

Contrary to the previous case, for fixed R_v the service probability decreases with R_p, which is easy to understand since larger sampling region corresponds to higher cache miss probability for users being sampled, while the number of users that can be served does not increase with the sampling range due to the service range constraint. Combining these two cases we have:

$$\begin{aligned} P_{sv} &= \mathbb{P}\left[\text{user in service range} \cap \text{content being cached}\right] \\ &= 1 - e^{-\lambda_c \pi \min\left\{R_p, R_v\right\}^2} - \sum_{k=1}^{\infty} e^{-\lambda_c \pi \min\left\{R_p, R_v\right\}^2} \cdot \\ &\quad \frac{\left(\lambda_c \pi \min\left\{R_p, R_v\right\}^2\right)^k}{k!} \prod_{j=1}^{k} P_m^j. \end{aligned} \tag{7.14}$$

We observe that for a given constraint on the service range of an SBS, the service probability increases with R_p till it reaches the service range limit R_v, then it decreases monotonically with R_p.

Sampling Cost Analysis. The regular content sampling allows the SBSs to learn the interests of local users, thus the service offered by small cell cachers can be better adapted to the local request pattern. On the other hand, the power consumption for the sampling procedure must be taken into consideration for energy efficiency reasons. Denote by C_s the sampling cost of the typical SBS at the origin with sampling region radius R_p, we introduce a cost function for the sampling process, which depends on the distance from the sampled users to the SBS, i.e.:

$$C_s = \sum_{j \in \Phi_u \cap \mathcal{B}(0,R_p)} f(\| x_j - Y_0 \|) \cdot \eta \tag{7.15}$$

where $\| x_j - Y_0 \|$ is the Euclidean distance between the j-th user in the sampling region and the typical SBS at the origin, f is a distance-dependent cost function, and η is a constant cost factor. For exposition convenience, we choose $\eta = 1$ and inverse pathloss model for f as $f(x) = x^\alpha$ where α is the pathloss exponent.

From Campbell's theorem [40], we have the mean cost when averaging over different realizations of PPP as

$$\begin{aligned} \mathbb{E}[C_s] &= \mathbb{E}\left[\sum_{j \in \Phi_u \cap \mathcal{B}(0,R_p)} \| x_j - Y_0 \|^\alpha \right] \\ &= 2\pi\lambda_u \int_0^{R_p} r^\alpha \cdot r\mathrm{d}r \\ &= \frac{2\pi\lambda_u}{\alpha+2}\left(R_p^{\alpha+2} - 1\right). \end{aligned} \tag{7.16}$$

We see that the sampling cost increases linearly with the user density and follows power law of order $\alpha + 2$ with the sampling region radius. Note that in dense network deployments, larger sampling distance results in improved learning of the request pattern; however, the sampling cost may be prohibitively high and should be taken into account when designing the optimal sampling distance.

Optimal Sampling Range under Cost Constraint. Figure 7.4 shows an example of the service probability as a function of the sampling region radius when there are no constraint on the service range. We see that increasing sampling range gives higher service probability, but the increment tends to saturate when R_v is relatively large. By taking into consideration the limited service range while ignoring the sampling cost, the optimal sampling region size is equal to R_v since it gives the highest service probability. We now search for the optimal sampling range that provides sufficiently good service probability while assuring the sampling cost is below a given constraint.

We formulate the problem as follows:

$$R_p^* = \arg\max_{R_p} \; P_{sv} \tag{7.17}$$

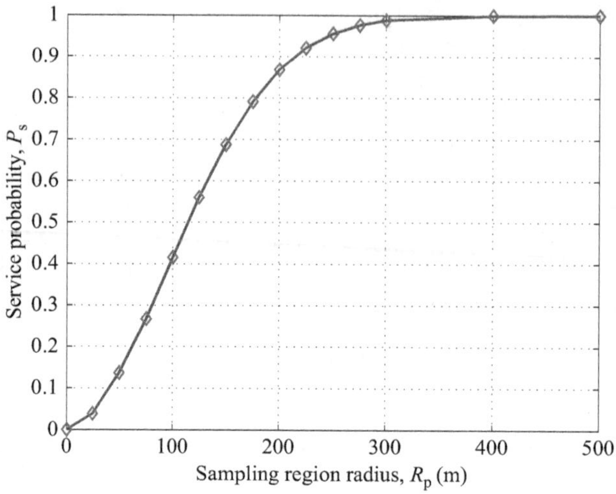

Figure 7.4 Service probability vs. sampling range with infinite service range. $q = 0$, $\gamma = 1$

Subject to:

$$R_p \in [0, R_v], \tag{7.18}$$

$$C_s(R_p^*) < C_{\max}, \tag{7.19}$$

where $C_{\max}$ is the predefined cost constraint. The first condition comes from the fact that when $R_p > R_v$, the service probability decreases monotonically with R_p. The second condition is due to the sampling cost constraint.

With the help of (7.16), we can derive the constraint on R_p to satisfy (7.19) as

$$R_p^* < \left(\frac{C_{\max}(\alpha + 2)}{2\pi\lambda_u + 1} \right)^{\frac{1}{\alpha+2}}. \tag{7.20}$$

Since P_{sv} is a monotonically increasing function of R_p for $R_p \in [0, R_v]$, combining these two constraints we have the optimal sampling distance given by

$$R_p^* = \min\left(R_v, \left(\frac{C_{\max}(\alpha + 2)}{2\pi\lambda_u + 1} \right)^{\frac{1}{\alpha+2}} \right).$$

7.3.1.3 Numerical results

We validate our analysis on cache-enabled small cell networks with local user interest sampling, and we show the trade-off between the small cell cache service probability and the sampling cost for different levels of local user interest correlation and user densities.

We set $\lambda_c = 2 \times 10^{-5}$ SBS/m^2 and $\lambda_u = 8 \times 10^{-4}$ users/m^2 as the density of SBSs and of users to be served. The service region of each SBS has radius

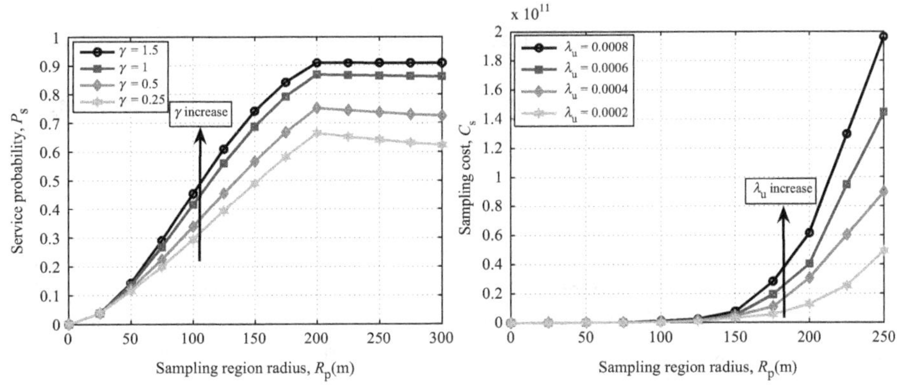

Figure 7.5 (Left) Service probability under different levels of user interest correlation. (Right) Sampling cost vs. sampling range with distance-dependent cost function

$R_v = 200$ m. The regular content library size of each user is $J = 10$. The maximum cache storage of an SBS is $M = 50$. We use $S = \min(JN, \lceil J(1 + 3\log N)\rceil)$ as the mapping function from the number of users to the overall local content library size. The popularity distribution is assumed to follow the Zipf-Mandelbrot law with $q = 0$. We assume the same shape parameter for the content popularity distribution in the sampling region of all the SBSs, meaning that $\gamma_j = \gamma$ for all j. Note that considering uniform or different shape parameters for the popularity distribution does not change the general trend of how the sampling distance affects the service probability P_{sv}. Numerical results are presented with different values of the shape parameter $\gamma \in [0.25, 1.5]$, which correspond to different user interest correlation levels, in order to see how the cache service performance scales with local user interest pattern.

In Figure 7.5(left) we plot the service probability P_{sv} as a function of the sampling range R_p for $\gamma = 0.25$, 0.5, 1, and 1.5, respectively. We see that P_{sv} increases with R_p before it reaches $R_v = 200$ m then decreases when R_p continues to grow. However, the speed of decrease is relatively small compared to the growth in $R_p \in [0, R_v]$, especially for higher values of γ. This is because higher γ corresponds to higher concentration of content popularity. Although increasing library size increases the number of files not being cached, the most popular files are cached with higher probability, which still leads to lower cache miss probability in the sampling region.

Figure 7.5(right) shows the simulated sampling cost per SBS as a function of the maximum sampling distance with the adopted distance-dependent cost function. The simulation results are obtained under different user density settings and by averaging over 5,000 PPP realizations. It shows the significant growth of the cost of local regular content sampling when increasing the sampling distance R_p. In a high user density scenario, the cost issue is of growing importance, showing the need for taking into consideration the sampling cost when searching for the optimal sampling distance.

To summarize, we investigated cache-enabled SCNs with local user interest sampling managed by SBSs for local request pattern learning. The main takeaway of this paper is the idea of local regular content sampling and the influence of the sampling distance on the cache service performance, which will be beneficial for designing the sampling procedure of cache-enabled SCNs.

7.3.2 D2D caching vs. small cell caching

In this section, we study the following question: given the storage capabilities of both SBS and mobile devices, where is it more efficient to cache the content? In other words, we would like to compare (to the extent possible) small-cell caching versus D2D caching. Although such comparison may be artificial or partially fair, being particularly sensitive to models and assumptions, some first conclusions can be drawn.

At first sight, D2D caching and SC caching may bring comparable gains in terms of network throughput, area spectral efficiency and energy efficiency. However, they are quite different in the following aspects:

- **Cache capacity**. Caching entities deployed in SBSs can have very large storage capability thanks to the low cost of storage units. In contrast, user devices, such as cellphones and tablets, have relatively small storage capacity and can only serve a rather small amount of requests generated by devices in their proximity.
- **Transmit power and coverage**. User devices normally transmit with much less power (e.g., 23 dBm) than SBSs (e.g., 24–30 dBm), which in turn corresponds to smaller covering range. As a result, the success probabilities of a cache-assisted transmission in the above two cases may be different.
- **Density of cache-served requests**. D2D communication usually involves small transmission distances. In networks with high user density, in the case of D2D caching, more simultaneous links are allowed to coexist in the same region sharing the same spectrum resources as compared to the case of SC caching. Moreover, a special case in D2D caching is when a user finds its requested file stored in its own device. In that case, the user request is satisfied without any delay and cost.
- **Power consumption**. In addition to the power consumption of devices or SBS, the power cost from fetching the content via the backhaul (in SC caching) can be significant. When a user request cannot be satisfied locally, either through D2D communication or from the SBS caches, the requested content will be retrieved from the core network through the small cell backhaul. In general the backhaul power consumption is much higher compared to the transmission power cost.

In this chapter, we take into account the aforementioned differences and compare the performance of caching in user devices (i.e., D2D caching) and caching in the SBSs (i.e., SC caching), in random topologies. We provide analytical expressions for key performance metrics, including the cache hit probability, the density of cache-served requests and the average power consumption.

Our results reveal that the performance of cache-enabled networks with either D2D caching or small cell caching heavily depends on the user density and the content popularity distribution. The presented analysis is a summary of [41], which has been extended in [42].

7.3.2.1 Network model

We consider a small cell network (SCN) where the SBSs are distributed in the two-dimensional Euclidean plane $\mathbb{R}^2$ according to a homogeneous PPP $\mathbf{\Phi}_\mathrm{s}$ with intensity λ_s. Mobile users are distributed according to another independent homogeneous PPP $\mathbf{\Phi}_\mathrm{u}$ with intensity λ_u [39]. Caching capabilities can be enabled either on user devices for potential D2D communication, referred to as D2D caching, or by installing storage units at the SBSs, coined as SC caching.

Each mobile user makes a random request with probability $\rho \in [0, 1]$. As a result, the active users to be served form a homogeneous PPP $\mathbf{\Phi}_\mathrm{u}^\mathrm{r}$ with intensity $\rho\lambda_\mathrm{u}$ (independent thinning). The inactive users form another homogeneous PPP $\mathbf{\Phi}_\mathrm{u}^\mathrm{t}$ with intensity $(1-\rho)\lambda_\mathrm{u}$. They can serve as potential D2D transmitters in the case with D2D caching mode, or remain silent if D2D communication is not enabled.

Depending on whether cache capability is enabled at the devices or at the edge/SBSs, when an active user requests for a file, the following cases may happen:

- with only D2D caching, if the requested file is not cached in its own device, the user searches for the file in the devices in its proximity within a certain distance. If there is at least one potential D2D transmitter that has the requested file cached, the file is transmitted from the nearest one. Otherwise, the user connects to the nearest SBS in order to download the file from the core network through the backhaul;
- with only SC caching, the active user always connects to the nearest SBS. If its associated SBS has the file cached inside, the SBS transmits the file directly to the user. Otherwise the file is downloaded from the core network through the backhaul and then transmitted to the user.

We assume spectrum sharing among concurrent transmissions in such network, i.e., both D2D and small cell communication links receive interference from coexisting transmitters.

We consider a finite content library $\mathscr{F} = \{f_1, \ldots f_N\}$ for the user requests, where f_i is the i-th most popular file and N is the library size. All files are assumed to have equal size, which is normalized to one. We use the standard Zipf law for the popularity distribution, meaning that the request probability of the i-th most popular file is

$$p_i = \frac{\Omega}{i^\gamma}, \tag{7.21}$$

where $\Omega = \left(\sum_{j=1}^{N} j^{-\gamma}\right)^{-1}$ is the normalization factor and γ is the shape parameter of Zipf law, which defines the correlation level of user requests. High values of γ means that most of the requests are generated from a few most popular files. For a user making a random request, p_i can be seen as the probability that the requested file is f_i.

Given knowledge of the content popularity distribution, depending on whether D2D caching or SC caching is adopted, we apply the following caching policies:

- with only D2D caching, since a random active user will most likely have multiple potential D2D transmitters, each device will independently cache files subject to its capacity-limited storage according to a common probability distribution, in order to increase the content diversity within the search distance [34]. The optimal caching probabilities are determined by minimizing the average *cache miss* probability;
- with only SC caching, since an active user always connects to the nearest SBS, there are no overlapping coverage areas of different SBSs. Thus, we apply the conventional "cache the most popular content" (MPC) policy, meaning that all SBSs cache the same most popular files within their cache capacity.

D2D Caching with Probabilistic Cache Placement. Denote by M_{d} the cache storage size of user devices, where with a random D2D caching policy, each user stores f_i with probability q_i. Let R_{d} be the maximum search/discovery distance of a user device for establishing D2D communication. The probability that no potential D2D transmitters are found when f_i is being requested is equivalent to the probability of having no points from $\mathbf{\Phi}_{\mathrm{u}}^{\mathrm{t}}$ with independent thinning probability q_i in the searching area (void probability). The cache mis-probability of f_i within the discovery distance R_{d} is given by

$$p_m^i(R_{\mathrm{d}}) = e^{-\pi\left(1-\rho\right)\lambda_{\mathrm{u}} q_i R_{\mathrm{d}}^2}. \tag{7.22}$$

The optimal caching probability vector $\mathbf{q} = [q_1, \ldots, q_N]$ for minimizing the average cache miss probability over $\mathscr{F}$ can be found by solving the following optimization problem:

$$\min_{\mathbf{q}} \sum_{i=1}^{N} p_i e^{-\pi\left(1-\rho\right)\lambda_{\mathrm{u}} q_i R_{\mathrm{d}}^2} \tag{7.23}$$

subject to:

$$\sum_{i=1}^{N} q_i - M_{\mathrm{d}} \leq 0 \tag{7.24}$$

$$q_i \in [0, 1]. \tag{7.25}$$

We use the optimal dual-solution searching (ODSA) algorithm proposed in [36] in order to find the optimal $\mathbf{q}$. In the remainder of the paper, it is assumed that all the results related to D2D caching are obtained using the optimal caching probabilities.

SC Caching with MPC Policy Denote by M_{s} the cache capacity of an SBS. According to the MPC caching policy, only files f_i with popularity order $i \in [1, M_{\mathrm{s}}]$ would be cached in each SBS.

7.3.2.2 Performance analysis

The potential gain of wireless content caching is mainly captured by the cache hit probability, which gives opportunity to handle user requests and to deliver content without having to retrieve it from the core network. Furthermore, there are potential spatial reuse gains by establishing proximity-based cache-assisted communication links sharing the same spectrum. In this section, we provide analytical results on several key performance metrics for cache-enabled cellular networks with either D2D caching or SC caching.

Cache hit probability

The cache hit probability is the probability for a random active user to find its requested file in local caches.

D2D caching

With D2D caching, a cache hit request may happen in two cases:

- a user requesting for a file may find it get stored in its own cache, we call this "self-request";
- when the requested file is not cached in its own device, the user finds it in the cache space of its nearby potential D2D transmitters within a distance R_d.

Denoting by $p^\mathrm{d}_\mathrm{self}$ the self-request probability of a random user, we have:

$$p^\mathrm{d}_\mathrm{self} = \sum_{i=1}^{N} p_i q_i. \tag{7.26}$$

The probability of an active user being served by a nearby potential D2D transmitter is given by

$$p^\mathrm{d}_\mathrm{r} = \sum_{i=1}^{N} p_i(1-q_i)\left(1 - e^{-\pi(1-\rho)\lambda_\mathrm{u} q_i R_\mathrm{d}^2}\right) \tag{7.27}$$

where $p^\mathrm{d}_{\mathrm{hit},i} = 1 - e^{-\pi(1-\rho)\lambda_\mathrm{u} q_i R_d^2}$ is the probability to have at least one potential D2D transmitters within distance R_d with file f_i cached.

Therefore, the cache hit probability is the sum of (7.26) and (7.27), given by

$$p^\mathrm{d}_\mathrm{hit} = p^\mathrm{d}_\mathrm{self} + p^\mathrm{d}_\mathrm{r} = 1 - \sum_{i=1}^{N} p_i(1-q_i)e^{-\pi(1-\rho)\lambda_\mathrm{u} q_i R_\mathrm{d}^2}. \tag{7.28}$$

SC Caching The cache hit probability in the case with SC caching is simply the probability that a user finds its requested file stored in the cache of its associated SBS. According to the MPC caching policy, we have that:

$$p^\mathrm{s}_\mathrm{hit} = \sum_{i=1}^{M_\mathrm{s}} p_i. \tag{7.29}$$

When an active user finds its requested file cached in either the devices in proximity or the nearest SBS, it is not guaranteed that the cache-assisted transmission of the file will be successful. We calculate here the success probability of a typical cache-assisted transmission conditioning on having a receiver at the origin.

For a given realization of the network, we assume having K established cache-assisted communication links with $\mathbb{E}[K] \leq \rho\lambda_{\mathrm{u}}$. For a random link $i \in [1, K]$, the received signal-to-interference-plus-noise ratio (SINR) is given by

$$\mathrm{SINR}_i = \frac{P_i |h_{i,i}|^2 d_{i,i}^{-\alpha}}{\sigma^2 + \sum_{j \in T \backslash \{i\}} P_j |h_{j,i}|^2 d_{j,i}^{-\alpha}}$$

where $P_i = \{P_{\mathrm{d}}, P_{\mathrm{s}}\}$ denotes the transmit power of either a D2D transmitter or an SBS, depending on whether caching capabilities are enabled on the mobile devices or at the SBSs; $h_{j,i}$ denotes the small-scale channel fading from the transmitter j to the receiver i, which follows $\mathcal{CN}(0, 1)$ (Rayleigh fading); $d_{j,i}$ denotes the distance between the transmitter j and the receiver i; σ^2 denotes the background noise power. Note that interference comes not only from cache-assisted transmissions, but also from the SBS-user links when the requested file of the user is not locally cached. Thus, in the D2D caching case, the set of active transmitters belongs to $T \subseteq \{\mathbf{\Phi}_{\mathrm{s}} \cup \mathbf{\Phi}_{\mathrm{u}}^{\mathrm{t}}\}$. In the SC caching case, $T \subseteq \mathbf{\Phi}_{\mathrm{s}}$, because active users can only be served by the SBSs. We assume interference-limited network, in which the background thermal noise is negligible, and the metric of interest is the signal-to-interference ratio (SIR).

D2D Caching. For a random active user requesting for a file, as given in (7.27), with probability $p_{\mathrm{r}}^{\mathrm{d}}$ the requested file is not cached in its own device, but in its nearby devices. Therefore, the density of cache-assisted communication links is $\rho\lambda_{\mathrm{u}}p_{\mathrm{r}}^{\mathrm{d}}$. Note that multiple users might find the same nearest D2D transmitter. In that case only one user can connect to this device, others have to search for another D2D transmitter. Denote by $\mathbf{\Phi}_{\mathrm{t}}^{\mathrm{d}}$ the set of active D2D transmitters; although the resulting set is not distributed according to a homogeneous PPP, the density of $\mathbf{\Phi}_{\mathrm{t}}^{\mathrm{d}}$ is given by

$$\lambda_{\mathrm{t}}^{\mathrm{d}} = \rho\lambda_{\mathrm{u}}p_{\mathrm{r}}^{\mathrm{d}}. \tag{7.30}$$

The set of users that cannot find their requested files in local caches will be served by the SBSs. The density of users to be served by the SBSs is $\lambda_{\mathrm{r}}^{\mathrm{s}} = \rho\lambda_{\mathrm{u}}(1 - p_{\mathrm{hit}}^{\mathrm{d}})$. According to the nearest SBS association, a Poisson–Voronoi tessellation is generated. From [43], the void probability of a typical Voronoi cell can be approximated as

$$p_{\mathrm{void}} \simeq \left(1 + \frac{\lambda_{\mathrm{r}}^{\mathrm{s}}}{3.5\lambda_{\mathrm{s}}}\right)^{-3.5}. \tag{7.31}$$

The density of active SBSs is thus given by

$$\begin{aligned} \lambda_{\mathrm{t}}^{\mathrm{s}} &= \lambda_{\mathrm{s}}(1 - p_{\mathrm{void}}) \\ &\simeq \lambda_{\mathrm{s}}\left(1 - \left(1 + \frac{\rho\lambda_{\mathrm{u}}(1 - p_{\mathrm{hit}}^{\mathrm{d}})}{3.5\lambda_{\mathrm{s}}}\right)^{-3.5}\right). \end{aligned} \tag{7.32}$$

Conditioning on having a typical D2D receiver at the origin with its associated transmitter at distance d_x, and assuming a homogeneous PPP for both active D2D transmitters $\Phi_{\rm t}^{\rm d}$ and active SBSs $\Phi_{\rm t}^{\rm s}$, the success probability is given as [44]

$$
\begin{aligned}
p_{\rm suc}^{\rm d} &= \mathbb{P}\left[\frac{P_{\rm d}|h_{i,i}|^2 d_x^{-\alpha}}{\sum_{j\in\Phi_{\rm t}^{\rm d}\backslash\{i\}} P_{\rm d}|h_{j,i}|^2 d_{j,i}^{-\alpha} + \sum_{k\in\Phi_{\rm t}^{\rm s}} P_{\rm s}|h_{k,i}|^2 d_{k,i}^{-\alpha}} > \theta\right] \\
&= \mathbb{E}\left[\mathscr{L}_{I_{\rm d}}\left(\theta d_x^{\alpha}\right)\cdot \mathscr{L}_{I_{\rm s}}\left(\theta \frac{P_{\rm s}}{P_{\rm d}} d_x^{\alpha}\right)\right] \qquad (7.33)\\
&= \mathbb{E}\left[\exp\left(-\frac{\pi d_x^2 \theta^{\frac{2}{\alpha}}}{\mathrm{sinc}\,(2/\alpha)}\left(\lambda_{\rm t}^{\rm d} + \lambda_{\rm t}^{\rm s}(P_{\rm s}/P_{\rm d})^{\frac{2}{\alpha}}\right)\right)\right]
\end{aligned}
$$

where $\mathscr{L}_{I_{\rm x}}(s) = \mathbb{E}[\exp(-sI_{\rm x})]$ is the Laplace transform of interference $I_{\rm x}$ and θ is the SIR threshold for successful D2D transmission. The expectation is over the distribution of d_x and over the content library $\mathscr{F}$.

When a cache hit occurs, the distribution of the D2D link distance d_x depends on the popularity order of the requested file. If f_i is requested by the typical user, conditioning on having at least one potential D2D transmitter within distance $R_{\rm d}$, the pdf of the D2D link distance d_x is given by

$$
f_{d_x}^i(r) = \begin{cases} \dfrac{2\pi(1-\rho)\lambda_{\rm u} q_i r}{1-e^{-\pi(1-\rho)\lambda_{\rm u} q_i R_{\rm d}^2}} e^{-\pi(1-\rho)\lambda_{\rm u} q_i r^2} & 0 \le r \le R_{\rm d} \\ 0 & r > R_{\rm d}. \end{cases} \qquad (7.34)
$$

Then, the approximate success probability when a cache hit of file f_i happens, is given by

$$
p_{\mathrm{suc},i}^{\rm d} = \int_0^{R_{\rm d}} \exp\left[-\frac{\pi r^2 \theta^{\frac{2}{\alpha}}}{\mathrm{sinc}\,(2/\alpha)}\left(\rho\lambda_{\rm u} p_{\rm r}^{\rm d} + \lambda_{\rm t}^{\rm s}(P_{\rm s}/P_{\rm d})^{\frac{2}{\alpha}}\right)\right] \qquad (7.35)
$$

$$
\cdot \frac{2\pi(1-\rho)\lambda_{\rm u} q_i r}{1-e^{-\pi(1-\rho)\lambda_{\rm u} q_i R_{\rm d}^2}} e^{-\pi(1-\rho)\lambda_{\rm u} q_i r^2}\,\mathrm{d}r, \qquad (7.36)
$$

where $p_{\rm r}^{\rm d}$ and $\lambda_{\rm t}^{\rm s}$ are given in (7.27) and (7.32), respectively.

SC Caching. In the case with SC caching, users are always connected to the nearest SBSs in both cache hit and cache miss events. Denote by $\tilde{\Phi}_{\rm t}^{\rm s}$ the set of active SBSs, similarly, using the void probability of Voronoi cell in (7.31), the density of active SBSs is given by

$$
\tilde{\lambda}_{\rm t}^{\rm s} \simeq \lambda_{\rm s}\left(1-\left(1+\frac{\rho\lambda_{\rm u}}{3.5\lambda_{\rm s}}\right)^{-3.5}\right). \qquad (7.37)
$$

Conditioning on having the typical receiver at the origin with its associated SBS at distance d_s and using nearest SBS association, we have the pdf of d_s given as

$$f_{d_s}(r) = 2\pi\lambda_{\mathrm{s}} r \cdot e^{-\pi\lambda_{\mathrm{s}} r}. \tag{7.38}$$

For a given SIR threshold θ, the success probability of the cache-assisted small cell transmission is given by

$$\begin{aligned} p_{\mathrm{suc}}^{\mathrm{s}} &= \mathbb{P}\left[\frac{P_{\mathrm{s}}|h_{i,i}|^2 d_s^{-\alpha}}{\sum_{k\in\tilde{\Phi}_{\mathrm{t}}^{\mathrm{s}}\backslash\{i\}} P_{\mathrm{s}}|h_{k,i}|^2 d_{k,i}^{-\alpha}} > \theta\right] \\ &= \int_0^{\infty} 2\pi\lambda_{\mathrm{s}} r \cdot e^{-\pi\lambda_{\mathrm{s}} r^2} \exp\left(-\frac{\pi\tilde{\lambda}_{\mathrm{t}}^{\mathrm{s}} r^2 \theta^{\frac{2}{\alpha}}}{\mathrm{sinc}\,(2/\alpha)}\right) \mathrm{d}r, \end{aligned} \tag{7.39}$$

where $\tilde{\lambda}_{\mathrm{t}}^{\mathrm{s}}$ is given in (7.37).

Density of cache-served requests

A user request is said to be "served" if the requested file is found in local caches and if the transmission of the file is successful. Based on the above results, we calculate the density of cache-served requests, which is the average number of requests that can be successfully and simultaneously handled by the local cache per unit area.

D2D Caching. In the D2D caching case, a random user request can be served either by self-request or through proximal D2D communication. Denote by $\mu_{\mathrm{suc}}^{\mathrm{d}}$ the density of cache-served requests, we have:

$$\begin{aligned} \mu_{\mathrm{suc}}^{\mathrm{d}} &= \rho\lambda_u \left(p_{\mathrm{self}}^{\mathrm{d}} + \sum_{i=1}^{N} p_i (1 - q_i) p_{\mathrm{hit},i}^{\mathrm{d}} p_{\mathrm{suc,i}}^{\mathrm{d}} \right) \\ &= \rho\lambda_u \sum_{i=1}^{N} p_i \left[q_i + (1 - q_i)\left(1 - e^{-\pi(1-\rho)\lambda_{\mathrm{u}} q_i R_d^2}\right) p_{\mathrm{suc,i}}^{\mathrm{d}} \right], \end{aligned} \tag{7.40}$$

where $p_{\mathrm{suc,i}}^{\mathrm{d}}$ is given in (7.36).

SC Caching. In the SC caching case, the maximum number of cache-assisted transmissions in a given time slot is limited by the density of the SBSs. The probability for an SBS to have at least one active user in its cell requesting for files that are stored in its cache is given by

$$p_{\mathrm{t}}^{\mathrm{s}} = 1 - \left(1 + \frac{\rho\lambda_{\mathrm{u}} p_{\mathrm{hit}}^{\mathrm{s}}}{3.5\lambda_{\mathrm{s}}}\right)^{-3.5}. \tag{7.41}$$

The density of cache-assisted transmission is $\lambda_s p_t^s$. Then, the density of cache-served requests, which is the density of successful cache-served small cell transmission, is given by

$$\mu_{suc}^s = \lambda_s p_t^s p_{suc}^s = \lambda_s \left(1 - \left(1 + \frac{\rho\lambda_u p_{hit}^s}{3.5\lambda_s}\right)^{-3.5}\right) p_{suc}^s \tag{7.42}$$

where p_{suc}^s is given in (7.39).

7.3.2.3 Power consumption

For a random user request, in a cache hit event, the consumed power for content delivery contains only the transmit power of either the D2D transmitter or the associated SBS. In a cache miss event, the requested file is first fetched from the core network via the backhaul and then transmitted from the nearest SBS to the user. Thus, additional energy is consumed at the backhaul. Denote P_b the backhaul power consumption required to handle a user request at a single SBS; we study below the power consumption per user request with either D2D caching or SC caching.

For the case with D2D caching, recall that when self-request occurs, no energy is consumed to serve the request. A random user request has probability p_r^d to be served by a nearby D2D transmitter, and probability $1 - p_{hit}^d$ to be served by the nearest SBS. The consumed power per user request is given as

$$P_{avg}^d = p_r^d P_d + \left(1 - p_{hit}^d\right)(P_s + P_b). \tag{7.43}$$

For the case with SC caching, when a random user request occurs, the transmission power of its nearest SBS is always consumed for both cache hit and cache miss events. The backhaul power is additionally consumed with probability $1 - p_{hit}^s$. Thus we have:

$$P_{avg}^s = P_s + \left(1 - p_{hit}^s\right)P_b. \tag{7.44}$$

7.3.2.4 Numerical results

For numerical evaluation, we set $\lambda_s = 10^{-5}$ SBS/m^2 and $\lambda_s = K \times 10^{-4}$ users/m^2, where K is a proportion factor. We choose $\rho = 0.2$, meaning that 20% of the users will request for a file in the content library $\mathscr{F}$. The cache storage size at an SBS and at a user device are $M_s = 10^4$ and $M_d = 10$, respectively. The content library has size $N = 10^5$, with popularity distribution given by Zipf law with shape parameter $\gamma = 0.7$ for the case with low popularity skewness, and $\gamma = 1.2$ for the other case. The searching distance of a user device to establish D2D link is $R_d = 75$ m. The transmit power of an SBS and a user device are $P_s = 100$ mW and $P_d = 2$ mW, respectively. Backhaul power consumption at the SBS to handle a user request is $P_b = 10$ W. The target SIR of successful transmissions is chosen as $\theta = 0$ dB.

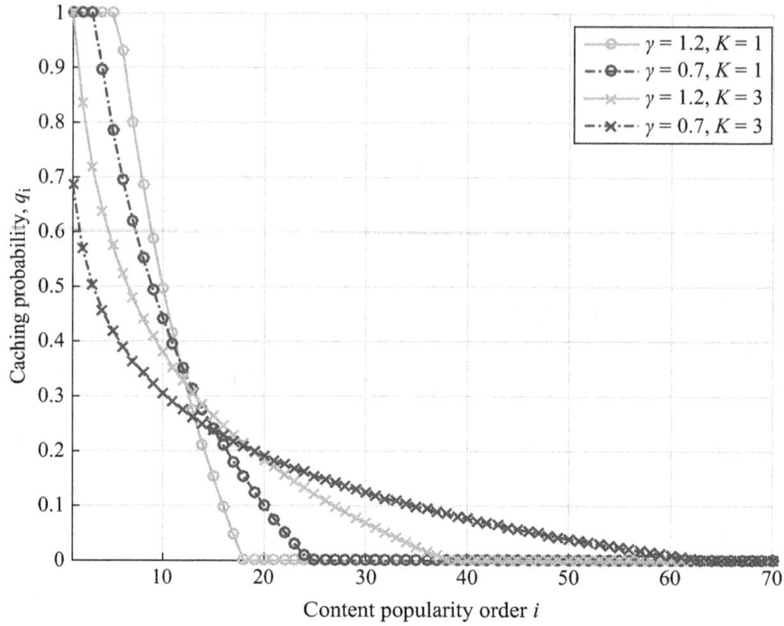

Figure 7.6 Optimal caching probability vector $\boldsymbol{q} = [q_1, \ldots, q_N]$

We present numerical results for $K \in [1, 9]$, in order to compare the performance between caching at the user devices and caching at the SBSs for different user density regimes.

In Figure 7.6, we plot the optimal caching probability q_i of file f_i, $i \in [1, N]$, obtained by solving the optimization problem in (7.23). Comparing the results with $\gamma = 1.2$ and with $\gamma = 0.7$, we see that for lower γ, users tend to cache more different files, because the user requests are more diverse due to the low popularity concentration level. We observe the same trend when user density is higher, e.g., $K = 3$. This is reasonable because with higher user density the probability to establish D2D communication is higher. Thus, more different files should be cached in user devices in order to serve more requests by cache-assisted D2D transmission.

Figure 7.7 shows the cache hit probability obtained with (7.28) and (7.29) for the case with D2D caching and with SC caching, respectively. As expected, the cache hit probability for both cases are higher with lower λ_s Furthermore, we see that caching at the SBSs results in much higher cache hit probability than caching at the mobile devices, as a result of the larger cache capacity. The density of cache-served requests measures how many requests can be successfully handled simultaneously using the local caches. From Figure 7.8, we see that D2D caching outperforms SC caching for higher γ, especially in the high user density regime. In the case with lower γ, the performance of SC caching is slightly better in the

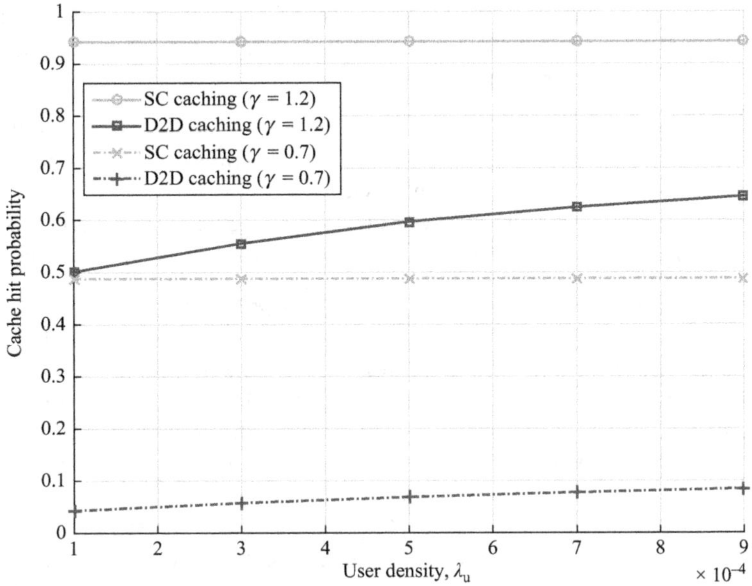

Figure 7.7 Cache hit probability vs. user density λ_s

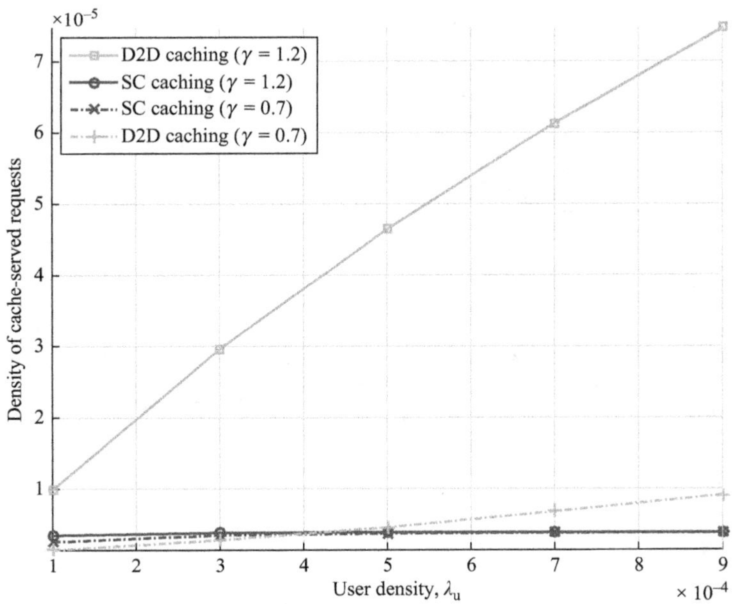

Figure 7.8 Density of cache-served requests vs. user density λ_s

sparse user regime, whereas the performance of D2D caching outperforms SC caching when the user density increases. The advantage of D2D caching is mainly due to the following two reasons:

- when the user density is high, the number of potential D2D transmitters are in general larger than the number of SBSs, which allows to have more simultaneous cache-assisted transmission links;
- self-request of D2D caching gives opportunity to handle a large amount of requests made by the users when the content popularity is highly concentrated.

Figure 7.9 plots the average power consumption per user request, showing that SC caching is expectedly more energy efficient than the case with D2D caching, which is mainly due to its high cache hit probability. Compared to the consumed power for fetching the content through the backhaul, the transmit power of both user devices and SBSs are relatively low. Hence, the higher the probability to serve a user request locally, the less power is needed. We also observe that the power consumption per user request with D2D caching decreases with the user density. This is because the number of potential D2D transmitters within the discovery distance of a user increases when the user density increases, thus giving higher probability to serve the user request by cache-assisted D2D transmission.

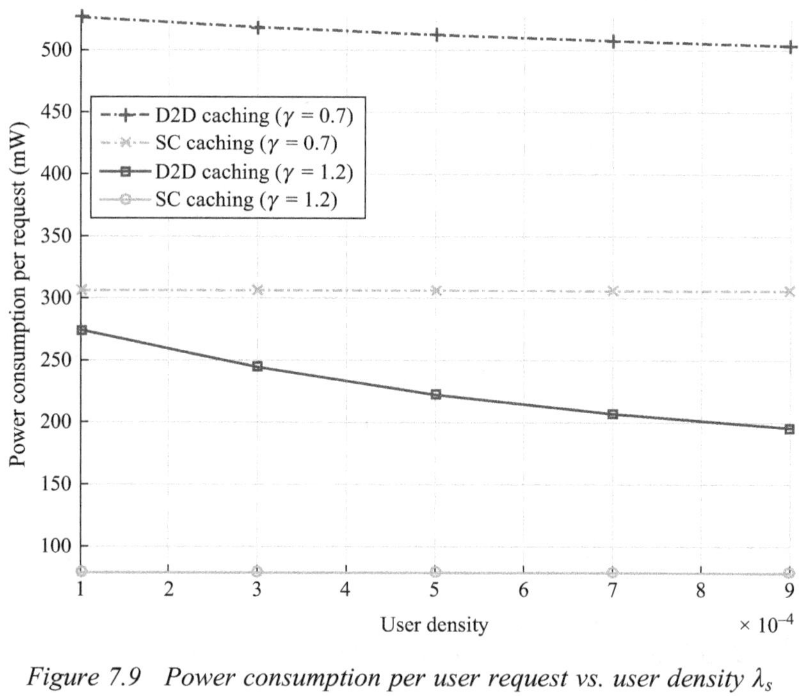

Figure 7.9 Power consumption per user request vs. user density λ_s

Combining these results, we have the following takeaway messages:

- in networks with high user density, D2D caching gives the opportunity to serve more user requests simultaneously through short-distance cache-assisted D2D communication;
- edge (SBSs) caching results in much higher cache hit probability because storage units at the SBSs have much larger capacity than at the mobile devices. As a result, SC caching is also more energy efficient because less power is consumed on average at the backhaul in order to download a file from the core network.

7.3.2.5 Discussion

We compared the performance of D2D caching and small cell caching in a spatial wireless network. The main takeaway message is that in wireless networks with high user density, it is beneficial in terms of the spatial reuse of communication resources to perform caching at the mobile devices and to benefit from proximal D2D communication. On the other hand, deploying large-capacity storage units for caching at the edge (i.e., in small cells) is more advantageous in terms of energy efficiency, as a result of the high cache hit probability in such case.

7.4 Research challenges for wireless caching

In this chapter we gave mathematical evidence of the caching benefits for the operation of future wireless networks. We complete the chapter by briefly summarizing the most challenging problems and the most interesting future directions of research for caching in wireless networks. The interested reader is also referred to [3,4,45].

7.4.1 Modelling the requests in wireless networks

The standard model for performance analysis of web caching is the Independence Reference Model (IRM): content m is requested according to an independent Poisson process with rate λp_m, where p_m refers to the content popularity, modelled by a power law, i.e., $p_m \propto m^{-\tau}$. This well-established model thrives due to its simplicity; it only has two parameters, namely λ to control the rate of requests, and τ to control the skew of the popularity.

The IRM assumes that the content popularity is static, which is of course not true. Breaking news, and the episodes of TV series are examples of ephemeral content with rapidly changing popularity; they appear, they become increasingly popular, and they gradually become unpopular again. Recent works [46,47] propose novel mathematical tools to model time-varying content popularity. The studies consider large samples of YouTube and VoD applications and discover: content popularity is time-varying and in addition has a consistent effect on analysing the caching performance. In the inhomogeneous Poisson model proposed in [46] each content is associated to a shot whose duration reflects the content lifespan and whose height denotes its instantaneous popularity. The model is called the Shot Noise Model (SNM), mirroring the Poisson noise from electronics. While the shape of the pulse is not important, the study [47] observes strong correlations between popularity and duration; apparently popular contents prosper longer. Finally, a class-based model [46] can conveniently capture spatio-temporal correlations while allowing analytical tractability. Mobile users are profound downloaders of ephemeral content, thus it is expected that a similar analysis for the case of wireless content will show stronger effects. Additionally, the time variations require perpetual popularity estimation, which is harder in wireless networks since the caches are used by smaller populations and hence "see" a sparser request sequence [4,48].

In fact, the concept of time-varying popularity is not entirely new to caching engineers. Older analyses of caching systems refer to it with the term "temporal locality," where bursts of requests occur for certain contents. Currently, web caching systems operate with dynamic eviction policies like Least-Recently-Used (LRU) and Least-Frequently-Used (LFU) which aim to capture temporal locality in a heuristic manner. The performance of LRU under the SNM model is analysed in [49]. A recent study [48] analyses the SNM model and gives the optimal policy for joint caching and popularity estimation. The result shows that (LFU-like) frequency estimation is optimal for contents that exist in the system for a while and have already received a few requests. Continuously estimating the popularity is more difficult in wireless systems where the requests of the population are split over multiple caches and those that reach any given cache are less. This development motivates novel caching techniques that employ learning methodologies to accurately track the evolution of content popularity over time.

7.4.2 Coded caching for broadcast medium exploitation

The fundamental work of Maddah-Ali and Niesen [50] proposed the use of caching to improve the efficiency of wireless transmissions in a cell. While traditional techniques cache entire contents at mobile terminals to eliminate retransmissions of popular content, the coded caching technique stores coded combinations of content chunks. Then, the BS can broadcast a coded sequence of messages so that all receivers in the group can combine it with the cached chunks to decode the contents. This is shown to yield resource blocks equal to $N(1 - K/M)/(1 + NK/M)$, where N is the number of users, K the cache size, and M the catalogue size. Hence, if the cacheable fraction of the catalogue K/M is kept fixed, then the required number of resource blocks does not increase with the number of users N, this can be verified by taking the limit $N \to \infty$ whereby the above quantity converges to a constant. In other words, in a scenario of a wireless BS with N users watching individual content, the result shows that we may increase N indefinitely while maintaining a fixed service quality.

This revolutionary idea has sprung a wealth of research efforts, such as device-to-device (D2D) networks [30,51], non-uniform content popularities [52], online caching policies [53], multi-servers [54], multi-library [55], and combination with CSI [56]. From the implementation point of view, promising research directions include extensions to capture system aspects such as (i) popularity skew,

(ii) asynchronous requests, (iii) finite code lengths, and (iv) cache sizes that scale slower than M. Assuming that these practical challenges are resolved, caching for wireless systems will become intertwined with physical layer techniques employed at the BS and the handheld.

7.4.3 Cooperative caching models

In the context of heterogeneous wireless networks, a user can connect in a variety of ways to the network in order to obtain the requested content. For example, contemporary smart phones receive the signal of more than 10 BSs simultaneously. In future densified cellular networks, the mobile will be connected to several femto-, pico-, or nano-cells. This can be exploited by cooperative caching techniques [32]. Since a user can retrieve the requested content from many network endpoints, neighbouring caches should collaborate and avoid storing the same objects multiple times [57].

Most content placement optimizations of wireless caching boil down to a set cover problem in a bipartite graph connecting the users to the reachable caches. Therefore, finding what contents to store at each cache is a difficult problem even if the popularities are assumed known [32]. It is possible to relax the problem to convex optimization by the use of distributed storage codes, where each cache stores coded combinations of contents [32]. These ideas lead to several interesting algorithms in the literature of multi-access caching [58–61].

7.4.4 Advanced spatial models

Stochastic geometry and point process theory have been instrumental in characterizing the performance of large-scale wireless networks and understanding the effect of interference and random topology on various metrics, such as coverage, area spectral efficiency and hitting probability. PPP has been widely used to model the network entities, mainly due to its tractability. Although PPP provides a good approximation for the interference associated with repulsive point processes, network topologies, especially cache-enabled wireless networks, may include other complex interactions, topological structures and correlations among the network entities. Exploring new point processes, which balances a trade-off between tractability and practicality, for modelling cache-enabled wireless networks is a challenging yet relevant research direction. In this regard, there have been some recent first efforts to analyse SC and D2D caching using binomial point processes and clustered processes processes [62–64].

References

[1] Cisco, "Cisco visual networking index: Global mobile data traffic forecast update 2014–2019," White Paper, 2015. [Online]. Available: http://goo.gl/tZ6QMk

[2] P. Gupta and P. R. Kumar, "The capacity of wireless networks," *IEEE Transactions on Information Theory*, vol. 46, no. 2, pp. 388–404, Mar. 2000.

[3] X. Wang, M. Chen, T. Taleb, A. Ksentini, and V. Leung, "Cache in the air: exploiting content caching and delivery techniques for 5G systems," *IEEE Communications Magazine*, vol. 52, no. 2, pp. 131–139, Feb. 2014.

[4] G. S. Paschos, E. Bastug, I. Land, G. Caire, and M. Debbah, "Wireless caching: technical misconceptions and business barriers,"*IEEE Communications Magazine*, vol. 54, no. 8, pp. 16–22, Aug. 2016.

[5] M. Franceschetti and R. Meester, *Random Networks for Communication*. New York, NY: Cambridge University Press, Series: Cambridge Series in Statistical and Probabilistic Mathematics (No. 24), 2007.

[6] J. Roberts and N. Sbihi, "Exploring the memory-bandwidth tradeoff in an information-centric network," in *Proceedings of the 2013 25th International Teletraffic Congress (ITC)*, Shanghai, China, 2013, pp. 1–9.

[7] S. Gitzenis, G. S. Paschos, and L. Tassiulas, "Asymptotic laws for content replication and delivery in wireless networks," in *Proceedings of the 2012 IEEE INFOCOM conference on Computer Communications*, Orlando, FL, USA, Mar. 2012, pp. 126–134.

[8] S. Gitzenis, G. Paschos, and L. Tassiulas, "Asymptotic laws for joint content replication and delivery in wireless networks," *IEEE Transactions on Information Theory*, vol. 59, no. 5, pp. 2760–2776, 2013.

[9] G. S. Paschos, S. Gitzenis, and L. Tassiulas, "The effect of caching in sustainability of large wireless networks," in *Proceedings of the 10th International Symposium on Modeling and Optimization in Mobile, Ad Hoc and Wireless Networks (WiOpt 2012)*, Paderborn, Germany, May 2012, pp. 355–360.

[10] S. Gitzenis, G. S. Paschos, and L. Tassiulas, "Enhancing wireless networks with caching: asymptotic laws, sustainability & trade-offs," *Computer Networks*, vol. 64, pp. 353–368, May 2014.

[11] M. Faloutsos, P. Faloutsos, and C. Faloutsos, "On power-law relationships of the internet topology," *ACM SIGCOMM Computer Communication Review*, vol. 29, no. 4, pp. 251–262, 1999.

[12] G. Alfano, M. Garetto, and E. Leonardi, "Content-centric wireless networks with limited buffers: when mobility hurts", *IEEE/ACM Transactions on Networking*, vol. 24, no. 1, pp. 299–311, Feb. 2016.

[13] U. Niesen, D. Shah, and G. Wornell, "Caching in wireless networks," in *Proceedings IEEE International Symposium on Information Theory*, Seoul, Korea, Jun. 2009, pp. 2111–2115.

[14] C. Fricker, P. Robert, J. Roberts, and N. Sbihi, "Impact of traffic mix on caching performance in a content-centric network," in *Proceedings of IEEE INFOCOM Workshops*, Orlando, FL, 2012, pp. 310–315.

[15] T. Bektas, J.-F. Cordeau, E. Erkut, and G. Laporte, "Exact algorithms for the joint object placement and request routing problem in content distribution networks," *Computers & Operations Research*, vol. 35, no. 12, pp. 3860–3884, 2008.

[16] J. Munoz-Gea, S. Traverso, and E. Leonardi, "Modeling and evaluation of multisource streaming strategies in P2P VoD systems," *IEEE Transactions on Consumer Electronics*, vol. 58, no. 4, pp. 1202–1210, 2012.

[17] V. Lenders, G. Karlsson, and M. May, "Wireless ad hoc podcasting," in *Proceedings of the Fourth Annual IEEE Communications Society Conference on Sensor, Mesh and Ad Hoc Communications and Networks (SECON)*, San Diego, CA, pp. 273–283.

[18] N. Golrezaei, A. G. Dimakis, and A. F. Molisch, "Wireless device-to-device communications with distributed caching," in *Proceedings of the IEEE International Symposium on Information Theory (ISIT)*, Cambridge, MA, 2012, pp. 2781–2785.

[19] X. Jiang, P. Gao, Y. Zhao, and Y. Shi, "Caching scheme research based on unstructured peer-to-peer network," *Physics Procedia*, vol. 25, pp. 1076–1083, 2012.

[20] C. Liaskos, S. Petridou, G. Papadimitriou, P. Nicopolitidis, M. Obaidat, and A. Pomportsis, "A novel clustering-driven approach to wireless data broadcasting," in *15th Annual Symposium of the IEEE/CVT*, Antwerp, Belgium, Nov. 2008.

[21] C. R. Cunha, A. Bestavros, and M. E. Crovella, "Characteristics of WWW client-based traces," in *View on NCSTRL*, Boston University, MA, USA, Jul. 1995.

[22] L. Breslau, P. Cue, P. Cao, L. Fan, G. Phillips, and S. Shenker, "Web caching and Zipf-like distributions: evidence and implications," in *Proceedings of the 1999 IEEE INFOCOM conference on Computer Communications*, New York, NY, Mar. 1999, pp. 126–134.

[23] L. A. Adamic and B. A. Huberman, "Zipf's law and the Internet," *Glottometrics*, vol. 3, pp. 143–150, 2002.

[24] V. N. Padmanabhan and L. Qiu, "The content and access dynamics of a busy web site: findings and implications," *ACM SIGCOMM Computer Communication Review*, vol. 30, no. 4, pp. 111–123, 2000.

[25] T. Yamakami, "A Zipf-like distribution of popularity and hits in the mobile web pages with short life time," in *Proceedings of Parallel and Distributed Computing, Applications and Technologies*, PDCAT'06, Taipei, ROC, Dec. 2006, pp. 240–243.

[26] M. Cha, H. Kwak, P. Rodriguez, Y.-Y. Ahn, and S. Moon, "I tube, you tube, everybody tubes: analyzing the world's largest user generated content video system," in *Proceedings of the Seventh ACM SIGCOMM Conference on Internet Measurement*, ACM, New York, NY, USA, 2007, pp. 1–14.

[27] G. Dán and N. Carlsson, "Power-law revisited: large scale measurement study of P2P content popularity," in *Proceedings of the 9th international conference on Peer-to-peer systems (IPTPS'10), USENIX Association*, Berkeley, CA, USA, 2010, pp. 1–5.

[28] H. Yu, D. Zheng, B. Y. Zhao, and W. Zheng, "Understanding user behavior in large-scale video-on-demand systems," in *ACM SIGOPS Operating Systems Review*, vol. 40, no. 4, 2006, pp. 333–344.

[29] M. A. Maddah-Ali and U. Niesen, "Fundamental limits of caching," *IEEE Transactions on Information Theory*, vol. 60, no. 5, pp. 2856–2867, May 2014.

[30] M. Ji, G. Caire, and A. F. Molisch, "Fundamental limits of caching in wireless D2D networks," *IEEE Transactions on Information Theory*, vol. 62, no. 2, pp. 849–869, Feb. 2016.

[31] E. Bastug, M. Bennis, M. Kountouris, and M. Debbah, "Cache-enabled small cell networks: modeling and tradeoffs," *EURASIP Journal on Wireless Communication and Networking*, vol. 2015, no. 1, pp. 1–11, Feb. 2015.

[32] N. Golrezaei, K. Shanmugam, A. Dimakis, A. Molisch, and G. Caire, "Femtocaching: wireless video content delivery through distributed caching helpers," in *Proceedings of the IEEE INFOCOM*, Mar. 2012, pp. 1107–1115.

[33] M. S. ElBamby, M. Bennis, W. Saad, and M. Latva-aho, "Content-aware user clustering and caching in wireless small cell networks," in *IEEE International Symposium on Wireless Communications Systems (ISWCS)*, Barcelona, Spain, Aug. 2014, pp. 945–949.

[34] M. Ji, G. Caire, and A. Molisch, "Optimal throughput-outage trade-off in wireless one-hop caching networks," in *Proceedings of the IEEE International. Symposium on Information Theory (ISIT)*, Jul. 2013, pp. 1461–1465.

[35] N. Golrezaei, A. Dimakis, and A. Molisch, "Scaling behavior for device-to-device communications with distributed caching," *IEEE Transactions on Information Theory*, vol. 60, no. 7, pp. 4286–4298, Jul. 2014.

[36] H. Kang, K. Park, K. Cho, and C. Kang, "Mobile caching policies for device-to-device (D2D) content delivery networking," in *Proceedings of the IEEE Conference on Computer Commun. Workshops (INFOCOM WKSHPS)*, Toronto, ON, Apr. 2014, pp. 299–304.

[37] M. Ji, G. Caire, and A. Molisch, "Wireless device-to-device caching networks: basic principles and system performance," *IEEE Journal on Selected Areas in Communication*, vol. 34, no. 1, pp. 176–189, Jan. 2016.

[38] Z. Chen and M. Kountouris, "Cache-enabled small cell networks with local user interest correlation," in *IEEE Sixteenth International Workshop on Signal Processing Advances in Wireless Communications (SPAWC)*, Stockholm, 2015, pp. 680–684.

[39] A. Baddeley, *Spatial Point Processes and Their Applications*. Berlin: Springer Verlag, 2007, vol. 1892, pp. 1–75.

[40] F. Baccelli and B. Błaszczyszyn, "Stochastic Geometry and Wireless Networks: Volume I Theory", *Foundations and Trends® in Networking*, vol. 3, no. 3–4, pp. 249–449, 2010.

[41] Z. Chen and M. Kountouris, "D2D caching vs. small cell caching: Where to cache content in a wireless network?," in *IEEE Seventeenth International Workshop on Signal Processing Advances in Wireless Communications (SPAWC)*, Edinburgh, 2016, pp. 1–6.

[42] Z. Chen, N. Pappas, and M. Kountouris, "Probabilistic Caching in Wireless D2D Networks: Cache Hit Optimal Versus Throughput Optimal," *IEEE Communications Letters*, vol. 21, no. 3, pp. 584–587, Mar. 2017.

[43] S. Lee and K. Huang, "Coverage and economy of cellular networks with many base stations," *IEEE Communication Letters*, vol. 16, no. 7, pp. 1038–1040, Jul. 2012.

[44] M. Haenggi and R. K. Ganti, "Interference in large wireless networks," *Foundation and Trends in Networking*, vol. 3, no. 2, pp. 127–248, Feb. 2009.

[45] E. Baştuğ, M. Bennis, M. Kountouris, and M. Debbah, "Cache-enabled small cell networks: modeling and tradeoffs," *EURASIP Journal on Wireless Communications and Networking*, no. 1, p. 41, Feb. 2015.

[46] S. Traverso, M. Ahmed, M. Garetto, P. Giaccone, E. Leonardi, and S. Niccolini, "Temporal locality in today's content caching: why it matters and how to model it," *ACM SIGCOMM Computer Communication Review*, vol. 43, no. 5, pp. 5–12, Nov. 2013.

[47] F. Olmos, B. Kauffmann, A. Simonian, and Y. Carlinet, "Catalog dynamics: impact of content publishing and perishing on the performance of a LRU cache," in *Proceedings of the 26th International Teletraffic Congress (ITC)*, Karlskrona, 2014, pp. 1–9.

[48] M. Leconte, G. S. Paschos, L. Gkatzikis, M. Draief, S. Vassilaras, and S. Chouvardas, "Placing dynamic content in caches with small population," in *Proceedings of the IEEE INFOCOM 2016 – The 35th Annual IEEE International Conference on Computer Communications*, San Francisco, CA, 2016, pp. 1–9.

[49] E. Leonardi and G. L. Torrisi, "Least recently used caches under the shot noise mode," in *IEEE Conference on Computer Communications (INFOCOM)*, Kowloon, 2015, pp. 2281–2289.

[50] M. Maddah-Ali and U. Niesen, "Fundamental limits of caching," *IEEE Transactions on Information Theory*, vol. 60, no. 5, pp. 2856–2867, May 2014.

[51] S.-W. Jeon, S.-N. Hong, M. Ji, and G. Caire, "On the capacity of multihop device-to-device caching networks," in *Proceedings of the IEEE Information Theory Workshop (ITW)*, Jerusalem, 2015, pp. 1–5.

[52] M. Ji, A. M. Tulino, J. Llorca, and G. Caire, "Order-optimal rate of caching and coded multicasting with random demands," *IEEE Transactions on Information Theory*, April 2017, doi:10.1109/TIT.2017.2695611

[53] R. Pedarsani, M. A. Maddah-Ali, and U. Niesen, "Online coded caching," *IEEE/ACM Transactions on Networking*, vol. 24, no. 2, pp. 836–845, April 2016.

[54] S. P. Shariatpanahi, S. A. Motahari, and B. H. Khalaj, "Multi-server coded caching," *IEEE Transactions on Information Theory*, vol. 62, no. 12, pp. 7253–7271, Dec. 2016.

[55] S. Sahraei and M. Gastpar, "Multi-library coded caching," in *Proceedings of the 54th Annual Allerton Conference on Communication, Control, and Computing (Allerton)*, Monticello IL, 2016, pp. 1012–1017.

[56] J. Zhang and P. Elia, "Fundamental limits of cache-aided wireless BC: interplay of coded-caching and CSIT feedback," *IEEE Transactions on Information Theory*, Feb. 2017, 10.1109/TIT.2017.2674668.

[57] Z. Chen, J. Lee, T. Q. S. Quek, and M. Kountouris, "Cooperative Caching and Transmission Design in Cluster-Centric Small Cell Networks," *IEEE Transactions on Wireless Communications*, vol. 16, no. 5, pp. 3401–3415, May 2017.

[58] K. Poularakis, G. Iosifidis, and L. Tassiulas, "Approximation algorithms for mobile data caching in small cell networks," *IEEE Transactions on Communications*, vol. 62, no. 10, pp. 3665–3677, 2014.

[59] K. Naveen, L. Massoulie, E. Baccelli, A. C. Viana, and D. Towsley, "On the interaction between content caching and request assignment in cellular cache networks," in *Proceedings of the Fifth Workshop on All Things Cellular: Operations, Applications and Challenges*. ACM, New York, NY, USA, 2015, pp. 37–42.

[60] M. Dehghan, A. Seetharam, B. Jiang, *et al.*, "On the complexity of optimal routing and content caching in heterogeneous networks," in *Proceedings of the IEEE Conference on Computer Communications (INFOCOM)*, Kowloon, Hong Kong, 2015, pp. 936–944.

[61] B. Blaszczyszyn and A. Giovanidis, "Optimal geographic caching in cellular networks," in *Proceedings of the IEEE International Conference on Communications (ICC)*, London, UK, Jun. 2015, pp. 3358–3363.

[62] M. Afshang, H. S. Dhillon, and P. H. J. Chong, "Fundamentals of Cluster-Centric Content Placement in Cache-Enabled Device-to-Device Networks," *IEEE Transactions on Communications*, vol. 64, no. 6, pp. 2511–2526, June 2016.

[63] E. Baştuğ, M. Bennis, M. Kountouris, and M. Debbah, "Edge caching for coverage and capacity-aided heterogeneous networks," in *IEEE International Symposium on Information Theory (ISIT)*, Barcelona, 2016, pp. 285–289.

[64] M. Afshang and H. S. Dhillon, "Optimal geographic caching in finite wireless networks," in *IEEE Seventeenth International Workshop on Signal Processing Advances in Wireless Communications (SPAWC)*, Edinburgh, 2016, pp. 1–5.

Chapter 8
Full duplexing

Kari Rikkinen[1], Markku Juntti[1], Visa Tapio[1], Ari Pouttu[1], Björn Debaillie[2], Cristina Lavin[3] and Mir Ghoraishi[4]

8.1 Introduction

The evolution of wireless communication networks with ever-increasing data traffic calls for air interface enhancements to exploit the finite radio resources more efficiently. Full-duplex (FD) transmission is considered as a promising air interface technique for future wireless systems as it tackles key issues such as spectral efficiency, link level capacity and latency. Therefore, wireless network operators include FD as potential technology candidate for 5G systems in their joint 5G vision statement [1].

The FD, or more explicitly in-band FD, transmission concept differs fundamentally from the conventional duplexing schemes, i.e., frequency division duplexing (FDD) and time division duplexing (TDD), which are used in legacy wireless communication systems. While the conventional duplexing schemes are based on separation of transmit and receive signals either in frequency or time domain, the FD concept enables a wireless transceiver to transmit and receive simultaneously in the same carrier frequency. Figure 8.1 illustrates the fundamental difference between the FD and the conventional half-duplex transmissions. The FD concept sets high requirements to the transceiver implementation due to simultaneous transmission and reception in the same frequency band, but can provide significant improvements as a return.

Until recently, the use of FD transmission in wireless communications systems was believed to be infeasible due to self-interference phenomenon which occurs as part of the transmitted signal leaks or couples to the co-located receiver. The self-interference deteriorates the receiver performance unless it can be significantly suppressed. In recent years, FD communications have attracted significant attention as a number of groups in industry and academia have proposed various designs to

[1]Pentti Kaiteran katu 1, Linnanmaa, PL 8000, 90014 Oulun yliopisto, Finland
[2]Interuniversity Micro-Electronic Center (IMEC), Kapeldreef 75, B-3001 Leuven, Belgium
[3]Information and Communication Technologies (TTI), Calle Albert Einstein 14, 39011 Santander, Spain
[4]5G Innovation Centre, Institute for Communication Systems, University of Surrey, Guildford, GU2 7XH, UK

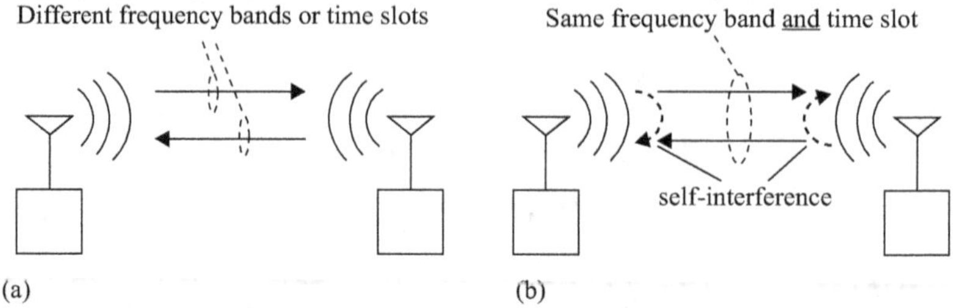

Figure 8.1 Transmission schemes: half-duplex transmission and in-band FD transmission

build FD transceivers. The proposed solutions include different antenna designs, radio frequency (RF) and analogue baseband circuit designs, digital baseband algorithms and their combinations to achieve high self-interference cancellation capabilities with different types of radio devices [2–8]. Published results from several research teams indicate that FD wireless technology is approaching the performance requirements of practical systems [6–8].

Enabling wireless devices to operate in FD mode offers potential for significant improvements in wireless systems operation. FD can double the two-way link capacity when compared to half-duplex transmission. It can enable more flexibility in spectrum usage. Same frequency resources can be used for one directional or bi-directional transmission. FD can further complement the legacy half-duplex systems based on TDD or FDD. Beyond spectral efficiency and physical layer, FD concepts can be advantageously utilised in higher layers, such as the medium access layer. FD transmission can reduce the air interface delays due to simultaneous reception of feedback information (control channels, signalling related to error correction protocol, etc.) and facilitate improved collision detection/avoidance mechanisms on contention based networks [9]. FD transceivers have also been proposed to be used to improve physical layer secrecy in wireless communications [10].

The in-band FD wireless technology has a long history, and the concept has been utilised already since 1940s in the design of continuous wave (CW) radar systems [11]. Terrestrial wireless communication systems like cellular mobile and Wi-Fi have not yet largely utilised FD transmission. The first known application of FD to wireless communications was relaying, where in-band repeaters were used to increase the coverage by receiving, amplifying and re-transmitting the wireless signal in the same frequency band [12]. However, interest in applying FD to future evolution of wireless systems has risen significantly during the past few years, and it is currently considered as a potential technology component for 5G radio systems [1,13]. The intended use cases proposed for FD in the literature include relaying, self-backhauling, microwave backhaul links, mesh networks, cognitive radio, access points/base stations and devices in Wi-Fi and mobile systems [10,14–20]. Short-range systems (e.g., femto or pico cells in mobile wireless systems and Wi-Fi systems) are especially promising operation environments for FD because the self-interference requirements are much more relaxed compared to large cell systems. This facilitates the implementation of FD technology to compact commercially attractive wireless devices.

This chapter discusses the FD transceiver and system solutions for 5G wireless networks. The focus is on FD transceiver solutions for compact form factor wireless devices, such as handheld ones and small access points/base stations (e.g., femto cell). We further concentrate on FD transmission solutions in small area radio communication systems. In Section 8.2, the challenges with FD transceiver implementation and FD transmission in small area radio communication systems are introduced. The FD transceiver architecture options and transceiver solutions for compact form factor devices are discussed, including examples on implemented hardware prototypes developed in the EU FP7 research project DUPLO [21] in Section 8.3. In Section 8.4, FD system aspects and potential performance gains of using FD transmission in small area radio systems are discussed. Finally, conclusions and proposals for further work are given in Section 8.5.

8.2 FD wireless transmission in 5G—scenarios and challenges

8.2.1 Full-duplex deployment scenarios

Densification of network deployments is a trend in 4G, and the development continues with 5G. Small cells improve spatial re-use of spectral resources and facilitate low transmit powers even with high bit rates. Small area systems are attractive application areas for the FD transmission, because the self-interference reduction requirements are much more relaxed compared to macro cells. This facilitates implementation of FD technology to compact commercially attractive wireless devices.

Figure 8.2 shows basic network set-ups related to the potential FD use cases and illustrates the FD implementation challenges. The first set-up in Figure 8.2(a) is the point-to-point connection between two nodes, either base station (BS) or user equipment (UE) types, or both. A typical use case example could be, e.g., a backhaul link or a single BS–UE connection in an isolated small cell. Also a Wi-Fi connection in a single cell falls into this category. In this set-up, the specific challenge with FD transmission is the self-interference (SI) in each node. In principle, the SI level should be reduced to noise floor level in order not to degrade the desired signal reception quality. Ideally, in this case, the FD transmission can double the link capacity over half-duplex connection.

Another potential use case for FD transmission is shown in Figure 8.2(b) where the same spectral resource is simultaneously used for two connections; i.e., the BS is transmitting a signal to one UE while receiving another signal from another UE. In this case, only the BS needs to be an FD capable device, while the UEs can be legacy half-duplex ones. An additional challenge is the inter-node interference (between the transmitting and receiving UE). In the relaying scenario shown in Figure 8.2(c), the relay node needs to be FD capable device. The direct link (UE to UE) can be considered either interference or desired signal channel depending on the relaying scheme.

In the multiple cell case, FD operation creates interference paths between all active nodes as illustrated in Figure 8.2(d). Thus, in the network level, the FD

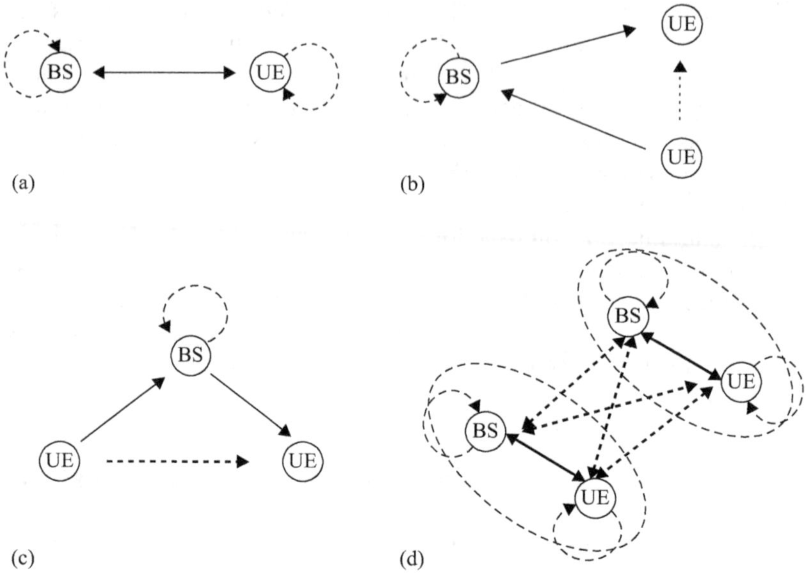

Figure 8.2 *FD deployment scenarios. The solid lines denote the desired signal channels, the dashed lines denote the self-interference channels, and the dotted lines denote the inter-node interference channels. The UE–UE link in relay case can be considered either interference or desired signal channel. (a) A point-to-point full duplex link, (b) a small cell with FD BS and half-duplex UEs, (c) an FD relay and (d) two neighbouring small cells with FD BS and UEs*

system implementation challenge is not only how to realise a compact FD device with high self-interference cancellation (SIC) capability, but also how to combat additional inter-node interference caused by the transmission taking place simultaneously with the reception. Furthermore, the network level gain of the FD transmission depends on multiple aspects, like traffic asymmetry, network architecture, propagation conditions, etc. The FD transmission in different network topologies is discussed in more detail in Section 8.4.

8.2.2 Full-duplex link budget and transceiver challenges

The main physical layer challenge in the FD wireless is self-interference: since transmission and reception occur on the same frequency simultaneously, the transmitted signal easily leaks into the receiver path, creating very strong in-band interference.

Figure 8.3 shows an FD link with a local and remote transceiver node. Each node has one antenna for the transmitter and one for the receiver. In this link, three sources of the SI limit the proper reception of the signal arriving from the remote node by the local RX. First, the leakage can occur on-chip or on-board (type A). Such direct cross-talk is likely to occur with dense integration. Second, the line-of-sight

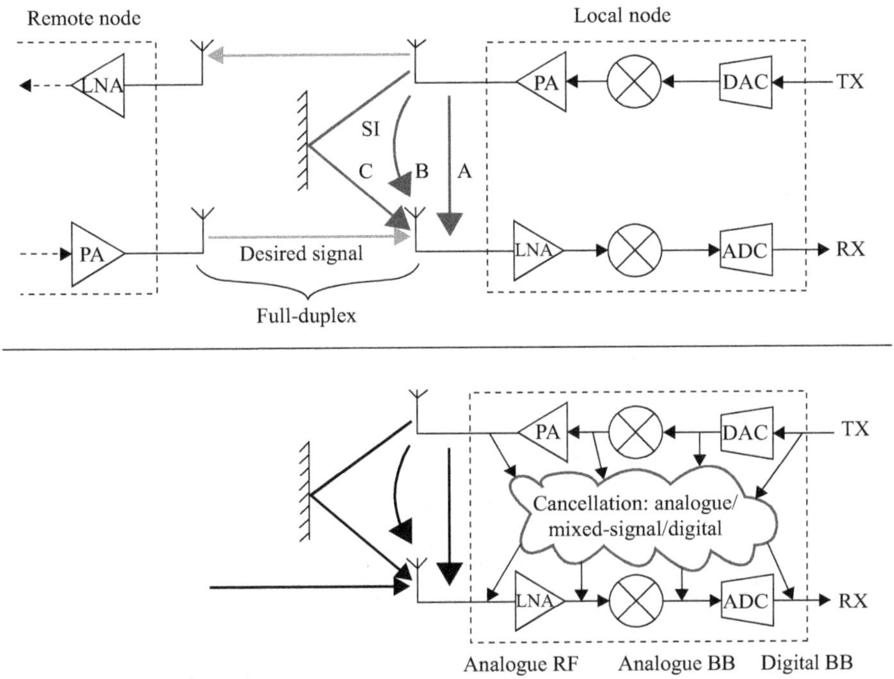

Figure 8.3 FD transceiver and system model. Top: Generic view of an FD link between a local and a remote node, subject to three types of self-interference: (a) Electrical crosstalk between TX and RX, (b) RF coupling due to limited antenna isolation and a varying antenna near-field, and (c) SI reflected by the environment. Bottom: Generic view of SI-cancellation in a single FD node, from various points in the TX chain to various points in the RX chain

leakage or spill-over between the two antennas may occur (type B). This SI is generally reduced by implementing antenna structures with specific radiation patterns or polarisations. Finally, TX signals may be reflected on nearby objects back into the receive antenna (type C). Such multipath reflections typically result in frequency-dependent SI.

The self-interference signal coupled into the receiver path consists of multiple components: the transmitted signal is corrupted by transmitter impairments, such as non-linearity, phase- and quantisation noise [22]. Some of these by-products are noisy, others are deterministic. This transmitted signal, including its by-products, is coupled into the receiver through various paths as illustrated in Figure 8.3. In order to achieve very high total self-interference rejection, the self-interference cancellation needs to be implemented constructively in multiple stages across the FD transceiver chain (isolation in the antenna part and further cancellation in the RF and baseband processing), as also illustrated in Figure 8.3. Furthermore, these

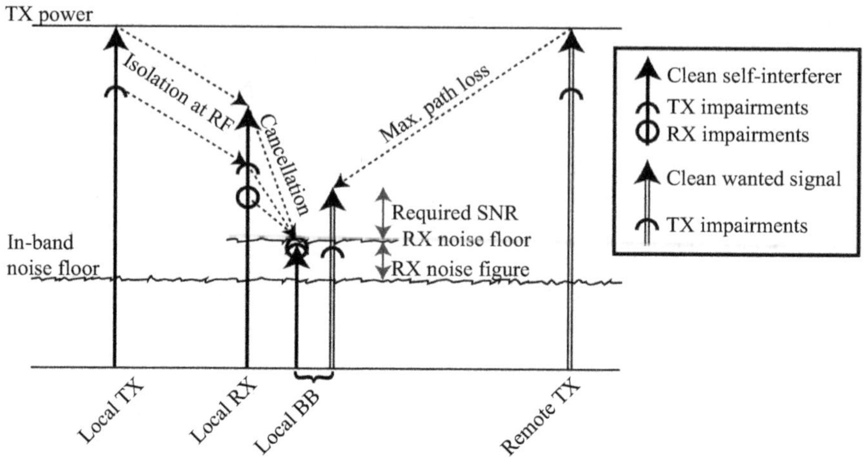

Figure 8.4 Visual representation of an FD link budget. The combination of isolation and cancellation techniques suppresses all self-interference components, preferably to below the receiver noise floor. © 2014 IEEE. Reprinted with permission from [22]

techniques should be self-adaptive to maintain the requirements over different propagation conditions and environments.

Figure 8.4 shows how the local and remote transmitter signals transfer to the local receiver baseband, where both nodes operate at equal average transmit power. The locally transmitted signal (Local TX) consists of a clean signal and its by-products due to transmitter impairments (half-circle). Antenna isolation will attenuate the self-interferences coupled to the receiver (Local RX), along with its transmitter impairments. Additional by-products will arise on this large signal due to receiver impairments (circle). Cancellation techniques are required to further reduce the self-interference and its by-products towards the receiver digital baseband (Local BB). If all self-interference components are suppressed to be below the receiver noise floor, the receiver sensitivity and tolerable path loss for the link is equal to its half-duplex equivalent, and the signal-to-noise ratio (SNR) is maintained to receive the signal from the remote node (Remote TX).

To quantify the link budget, the required self-interference rejection is calculated for a typical small distance wireless link with a 10 MHz channel bandwidth and a relaxed receiver noise figure of 14 dB. Then, the in-band receiver noise floor equals:

$$-174\,\text{dBm/Hz} + 10^{*}\log(10\,\text{MHz}) + 14\,\text{dB} = -90\,\text{dBm}.$$

With a realistic average transmission power of 10 dBm, the required rejection equals, thus, at least 100 dB to prevent sensitivity losses compared to a half-duplex link. In practical antenna solutions [3], the effective isolation is limited to approximately 40 dB due to reflections from the environment. Therefore, an additional 60 dB of cancellation is required to achieve the targeted 100 dB rejection and to suppress the self-interference below the receiver noise floor. In case of a single port antenna FD transceiver all the required self-interference rejection needs to be achieved with RF and baseband cancellation circuitries.

8.2.3 MAC protocol design challenges

To fully exploit the potential of the FD transmission in wireless networks, the design of medium-access control (MAC) and higher layer protocols is as important as that of the physical layer.

The FD transmission has the potential to significantly increase the overall throughput of a wireless network, beyond simply doubling the spectral efficiency of a point-to-point link. This is because the use of FD nodes in the network enables true real-time collision detection. When a wireless system operates in the half-duplex mode, it cannot transmit and detect collisions at the same time. Hence when a collision happens the entire packet is lost. As a result carrier-sense multiple-access with collision-avoidance (CSMA/CA) is commonly used. Therein the sender only transmits when it senses an idle channel. In an FD capable transceiver this limitation is removed, thus, offering the possibility to design MAC algorithms using real-time collision detection. This reduces the collision detection time and thereby increases the throughput. Furthermore, the FD transmission helps to solve some important problems in wireless ad hoc networks, such as the hidden node problem, large end-to-end delay, and loss of throughput due to congestion [15].

In infrastructure-based networks, such as cellular ones, scheduling (e.g., resource and power allocation) are important in optimising the performance. The simultaneous uplink and downlink transmission in the FD mode results in additional intra-cell interference. Intercell interference in dense networks can also be more severe. Therefore, to achieve the potential gain of the FD mode, an intelligent scheduler to pair the downlink and uplink users with corresponding transmission powers is necessary. Due to the combinatorial nature of pairing multiple uplink and downlink users on each subcarrier (in an OFDM based system), and also the difficulty of optimal power allocation to each subcarrier, resource allocation in FD systems is very challenging. Estimation of the internode interference channels is also needed.

Another challenge in the design and deployment of an efficient FD MAC is how to address the problem of asymmetric data flow in a network with symmetric link capacities. The traffic in multimedia communications, such as data and video, is generally asymmetric in uplink and downlink. The FD MAC needs to come up with solutions to benefit from FD operation with asymmetric data traffic.

- Furthermore, in order to include the FD technology in future cellular standards as well as exploring applications of the technology in the current wireless infrastructure, the coexistence of FD and half-duplex nodes should be addressed and managed appropriately. For this purpose MAC protocol needs to maintain fairness among the nodes in assigning the opportunities to access the medium.

8.3 Full-duplex transceiver

8.3.1 General

In the recent years, an increasing amount of research has been performed to solve the self-interference problem in FD transceivers; different techniques have been investigated to prevent and cancel self-interference leakage from the TX to the RX. To prevent direct coupling in the FD antenna, multi-antenna structures have been proposed to exploit the spacing and positioning of the antenna elements to create isolation in the antenna structure [3,23]. Such antenna structures can achieve up to 40 dB of isolation, but prevent dense integration due to the required physical distances between the antenna elements. The other techniques mainly cancel the self-interference in the receiver path based on the knowledge of the transmitted signal. This cancellation can be performed at different locations in the transceiver, namely in digital baseband, analogue baseband and/or analogue RF. Digital cancellation leverages the benefits of digital processing, but does not compensate for the noise and distortion generated in the analogue transceiver and cannot prevent the receiver to saturate due to strong self-interference signals. Analogue and RF cancellation enable cancelling the noise and distortion induced by the transmitter, and relaxing the receiver linearity requirements as the self-interference is reduced early in the receiver chain. Few mixed-signal cancellation architectures have been proposed such that the digital transmit signal is used as a reference to compensate for the self-interference at the receiver RF input [5,24]. This architecture requires a dedicated up-converter, which in practice introduces its own noise and distortion and limits the cancellation efficiency.

To achieve an overall self-interference rejection of more than 100 dB, a combination of self-interference isolation and cancellation techniques is required. A recent transceiver design has been described in [6] offering an attractive FD performance with a self-interference isolation of more than 100 dB over 80 MHz bandwidth. This design uses a single port antenna in combination with an RF circulator and applies both digital and analogue cancellation. This architecture is being used in the commercial products of Kumu Networks™ targeting wireless backhauling and fixed-location relaying. The transceiver design is however not suitable for form-factor critical radios applied in mobile devices and small-cell access points. The analogue cancellation structure implements a network of physical delay lines which cannot be substantially scaled for a given operation frequency. Therefore, other transceiver solutions are required for portable 5G applications.

We present different potential transceiver solutions which support integration in compact and portable radios. Section 8.3.2 discusses several RF/analogue solutions and antenna structure supporting FD operation and Section 8.3.3 presents digital cancellation solutions which complete the self-interference system cancellation. For the typical short-distance or indoor scenario, the total self-interference rejection should equal to about 100 dB over a 10 MHz bandwidth (Section 8.2.2). The presented analogue/RF cancellation solutions cancel the direct leakage (cross-talk in the transceiver and direct coupling in the antenna), whereas the digital cancellation also covers the multi-path reflections. Reflected signals are subject to a substantial delay and cause frequency-dependent self-interferences. As for the targeted short-range or indoor scenario the multi-path reflection is about 40–65 dB smaller than the direct leakage, the self-interference cancellation should equal to about 50 dB. Section 8.3.4 integrates the analogue/RF and digital solutions and a proof-of-concept validates the FD system performance. Finally, Section 8.3.5 discusses application of FD transmission to multiple-input multiple-output (MIMO) systems. The implementation examples in this chapter are based on hardware prototypes developed in the EU FP7 research project DUPLO [21].

8.3.2 RF/analogue and antenna solutions for compact form factor devices

The FD concept is contradictory to densely integrated radios, as integration promotes signal leakage within the transceiver circuits and between the antenna elements. In this section, disruptive analogue architectures are discussed which enable to minimise the self-interference leakage while sustaining the dense (co-) integration in compact radios for mobile and small-area access applications.

The three architectures depicted in Figure 8.5 are considered to minimise the direct leakage from the transmitter to its own receiver. The first architecture is a complete analogue front-end, which implements a novel self-interference cancellation mechanism including a dynamic downconversion. The analogue part indicated by the dashed line has been prototyped in the CMOS technology. It supports co-integration with legacy systems, enables low cost fabrication and operates at low power. The second architecture includes a single antenna structure with native isolation between its ports and an active cancellation network at the antenna ports. This antenna structure is compact, and, in combination with the cancellation network, offers a robust self-interference isolation over changing operation conditions. The third architecture implements an innovative duplexer which enables a legacy single-port antenna to be used for FD communications. This duplexer is prototyped in SoI CMOS technology as indicated with the dashed line.

These three architectures aim at integration in compact radios, and offer specific benefits and characteristics that can be exploited in different applications. Therefore, they are not intended to be combined.

8.3.2.1 Self-interference cancelling front-end

In compact FD radio devices, space does not permit sophisticated antenna solutions to provide high SI isolation. Also, a varying near-field environment (e.g., a person holding the device) constantly changes the magnitude and phase of the self-interference, requiring an adaptive solution. Experiments using two crossed WLAN dipole antennas as a simple FD antenna showed that 25 dB isolation is a reasonable assumption. This leaves room for 20–40 dB improvement in the analogue domain before the self-interference caused by reflection on nearby objects becomes the bottleneck [22]. Given this scenario, an innovative and novel front-end architecture

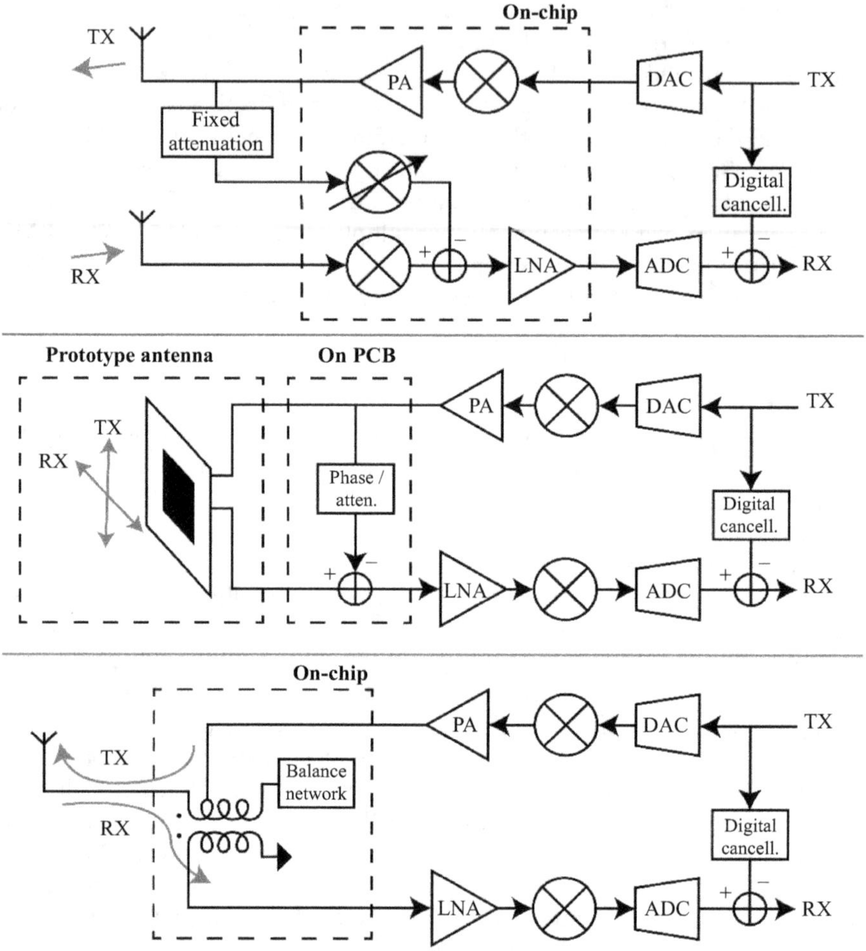

Figure 8.5 Three FD transceiver architectures supporting integration in compact radios. Top: front-end with self-interference cancelling receiver, middle: dual-polarised antenna and active RF cancellation, bottom: electrical balance duplexer

has been presented in [8,25,26]. This design achieves cancellation by subtracting the self-interference at analogue baseband of the receiver based on the transmitted RF signal. This topology is mainly suitable to actively boost the isolation of a moderate antenna solution, e.g., when the antenna provides a moderate 15–20 dB isolation, up to 30–45 dB before reflections become the dominant self-interference. These reflections are then cancelled in the digital baseband by a second cancellation path that accounts for delay in the digital domain.

A strong point of this architecture is that attenuation, phase shift and down-conversion can be combined into one functional component: a vector modulator mixer.

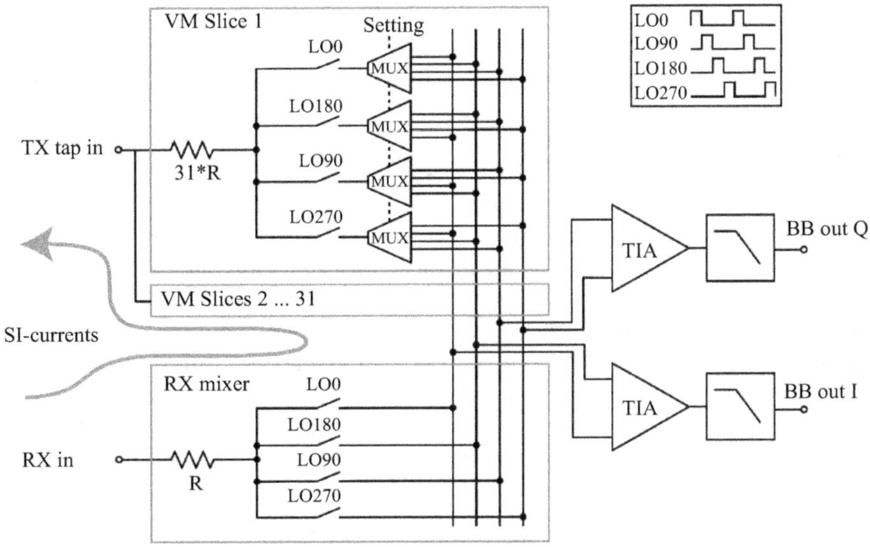

Figure 8.6 Simplified schematic of the self-interference cancelling receiver. One slice of the 31-slice vector modulator (VM) downmixer is shown in detail. Implementation details are shown in [25,26]

However, since the cancellation point is located in the analogue baseband processing, the strong self-interferer is fully present in the LNA input. To prevent drowning the weak underlying desired signal, high linearity is required in the first receive stages. This can be accomplished by using a mixer-first architecture for both the main mixer and the vector modulator. Circuit simulations confirmed the unique potential of this architecture and indicated that the drawbacks can be resolved by using appropriate circuit topologies.

Figure 8.6 shows a simplified schematic of the receiver with including the downmixer as presented in [8,25,26]. Essentially, both the VM and the main RX mixer are four-phase switched resistor mixers which maintain high linearity under large self-interference powers. The vector modulator is a sliced version of the main RX mixer, incorporating multiplexers in each slice that steer the four-phase output current to the appropriate baseband phases. The amount of slices of the VM determines the number of phase/amplitude constellation points that can be covered and thus the amount of cancellation that can be achieved due to quantisation effects. In [25,26], 31 slices are used allowing 27.1 dB SI-cancellation.

The prototyped front-end covers an area of 2 mm^2 in 65 nm CMOS and has been validated based on measurements [25,26]. The self-interference cancelling performance was measured with two crossed WLAN dipole antennas (~30 dB measured isolation) and 802.11a-like multi-tone signal over a bandwidth of 16.25 MHz, resulting in a total cancellation value of more than 45 dB. Also the phase noise immunity in FD mode has been investigated when using the same frequency synthesiser of up- and down-conversion. Such immunity is required as experiments show that different synthesisers generating −38 dBc uncorrelated phase noise in the band of interest induces a combined noise floor of −35 dBc, limiting the cancellation. These measurements showed a very high phase noise immunity in FD.

8.3.2.2 Dual-polarised antenna and active analogue cancellation

The design of an FD antenna structure targets a high intrinsic isolation and low cross-polarisation between the transmit and receive antenna ports. Such isolation is traditionally achieved with multi-antenna elements, implementing radiation shields and/or destructive radiation patterns to prevent leakage from the transmit to the receive antenna(s). When targeting a small form-factor antenna structure, an alternative architecture building on a single antenna element with orthogonal polarisation is attractive. Such antenna structure is depicted in Figure 8.5, where the transmitter and receiver are connected to the corresponding polarisation planes via the two antenna ports. Fulfilling both the isolation and cross-polarisation requirements simultaneously is very challenging, especially for compact antenna solutions where both excitation ports are closely spaced and the parasitic coupling between them is high.

Recently, a compact form-factor dual-polarised antenna design for FD has been presented in [8,27]. These designs combine different excitation mechanism, such as slot-coupled and multi-point excitation feeding networks in a stacked-patch antenna structure to achieve attractive isolation values. The stacked structure consists of two square patches and a symmetrical excitation network. Also, dual-linear polarisation is used because linear polarisation is generally not operation-frequency dependent (in contrast to circularly polarised antennas).

To enable a low profile, low cost and eased integration, these designs are implemented using microstrip technology. For a microstrip patch antenna, dual linearly polarised operation is obtained by using a pair of probe feeds to respectively excite two orthogonal fundamental modes from a single radiating patch. The discrimination between the two polarisations could however be degraded by undesirable higher-order modes, which can cause cross-polarisation. To suppress the cross-polarisation at each feeding port of the dual-polarised microstrip antenna, thus improving the isolation level between ports, the antenna structure incorporates a symmetrical dual-feed excitation network with a phase difference of 180° and a pair of coupling slots.

The design presented in [8] is an improved redesign of antenna described in [27]. The optimisation criteria were to reduce the form-factor, to increase the self-interference isolation and improve the impedance matching. The resulting antenna geometry of the second prototype measuring 60 × 60 × 8 mm is illustrated in Figure 8.7 and consists of two square patches and three substrate layers with the same thickness and dielectric constant. The horizontal plane (PORT 1) implements a 1/2 divider and the vertical polarisation (PORT 2) is excited by a 50 Ω microstrip line and an internal via. A rigid dielectric material with a dielectric constant of 2.5 has been included to separate the two square patches. The reported measured specifications are very attractive for FD, offering more than 50 dB of antenna

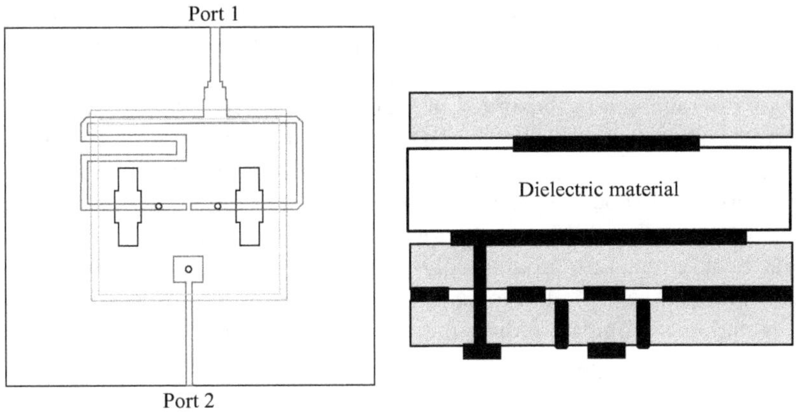

Figure 8.7 Geometry of the dual-polarised antenna. Left: top view of the antenna, right: multilayer antenna stack-up. © 2015 IEEE. Reprinted with permission from [8]

isolation, 6.5 dB of antenna gain and 75% antenna efficiency over a bandwidth of 10 MHz. These measurements are also performed over 80 MHz bandwidth, indicating only a small performance degradation.

The measurements indicate however a substantial degradation of the FD antenna performance when objects are placed or moved in the vicinity. To maintain a self-interference isolation at the antenna (before the LNA), implementing an active cancellation network operating in-between the antenna ports is advisable as indicated in Figure 8.5. This active RF cancellation network uses an attenuated and phase shifted copy of the transmitted RF signal to cancel the self-interference in the receiver path before the LNA. The results reported in [8] illustrate an improved suppression of the self-interference in addition to the antenna isolation, resulting in 62 dB cancellation. The antenna prototype combined with the active cancellation network has been included in the FD transceiver prototype described in Section 8.3.4 and has been validated over different operation conditions.

8.3.2.3 Electrical balance duplexer

The antenna size is one of the main bottlenecks for compact integration of the FD radios. The antenna technique described in the previous section uses a single antenna element, avoiding the physical distance between the traditional multi-antenna structures. This single antenna element is however specific for FD operation and comes with two ports. When targeting denser integration and co-integration with legacy systems, using a conventional off-the-shelf antenna is more attractive. Such single port antennas are available in extremely small sizes (e.g., SMD components) and they are already implemented in legacy systems. By sharing this antenna for consequent operation between the FD transceiver and the legacy system, a high implementation efficiency can be achieved when designing a multi-standard radio.

Using such conventional antenna for FD operation implies that there is no antenna isolation, and that a duplexer is required to prevent transmitter signals to couple into its own receiver. This duplexer is connecting the FD transceiver with the single port antenna as illustrated in Figure 8.5. Surface-acoustic wave (SAW)-type duplexers are commonly used in FDD systems to prevent the transmitter signal from leaking into the receiver by means of filtering. However, such duplexers cannot be used in FD because there is no frequency separation. Another option is to use a circulator to route the transmitter and receiver signals over the common antenna. Such circulators provide only a moderate isolation of about 20 dB and require additional cancellation loops at RF to be suitable for FD. When targeting implementation in compact radios, the combined circulator and complex RF circuitry are rather unattractive.

An alternative duplexer topology has been investigated namely the electrical balance duplexer (EBD). The EBD concept was initially proposed to achieve tunable duplexer filters for FDD, but is also very attractive for FD because of its potential for efficient implementation in CMOS technology, its high isolation factor to prevent leakage towards its own receiver and its tuning flexibility. The EBD operation principle is illustrated in Figure 8.5; it mainly comprises a hybrid transformer and a balance network. The electrical balance network is essentially a tunable dummy load impedance which is tuned to match the antenna impedance such that transmitted signals are not propagating through the transformer to its own receiver. The balance network is made tunable to support the usage of a wide range of antenna types (with different impedances) and to trace eventual antenna impedance variations due to, e.g., dynamic operation conditions or moving objects in the antenna proximity.

The EBD is an RF circuit which performs self-interference cancellation very close to the antenna. This protects the first receiver elements from saturation due to self-interference and from distortions caused by the transmitter impairments (e.g., non-linearity, phase noise) in the cancellation. Moreover, by virtue of the purely passive structure of the EBD, any noise and non-linearity products generated in the transmitter are also cancelled by the circuit itself.

Recently, the EBD topology has been introduced [22] and prototype-validated for FD application. The EBD prototype presented in [27] implements an octagonal transformer, an RC balance network with tunable capacitance and resistance, and a differential receiver buffer in 180 nm UMC CMOS technology. The chip covers an area of 1.5×1.5 mm and is included in the FD transceiver prototype described in Section 8.3.4. An intelligent tuning algorithm has been developed [28] to rapidly tune and track the optimal configuration of the balance network. The validation measurements indicated some critical issues as discussed in [27]:

- Implementation loss in both the transmitter and receiver RF path of about 6 dB: although this loss seems considerably high, it is comparable with legacy FDD systems.
- Severe antenna impedance dependency over frequency and operation condition (e.g., moving objects close to the antenna); dynamic (re-)tuning of the balance network is of essence.

- Limited isolation bandwidth: the EBD offers an extremely high isolation peak of >90 dB at balance condition, but its peaking spectral shape limits the average isolation over a wider bandwidth. This peaking shape is caused by the antenna impedance variation across (in-band) frequencies. As the balance network cannot be designed to trace the antenna impedance over the instantaneous bandwidth, a novel design approach is required to increase the operation bandwidth.
- Distortion in the balance network leaks to the receiver. The EBD transformer does not cancel distortions generated in the balance network to leak towards the receiver. This directly limits the transmit power to prevent non-linearity generated in the balance network. This finding indicates a strict linearity requirement on the balance network to support transit powers greater than −10 dBm.

This prototype has been redesigned as described in [8,29] to increase the FD performance. This prototype has been implemented in 180 nm Silicon-on-Insulator (SoI) CMOS. An alternative balance network topology as illustrated in Figure 8.8 was used to achieve high linearity and tuning freedom while maintaining the implementation efficiency. A four-dimensional tuning capability of the balance network enables to tune the real and imaginary impedance across the frequency and to trade-off the bandwidth with the average SIC, whereas the R/C network of the first prototype only supported tuning the single isolation peak across the frequency. Note that as oppose to the first prototype, a single-ended topology at the receiver side of the hybrid transformer is preferred to avoid a common-mode SI and to enable high-power operation.

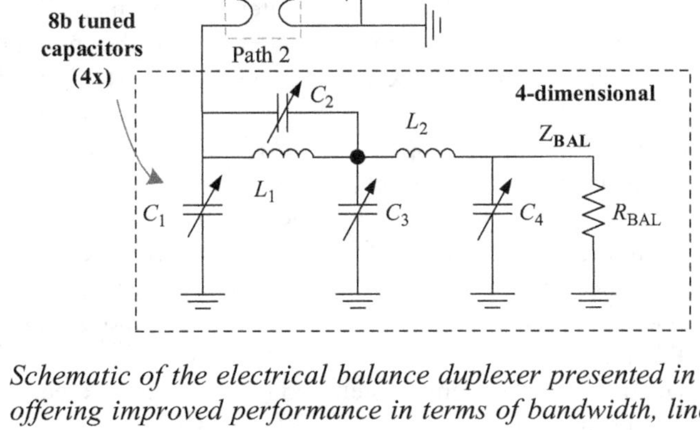

Figure 8.8 Schematic of the electrical balance duplexer presented in [8,29] offering improved performance in terms of bandwidth, linearity and transmit power with respect to [27]

8.3.3 Digital baseband solutions for full-duplex transceiver

The problem of self-interference cancellation in the digital domain resembles to echo-cancellation in digital subscriber line (DSL) modems, cellular relays and digital video broadcasting (DVB) repeaters [30–32]. Hence, same principles used in the DSL modem's echo cancellers can be used in the FD radio systems to mitigate the self-interference. In FD wireless transmission, the requirements are more restrictive and demanding. Compared to the DSL echo cancellers, broad frequency band and nonlinear effects due to RF and analogue impairments make the self-interference in compact FD radios more complex and hard to emulate. In general, relays can enjoy relatively high isolation at antenna level by using their larger form-factor and possibility of using directional antennas due to their specific application scenarios. On the other hand, the self-interference channel in FD portable and mobile radios can be variable due to the dynamic environment in which the devices are used, while the echo channel in the DSL and the interference loop (self-interference channel) in relays and repeaters are more or less static.

In an FD receiver, pre-ADC cancellations, including transmit-receive path isolation (due to using separate antennas for transmission and reception or by using isolation circuits), and RF/analogue SI cancellations, remove only part of the self-interference. Current analogue solutions offer up to 60 dB SI cancellation capabilities [6]. Since the total SI cancellation requirement can be more than 100 dB, both RF/analogue and digital SI cancellation techniques are required. This means that the remaining SI needs to be cancelled at the digital baseband. The residual SI power received at the baseband can be up to 60 dB higher than the noise-level and the SI signal includes nonlinear components created at the power-amplifier, low-noise amplifier, etc.

A general block diagram for the digital cancellation is displayed in Figure 8.9. Known transmitted data is used for SI channel estimation, the accuracy of which depends on the self-interference-to-noise ratio. In a real scenario, SI channel includes multipath and it can change dynamically. Using the estimated linear and nonlinear components, an estimate of total self-interference is subtracted from the received signal. Maximum achievable self-interference cancellation depends on the channel estimation accuracy and the accuracy of the nonlinear models. Furthermore cancellation performance depends on the pre-ADC cancellation as it can cause noise-level elevation [32].

One of the main transmitter impairments affecting the SI cancellation performance is the non-linearity of power amplifiers. In order to suppress interference components caused by non-linearities a model for non-linearity is needed. One commonly used model for transmitters is the Hammerstein model, which is a cascade of static nonlinear part followed by a linear filter [33]. It is a commonly used method for modelling PAs with memory and a natural choice for modelling a cascade of a PA and frequency selective self-interference channel as illustrated in Figure 8.10. The usage of the Hammerstein model in self-interference cancellation has been introduced in [34].

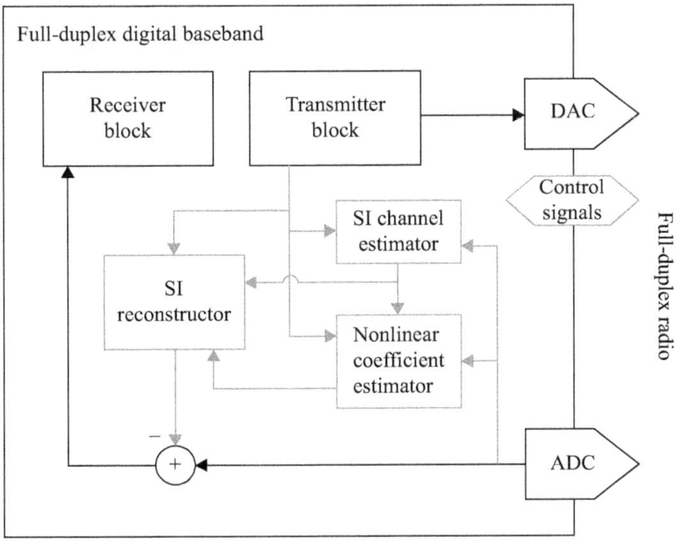

Figure 8.9 Block diagram of a general digital self-interference cancellation

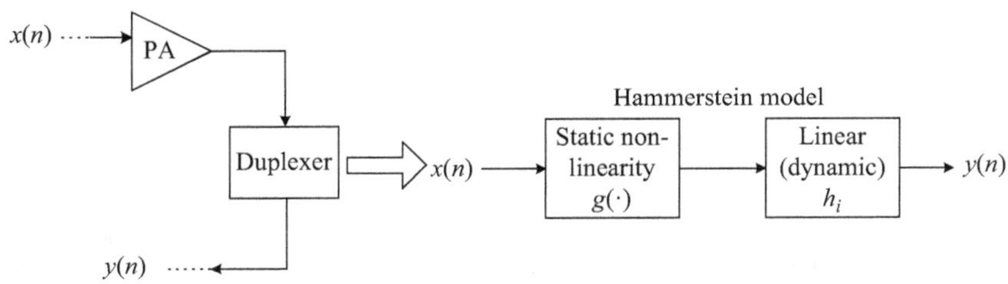

Figure 8.10 Hammerstein model

The performance of the Hammerstein model based self-interference cancellation is demonstrated with simulation results in Figure 8.11. The signal is an OFDM signal with 52 subcarriers. The power amplifier (PA) in the transmitter has been modelled using the Modified-Saleh I model [33]. Parameter OBO in Figure 8.11 indicates the power amplifier back-off value used in the simulations. The back-off is defined here as the difference between the saturation level of the PA model and the peak power of the PA input signal. Parameter INR is the self-interference-to-noise power ratio before the baseband cancellation. The duplexer in Figure 8.10 is an electrical balance based analogue self-interference cancellation circuitry [22]. In the receiver, the signal used for SI cancellation is generated using the Hammerstein model. The static non-linearity is modelled as third- and fifth-order polynomial functions and the linear part is a FIR filter. Coefficients of the FIR filter are estimated using a predetermined training signal.

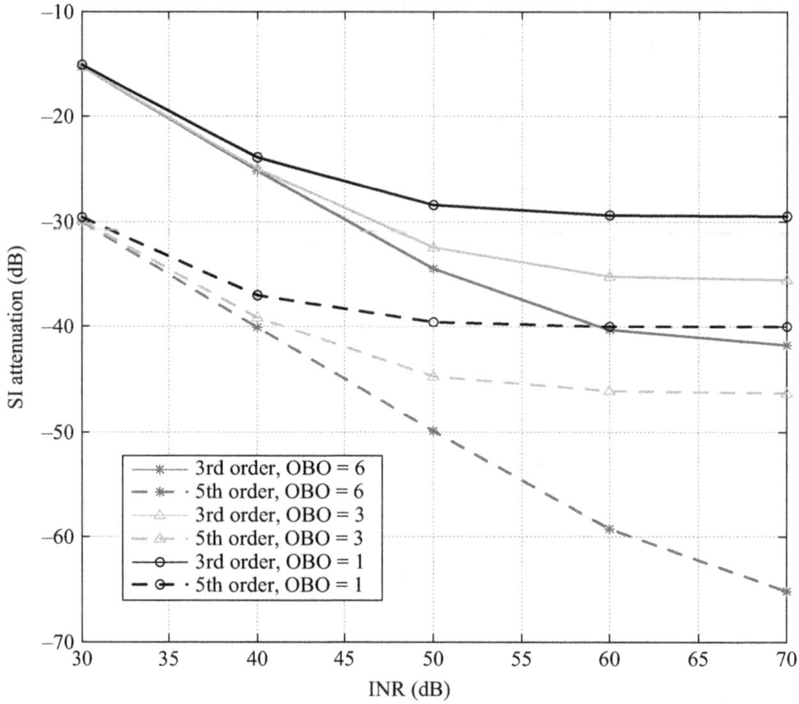

Figure 8.11 Performance of Hammerstein based SI cancellation

In a practical FD transceiver, the SI isolation for the wideband signal is not uniform over the band of interest. For instance, the electrical balance based analogue self-interference attenuation circuitry is tuned at a specific frequency (usually in the middle of the band) and then the isolation level degrades when going further from the tuned point [35]. This difference can reach 15 dB or more. The frequency selectivity of the SI isolation will increase the complexity of the Hammerstein model parameter estimation. For example, in the simulations presented above the FIR length was 52 taps.

One solution to mitigate the effects of the frequency selective transmit-receive path isolation could be an uneven SI cancellation over the band, i.e., the band can be divided into several subbands and then SI cancellation is performed at each subband independent from other subbands as presented in Figure 8.12.

Other non-idealities reducing the SI cancellation performance include IQ imbalance, digital-to-analogue and analogue-to-digital converters and phase noise [35]. Common to all these is that their impact on the received self-interference signal cannot be estimated using a simple channel estimator applied in the estimation of channel between nodes. However, all the phenomena are well understood and a number of techniques have been proposed in open literature for estimating and compensating for them, see e.g., [36] and references therein.

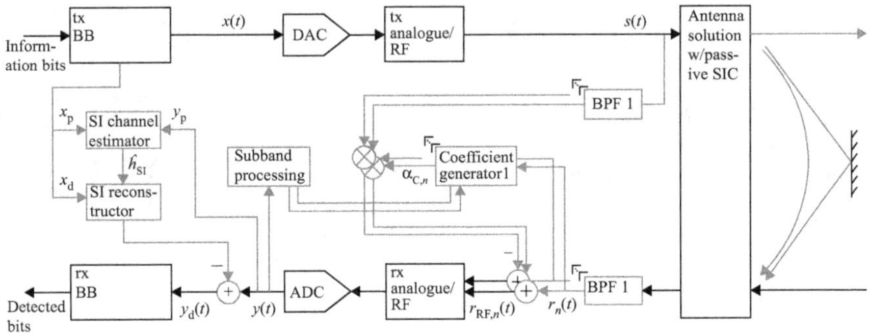

Figure 8.12 Block diagram for joint analogue/RF and digital baseband SI cancellation

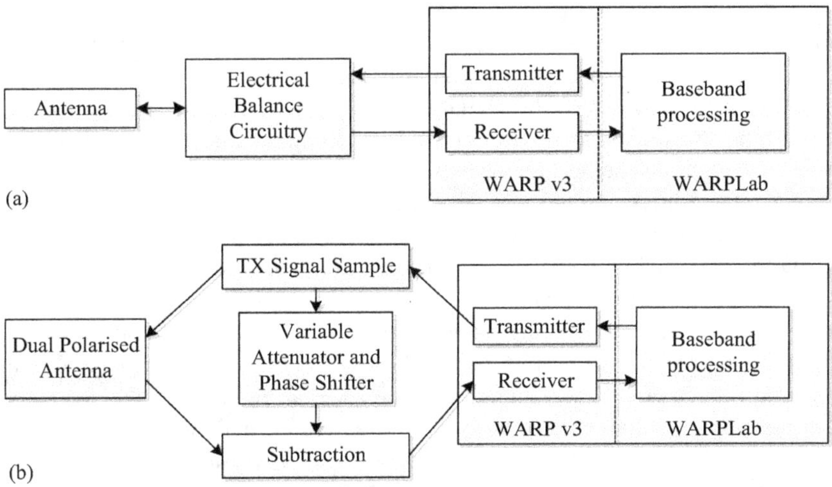

Figure 8.13 FD prototype transceiver architecture. (a) Single-port antenna solution with electrical balance duplexer and (b) dual-port antenna solution with a dual-polarised antenna and an active cancellation circuit

8.3.4 Prototype transceiver design and testing

Implementation of hardware prototype transceivers enables evaluation of FD communications in practical operation environments. For that purpose, the analogue and antenna cancellation solutions described in Section 8.3.2 have been integrated in two different FD transceiver prototypes.

The first prototype, the single-port antenna radio node, is illustrated in Figure 8.13(a). This transceiver integrates the electrical balance duplexer and baseband digital cancellation algorithms. WARP v3 radio board [37] and WARPLab 7 [38] framework have been used for integration. The electrical balance circuit

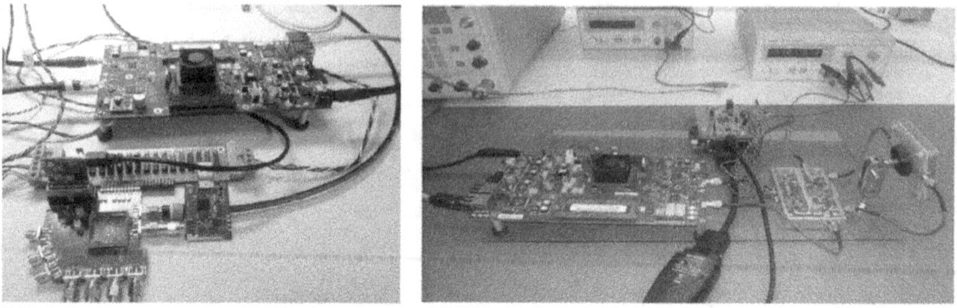

(a) (b)

Figure 8.14 FD prototype hardware implementations. (a) Single-port antenna solution and (b) dual-port antenna solution

interfaces between the RF ports of the WARP and the single port antenna, and isolates the received signals from the transmitted signals. The baseband functionality, including the test signals and digital baseband cancellation has been implemented in MATLAB®. OFDM signal with 64 subcarriers and 20 MHz signal bandwidth has been used as the test signal waveform. Two linear digital self-interference cancellation schemes, based on time domain and frequency domain cancellation, have been implemented into the demonstrator.

The second implemented FD prototype is the dual-port antenna radio node. This FD transceiver integrates the dual-polarised antenna and the active cancellation network previously described in Section 8.3.2. The architecture of the radio transceiver is illustrated in Figure 8.13(b). As in the case of the single-port antenna radio node, the WARP v3 radio board and WARPLab framework have been used for integration. The FD baseband and digital cancellation block is also the same as that in the single-port antenna prototype. The active cancellation network is tuned by means of using a gradient descent algorithm which is implemented in an external microcontroller. The implemented hardware transceiver prototypes for single-port antenna and dual-port antenna solution are shown in Figure 8.14(a) and (b), respectively.

The FD evaluation platform builds on two FD radio nodes which communicate through a wireless FD point-to-point link as Figure 8.15 illustrates for the single-port antenna solution. In principle, similar point-to-point link set-up is used with the dual-port antenna solution.

The evaluation measurements have been performed in an unshielded open lab environment with people and objects moving close to the FD radio nodes. Channels in the 2.4 GHz ISM band have been used and different transmit powers and modulation schemes have also been validated. Figure 8.16 illustrates the system performance with single-port antenna solution over different link distances when a BPSK signal with 0 dBm of WARP output power is transmitted. The error vector magnitude (EVM) of the constellation after demodulation has been measured at both radio nodes when the nodes are operating in FD and in half-duplex (no SI). The results show that the EVM performance during the FD operation closely

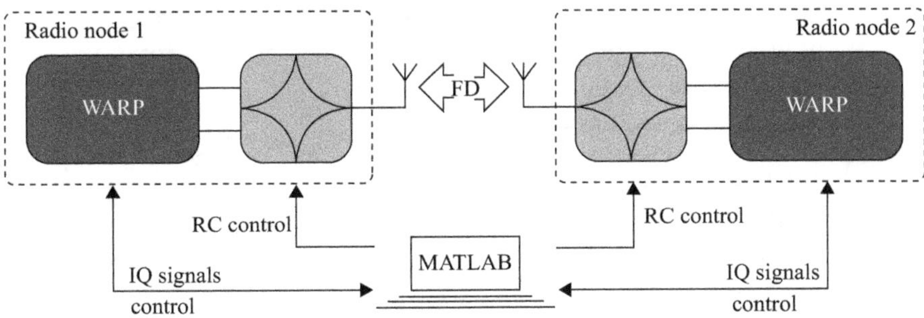

Figure 8.15 FD evaluation platform building on two single-port antenna radio nodes

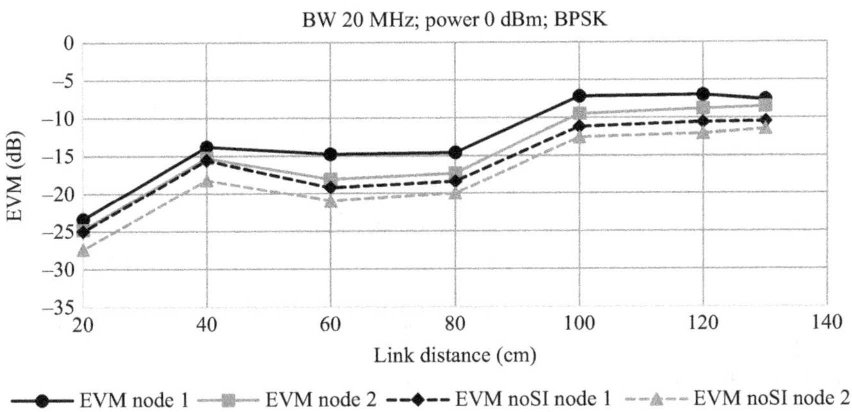

Figure 8.16 EVM performance versus the link distance (single-port antenna radio node)

matches with that of the half-duplex mode. This means that the SI is sufficiently reduced by the SIC solutions to maintain the SNR at the receiver. The main limitations in terms of the link distance are due to the linearity issues of the first prototype of the EBD already discussed in Section 8.3.2. More advanced experimental measurements of the FD wireless link between two single-port antenna radio nodes have been presented in [39]. The total self-interference cancellation provided by the single-port antenna radio node is 65 dB in 20 MHz bandwidth.

Figure 8.17 illustrates the system performance with dual-port antenna solution over different link distances when a 64 QAM signal is transmitted with 0 dBm of WARP transmit power. In this case, FDAC in Figure 8.17 represents the EVM when only analogue cancellation is applied to reduce the SI (the analogue SIC is 60 dB in 20 MHz BW), FDADC represents the EVM when both analogue and digital cancellation are applied (digital cancellation provides 30 dB of additional SIC) and HD is the EVM performance of the half-duplex link. The obtained results shows how the FD link provides similar performance to the half-duplex link, and only a

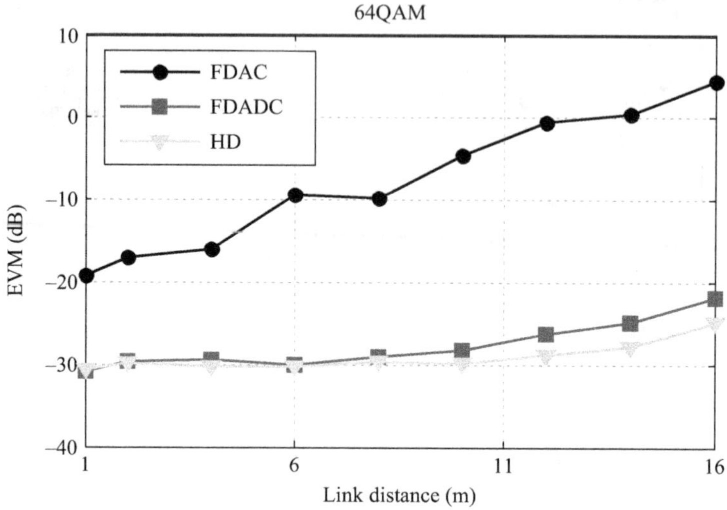

Figure 8.17 EVM performance versus the link distance (dual-port antenna radio node)

minor degradation in the EVM of 2% is observed at a maximum link distance of 16 meters. Extended measurements of this evaluation setup have been presented in [39].

In summary, the conducted performance measurements indicate that the achievable self-interference cancellation capability is between 65 dB and 90 dB in 20 MHz bandwidth, with single port antenna solution and dual port antenna solution, respectively. Although the targeted 100 dB self-interference cancellation could not be achieved with these hardware prototypes, the results are encouraging as there are already identified solutions to overcome the limitations of the first prototype implementations. For example, an improved balance network solution is reported in [29].

8.3.5 Multiple antenna full-duplex systems

It is known that achieving spatial gain of the wireless channel by using multiple antenna techniques in MIMO systems is one of the most important breakthroughs in wireless technology during the last couple of decades [40]. Thus, it is well recognised that if FD is going to be widely deployed, it needs to work well in MIMO systems. However, the extension of the FD scheme, or more specifically self-interference cancellation, to the MIMO set-up is by no means straightforward. In fact FD operation and the coexistence with the MIMO technology is now known as an important barrier in popularising FD technology. The problem in the design and implementation of an FD MIMO system is multifold as each transmit signal interferes with all local receive antenna/RF chains as illustrated in Figure 8.18, where it is assumed that each antenna can transmit and receive at the same time (single antenna full duplex operation) by using a circulator or duplexer circuit [22].

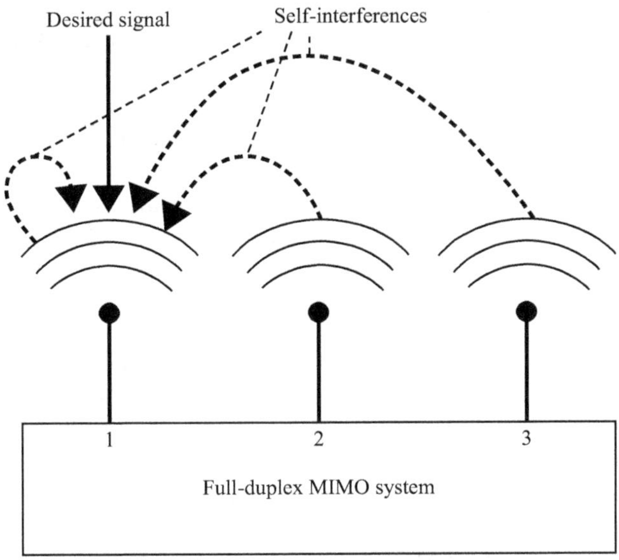

Figure 8.18 Schematic of the FD transceiver. Desired signal and self-interferences at antenna 1 only are illustrated for the sake of clarity

To gain a better understanding of the performance of the FD MIMO systems under practical constraints, the work presented in [41] derives achievable rate bounds using a realistic system model, including channel estimation errors and effects of the limited dynamic range. It has been shown that the FD MIMO spectral efficiency is uniformly better than its optimised half-duplex counterpart and nearly doubles when operating within the dynamic range constraints. A similar approach was used to obtain the achievable rates of a decode-and-forward based FD MIMO relay and a technique to increase the lower-bound for transmission rate is proposed [42].

When using multiple antennas at the FD transceiver, one way to suppress the self-interference would be by using extra spatial degrees-of-freedom, i.e., by employing techniques such as beamforming, null-steering and precoding to supress the self-interference [43–47]. One of the first proposals for an FD MIMO system requires $4M$ antennas for building an FD M antenna MIMO radio, and even then fails to provide the required self-interference cancellation for Wi-Fi systems (20 MHz bandwidth) to achieve the expected doubling of throughput [44]. In the context of FD MIMO relays, several analyses have been discussed ([42,45] and references therein). Notably beamforming, null-steering and MMSE filtering to suppress the self-interference in an FD MIMO relay is investigated in [45]. The self-interference cancellation based on null-steering and antenna polarisation diversity of FD MIMO system was investigated in [43]. On the other hand, assuming a certain level of FD MIMO self-interference cancellation is achieved at the RF and analogue circuits, solutions at the digital baseband to suppress the remaining residual self-interference and its nonlinear component is a common approach in the analysis of such systems [41,42,45,46,48], where the development of robust signal processing techniques to mitigate the residual self-interference is motivated. The nonlinear self-interference modelling and cancellation taking into account the effects of IQ imbalance and power amplifier non-linearities and memory was investigated in [48], whereas precoding algorithms at the digital baseband was used to mitigate the residual self-interference for an FD MIMO system [46,47]. These analyses show that on top of self-interference cancellations at RF and analogue circuits, employing precoding algorithms at the digital baseband can help to improve the system performance [46]. The performance of these solutions, as that of their single antenna counterparts, depends on the accuracy of the MIMO self-interference channel estimation as well on how precisely the non-linear components in the self-interference are estimated and modelled. In an alternate approach, precoding was used in a kind of mixed signal approach to cancel the FD MIMO self-interference while beamforming is used in the signal transmission [47]. In most of the above mentioned techniques it is assumed that the self-interference cancellation at RF is already achieved, which, together with a reasonable transmit-receive isolation, can cancel out significant amount of the self-interference. Other approaches propose to use extra receive/transmit chains to reproduce the self-interference component, which is not reasonable due to additional cost/complexity in MIMO transceiver design/implementation [47,49].

Little discussion exists on practical solutions to cancel/mitigate self-interference from multiple transmitters of the MIMO system at RF/analogue levels. A straightforward approach would be the extension of self-interference cancellation in single-input-single-output solutions to the MIMO system by using a dedicated cancellation circuit for each transmit signal at every receive chain of the MIMO transceiver. This however exponentially increases the complexity of the FD MIMO transceiver system. Recently a practical method has been proposed to overcome the self-interference cancellation challenge between all FD MIMO transceiver chains [50]. Instead of introducing a separate copy of the cancellation circuit for each pair of MIMO transceiver chains, and cancellation algorithm for each pair of chains that experiences the so called cross-talk, only one cancellation circuit per receive chain has been used. It has been further assumed that MIMO chains are collocated and share similar environment, so that the self-interference channels from all transmit chains to each receive chain are common. Moreover, a method to estimate self-interference channels from all transmit chains to each receive chains concurrently is used improving the overall performance that otherwise would degrade linearly with the number of MIMO antennas at each link end. The work has been prototyped and performance of the proposed schemes has been evaluated using off-the-shelf radios and test equipment [50]. The remaining question is whether the assumed common channel for the M transmit-chains to each receive chain is always a good estimate of the MIMO self-interference channels. Moreover the self-interference cancellation using delay and attenuation in RF domain has its own limitations in recreation of long delays. Overall, self-interference cancellation for MIMO systems still needs careful consideration and novel approaches. Moreover, to improve the confidence, these solutions need to be implemented on the real transceivers and the performance shall be compared to MIMO systems in terms of achievable throughput in various channels. Meanwhile digital solutions can help improving the overall self-interference cancellation of the FD MIMO systems.

8.4 Full-duplex transmission in wireless networks

8.4.1 General

FD transmission is an attractive air interface technology to improve system performance in future wireless systems, and it has gained considerable attention from academia and industry recently. Versatile overview of recent advances in FD transceiver technology and system solutions is presented, e.g., in [9,14,15]. This section introduces some of the system or network level results obtained in the DUPLO project as well as points out significant research results achieved in other research papers.

In infrastructure-based networks, such as cellular networks, scheduling plays pivotal role in harvesting the benefits of FD transmission. Simultaneous uplink and downlink transmission introduced with the FD mode results in additional intra-cell and inter-cell interference. Therefore, to achieve the potential gain of the FD mode, an intelligent scheduler to pair the downlink and uplink users with corresponding transmission powers is necessary. Due to the combinatorial nature of pairing multiple uplink and downlink users on each subcarrier, and also the difficulty of optimal power allocation to each subcarrier, resource allocation in FD systems to maximise the performance of the network is very challenging.

A resource allocation algorithm using matching theory to allocate the subcarriers among uplink and downlink users was proposed in [51]. A cell partitioning algorithm to allocate the frequency resources has been proposed in [52]. Both [51,52] consider a single cell FD OFDMA system, and the proposed algorithms in these papers cannot be directly applied for FD multi-cell systems. Moreover, in [52] a user can be allocated to only one subcarrier, and thus the transmission power of the uplink users is fixed, and cannot be adjusted. A suboptimal scheduling algorithm that selects the transmission direction of each user in a multi-cell scenario was proposed in [53]. In this system, interference from the users of neighbouring cells is ignored. A hybrid scheduler that can switch between FD and half-duplex mode has been proposed in [54] for the single small-cell scenario. The power allocation has not been considered in both [53,54]. A user pairing algorithm has been proposed in [55] for the multi-cell FD system without taking the interference from the neighbouring cell users into account, which makes the resource allocation problem easier. Moreover, although the transmission power of the uplink users can be adjusted, the transmission power of the base-stations is assumed to be fixed in [56].

In mobile ad hoc networks, FD transmission helps to solve hidden node problems and reduce collisions, and thus provides means to increase network throughput [3].

In an ideal two-hop FD relaying scenario where the self-interference at the relay node is perfectly cancelled, it has been shown that FD relay systems in general achieve an end-to-end ergodic capacity twice as high as that of the half-duplex systems [57]. Limited self-interference cancellation capability in the relaying node will reduce the FD gain. However, e.g., with infrastructure relays, the self-interference can be efficiently cancelled with antenna solutions. In multihop relays and mobile ad hoc networks the main challenges with FD transmission are related to increased inter-node interference as all the active nodes are transmitting simultaneously. One solution is to use directional antennas to reduce inter-node interference [58]. Also an efficient routing protocol can be an important component for increasing end-to-end throughput.

Different distributed MAC protocols have been proposed for the FD transmission, including Contraflow [59] and FD-MAC [60]. The reported throughput gain of the FD-MAC protocol over half-duplex transmission is 70% [60]. Another MAC protocol proposed for FD radio is Janus [61]. This protocol is based on centralised scheduling mechanism, and it has been reported to achieve 2.5 times the throughput of a CSMA/CA based half-duplex system [61]. Other references in the literature include, e.g., [55,62], which introduce FD MAC protocols for relay applications.

8.4.2 Full-duplex transmission in small area radio communication systems

This section discusses potential use cases for FD transmission in small area radio communication systems, and evaluates the achievable performance gains of FD transmission over half-duplex (HD) transmission in different network set-ups. The discussed examples are based on system performance investigations conducted in the DUPLO project. The scenarios covered in this section are listed here below. It is worth noticing that the performance and the algorithms and protocols related to the assessed study cases are treated in summary fashion in this section and more detailed results and assumptions are available in [63–82].

1. A *point-to-point link* is an essential unit and basic building block of many wireless networks. For example, in a wireless local area network (WLAN), every transmission between user devices and their access point (AP) can be regarded as a point-to-point link during the transmission period. Yet another example would be wireless backhauling in small cell deployments on mobile platforms such as trains, cars or airplanes.
2. *Standalone LTE* [83–86] *small cell* with an FD base station serving multiple UEs which operate (a) in half duplex mode or (b) in full duplex mode. A typical realisation of this use case could be a small cell deployed at home in, e.g., residential area.
3. *Multiple LTE* [83–86] *small cells* with an FD base station serving UEs which operate (a) in half duplex mode or (b) in full duplex mode. A typical realisation of this use case could be a small cell deployment in an office space, sports arena or shopping centre to name a few.

8.4.2.1 A point-to-point link

A point-to-point connection between two nodes forms the basic use case for FD transmission. The first results on the FD link performance are based on comparison of achievable rate region and sum power efficiency for bi-directional FD or half-duplex point-to-point connection utilising cyclic prefix (CP) assisted OFDM waveform for the signal transmission [63,64]. The impact of the transceiver imperfection on the FD link performance has been included in the system model. The non-ideality is quantified as Error Vector Magnitude (EVM) level, and it is modelled as Gaussian noise added to the original signal by the transmitter. The assumed FD transceiver suppresses the self-interference via a three-stage process. The stages are antenna isolation, RF cancellation and digital baseband cancellation.

Figure 8.19 shows the achievable rate region for half-duplex and FD links with different transmit power levels, assuming that the link distance between the nodes is 20 m. The results indicate, that with short transmission distances, the maximum sum rate achieved by the FD link is almost double as that of the half-duplex link

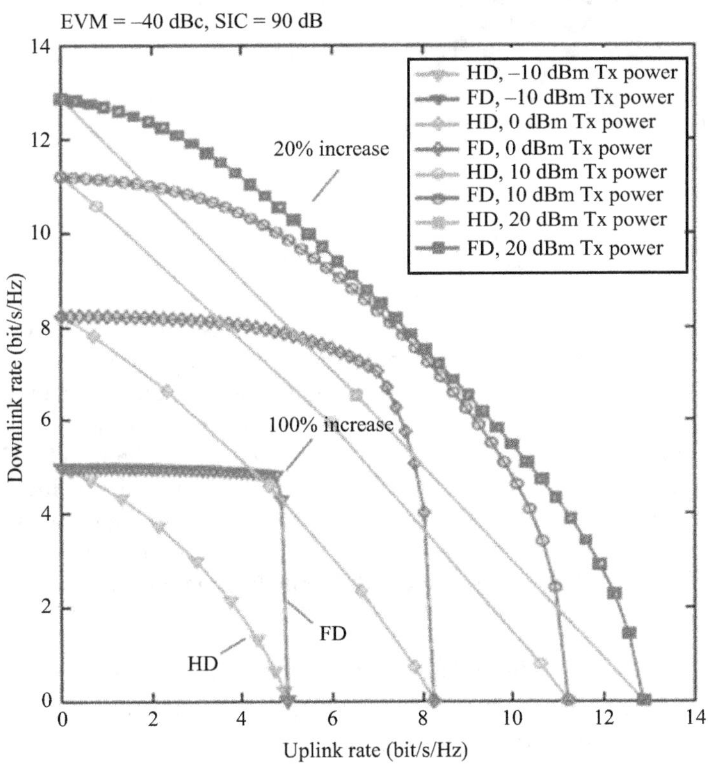

Figure 8.19 FD and half-duplex rate regions with different maximum transmission powers (−10 dBm, 0 dBm, 10 dBm, 20 dBm). The half-duplex and FD rate regions with −10 dBm transmit power marked with 'HD' and 'FD', respectively

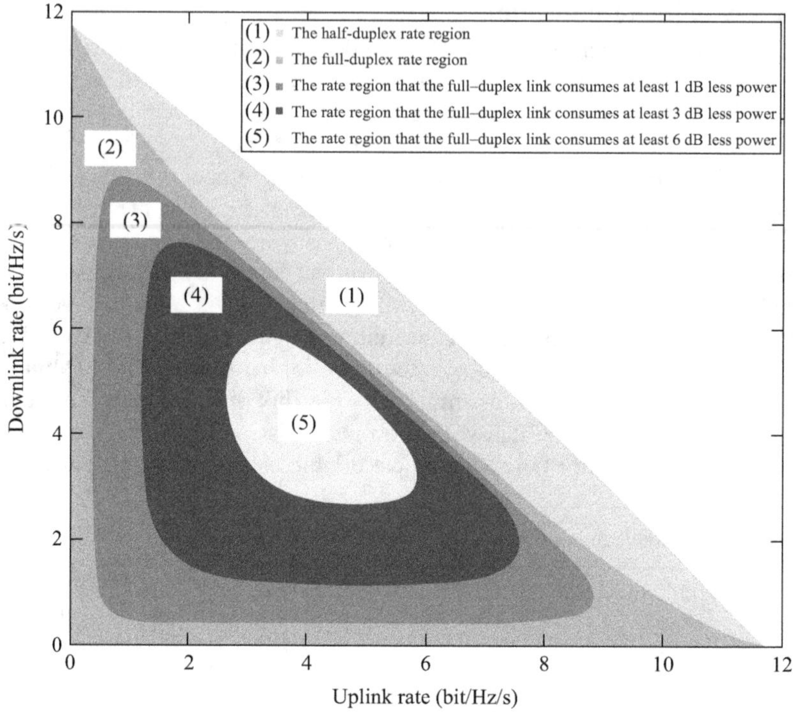

Figure 8.20 Sum transmission powers of using FD and half-duplex technologies. © 2014 IEEE. Reprinted with permission from [63]

when the transmission power is −10 dBm. As the transmission power is increased to 20 dBm, the maximum sum rate improvement is reduced to about 20% [63].

Energy efficiency is also important to be assessed in the comparative framework of FD and half-duplex communications. For that, the consumed transmission power to achieve a certain rate region is evaluated next. Figure 8.20 shows the achievable rate region contours with different power consumption values. It can be concluded that the FD link can achieve over 75% of its rate region with 1 dB less power consumption than that of the half-duplex link. Furthermore, in the area around downlink–uplink rate pair {4.5, 4.5} bit/Hz/s the power efficiency advantage is larger than 6 dB.

8.4.2.2 Single standalone full-duplex small cell

In the first study case in this section a single cell with 200 m cell radius is considered, where the users' locations are randomly generated and uniformly distributed within a cell. The FD base station has 30 dBm maximum transmit power, and the users each have 23 dBm transmit power. User devices are operating in half-duplex mode. The total number of users is 20, with 10 downlink users and 10 uplink users. Base station uses a game theory based iterative algorithm to jointly optimise uplink power allocations and downlink subcarrier and power allocations to maximise sum rate of the system [80].

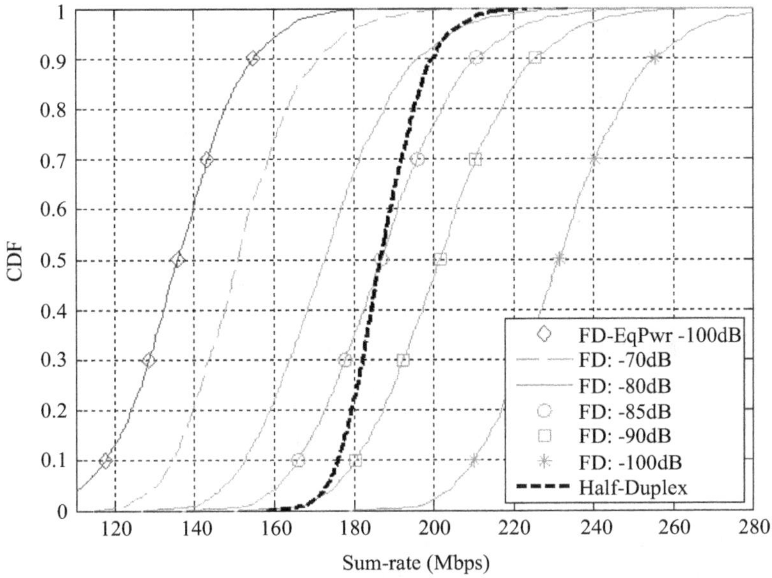

Figure 8.21 Sum-rate comparison for FD and half-duplex systems. © 2015 IEEE. Reprinted with permission from [80]

Figure 8.21 shows the cumulative distribution function (CDF) of the sum-rate for FD system, with different self-interference cancellation values. As benchmarks for comparison, FD with equal power allocation and half-duplex system (TDD) with optimal subcarrier and power allocation are used. In this case, it can be concluded that with 85 dB SIC, the FD system performance equals HD system performance but with SIC of 110 dB, 40% increase in performance is observed. Furthermore, the results show that the proposed algorithm significantly improved the FD performance (by about 70%) comparing to equal resource power allocation, emphasising the importance of radio resource management scheme in harvesting the gains of FD transmission. Another study case with picocell characteristics (cell radius 40 m, BS transmit power 23 dBm, UE transmit power 24 dBm) and smart scheduling and power allocation strategy in the base station concluded that the self-interference needs to cancel at least 80 dB to get performance gains with FD over half-duplex transmission [70].

In [81], the resource allocation algorithm [80] is extended to the FD-BS with multiple FD-UEs scenario, where the subcarriers are exclusively allocated to the users to ensure orthogonality among the users and eliminate the inter-user interference. Furthermore, the FD system performance is evaluated under different system settings, such as cell coverage area and the nodes' maximum transmission power. The system-level evaluation has shown that FD technique superiority to HD is more prominent in the case of cells with small coverage area and low transmission power nodes. For example, with 5 dBm maximum transmission power of BS and UEs, 39%, 50% and 70% gains can be achieved by FD with 90, 95 and 100 dB self-interference cancellation factors, respectively. Thus with proper radio resources allocation, considerable gain in spectral efficiency can be achieved by

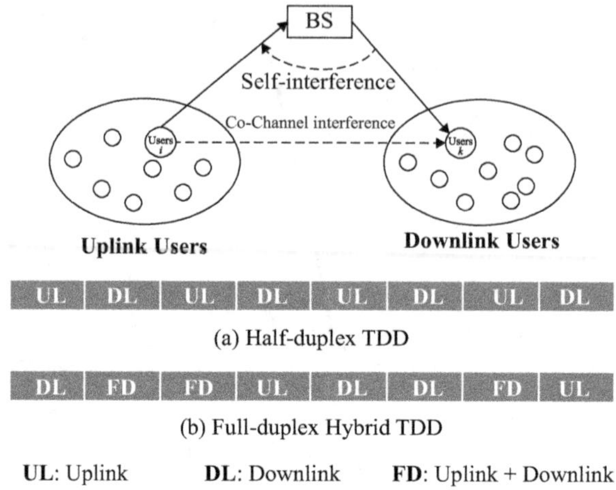

Simulation parameters

Parameter	**Settings**
Cell radius	40 m
Number of subcarriers	$N = 1024$
$[\lvert\mathcal{I}^{UL}\rvert, \lvert\mathcal{K}^{DL}\rvert]$	[10, 10]
Bandwidth	10 MHz
Maximum BS power	$P_T = 24$ dBm
Maximum user power	$P_i = 23$ dBm
Thermal noise density	–174 dBm/Hz
Noise figure	BS: 13 dB, User: 9 dB
Path loss (dB) between BS and users (*d* in km)	LOS: $103.8 + 20.9 \log_{10} d$ NLOS: $145.4 + 37.5 \log_{10} d$
Path loss (dB) between users (*d* in km)	LOS: $98.45 + 20 \log_{10} d$, $d \leq 50$ m NLOS: $175.78 + 40 \log_{10} d$, $d \leq 50$ m
Shadowing standard deviation	LOS: 10 dB NLOS: 12 dB

Figure 8.22 System model and simulation parameters used with hybrid FD/HD scheduling. © 2015 IEEE. Reprinted with permission from [71]

using the FD technique with the currently feasible self-interference cancellation capabilities.

Depending on the locations of the mobile users, propagation channels, the self-interference cancellation capability of the BS, transmission power of the mobile users and BS, etc., it might sometimes be better to use HD transmission mode instead of FD transmission. Therefore, a dynamic hybrid scheduler that can switch between HD uplink, HD downlink and FD mode opportunistically to maximise the sum-rate has been proposed in [71]. The system model is shown in Figure 8.22. In Figure 8.23, it is shown that at high self-interference cancellation values, FD

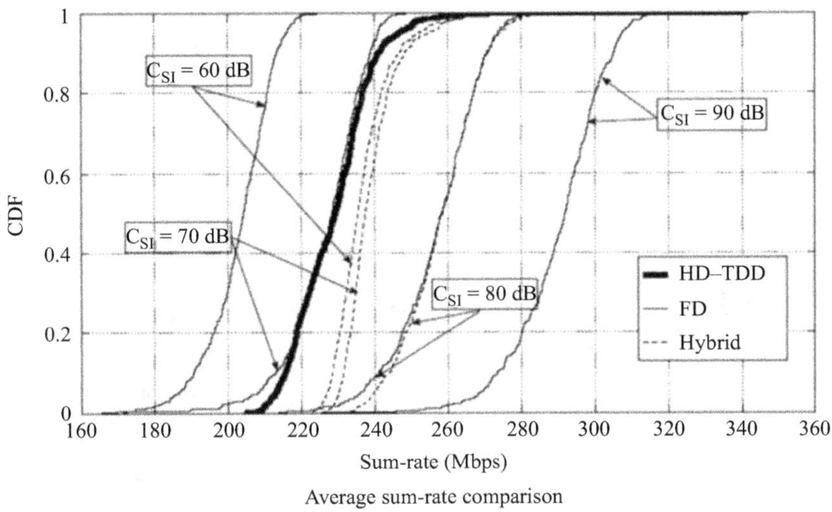

Figure 8.23 Average sum-rate comparison in single standalone cell for FD, HD and hybrid FD/HD schemes. © 2015 IEEE. Reprinted with permission from [71]

scheduling outperforms HD-TDD scheduling, and hybrid scheduling switches to the FD scheduling, so it is beneficial to allocate each time slot to simultaneous uplink and downlink transmission. On the other hand, at low self-interference cancellation values, HD-TDD scheduling outperforms FD scheduling, and hybrid scheduling starts switching to HD-TDD scheduling, so it is beneficial to allocate time slots either uplink or downlink transmission.

System performance studies with partial channel state information reported in [72] show how the average goodput performance of the FD system (uplink plus downlink) under different channel estimation errors the performance degrades as the estimation error increases. There can easily be a 10%–20% system performance decrease with moderate channel estimation errors, thus highlighting the importance of high quality channel estimators with FD operation.

8.4.2.3 Multiple full-duplex small cells

This section provides first system simulation results for FD operation in multiple LTE small cells environment, assuming that there is only single FD user equipment (UE) per cell. Therefore, the focus is on investigating the impact of inter-cell interference due to FD transmission. As a result, achievable system throughput with different self-interference cancellation values are provided. Half-duplex TDD operation mode is used as the reference scheme in comparisons. User scheduling and power control is not considered in the simulations. Two types of multiple cell deployment scenarios are considered, namely indoor and outdoor scenarios. The deployment scenarios follow the 3GPP femto and picocell models [84], and system simulation parameters are given in [82]. For both the indoor and outdoor scenarios, the average UE throughput results in the DL and UL directions are provided.

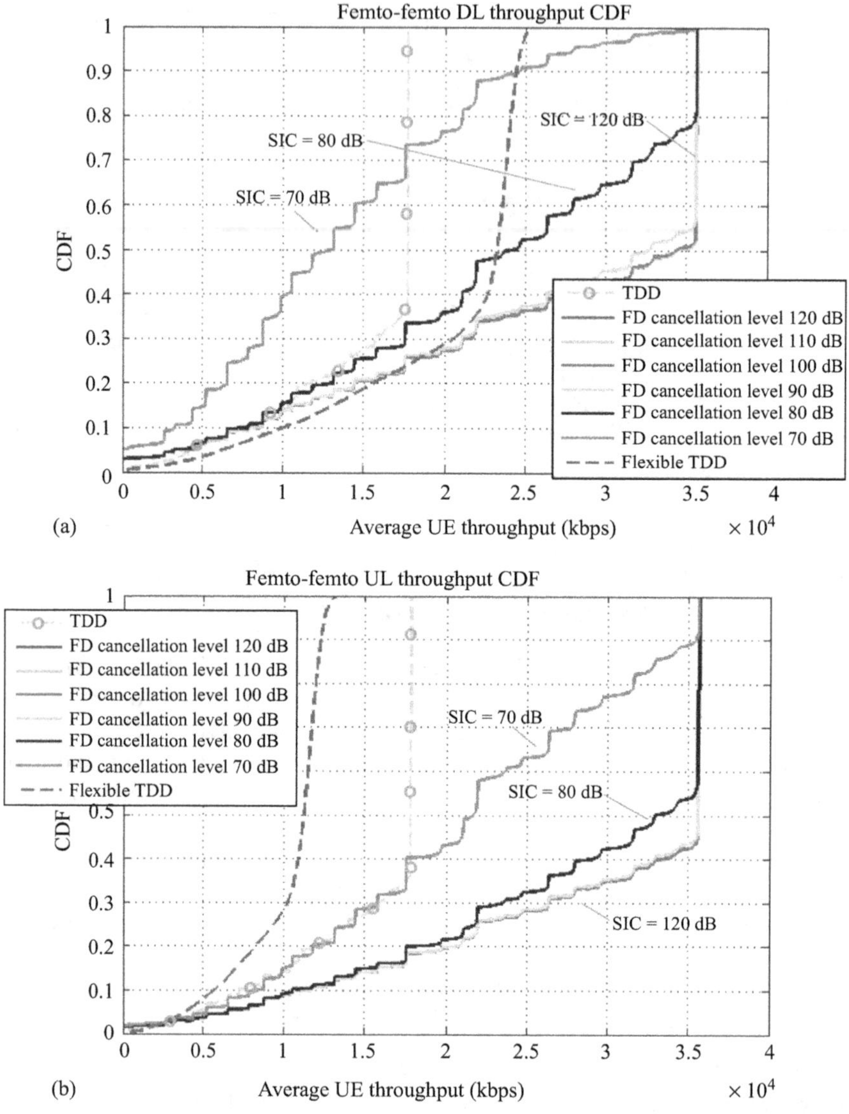

Figure 8.24 Indoor (a) DL and (b) UL average UE throughput

Figure 8.24(a) and (b) present the CDF of the average UE throughput in the DL and UL of the indoor scenarios, respectively. From these figures, the maximum throughput in the FD deployment becomes twice as much the one achieved in TDD mode, as expected. It is worth noting that the FD scheme not only outperforms the TDD in the low SINR regime, but also guarantees the maximum achievable throughput for about 50% of the FD UEs if the self-interference cancellation level is at least 80 dB. By comparing the FD and TDD coverage in this particular scenario, the former is only slightly worse than the latter in the DL, while both are

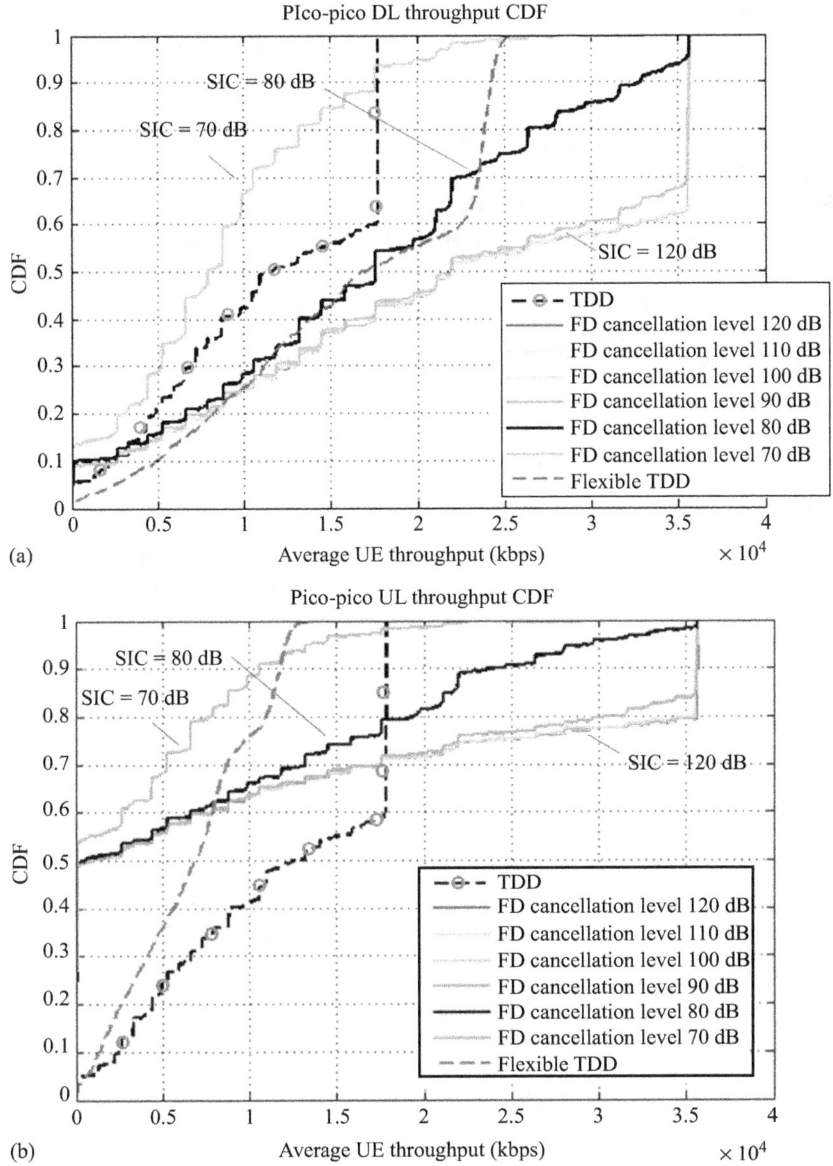

Figure 8.25 Outdoor (a) DL and (b) UL average UE throughput

equal in the UL. While the flexible TDD scheme [82] outperforms the standard TDD in the DL, it performs poorer in the UL. In all cases, excluding FD with 70 dB self-interference cancellation, the used modulation and coding scheme (MCS) limits the achievable maximum throughput.

Figure 8.25(a) and (b) illustrates the average UE throughput in the DL and UL of the outdoor scenarios, respectively. In the DL of these scenarios, the FD scheme with self-interference cancellation of at least 80 dB achieves higher throughput than the standard TDD while providing almost similar coverage (low values of CDF). Moreover, about 40% of the FD UEs achieve the maximum throughput in the DL direction. In the UL, while nearly 30% of the FD UEs with self-interference cancellation of at least 90 dB achieve throughput higher than that of the TDD users, only about 20% of them perform as well when the cancellation level is 80 dB. The coverage is actually the main limitation of this scenario: only 5% of the TDD UEs are in outage in contrast with nearly 50% of the FD ones. The reason is the additional inter-cell-interference in the FD deployments, more specifically the DL to UL cross interference and poor isolation between the outdoor BSs.

Based on these results, it is evident that in indoor scenarios with good isolation between BSs, the full duplex configuration can outperform the TDD schemes in terms of average UE throughput while providing similar coverage as long as the self-interference cancellation level is at least 80 dB. In addition, whenever the isolation between the BSs is poor, both the coverage and the achievable data rates of FD-enabled UEs decrease due to high DL to UL cross interference. However, these results are limited to scenarios with a single UE per cell site and further studies are needed to confirm if they still hold in multiuser deployments (more than one UE per cell). On the other hand, these results clearly point the need for device

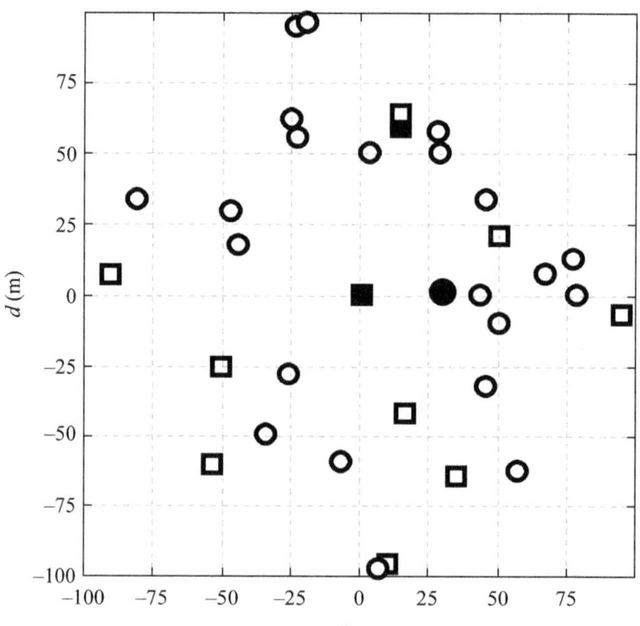

Figure 8.26 Illustration of a random deployment of small cells and user terminals over an arbitrary network area. Circles represent user terminals, while squares depict small cell base stations. At the centre is depicted the typical pair of BS and UE (filled circle and square). © 2014 IEEE. Reprinted with permission from [68]

interference mitigation techniques to reduce the DL to UL cross interference and, as a result, to enable full duplex communications in scenarios with poor isolation between nodes, especially BSs with high transmission power. For instance, power control and inter-cell-interference cancellation or coordination schemes are envisaged as promising for interference management solutions.

Another small cell study scenario considers a random network topology where UEs and small cell BSs are uniformly scattered over network area, see Figure 8.26. The devices are able to transmit and receive simultaneously at a cost of self-interference. Even though employing advanced interference cancellation and antenna attenuation techniques, there still remains a residual self-interference level, which can be modelled as a random variable that allows one to emulate various (non) line-of-sight configurations. Generally, this analytical framework is based on the stochastic geometry concepts [67,68]. The network topology is modelled using a Poisson point process, while wireless links follow a composite Log-Normal (LN) shadowing and Nakagami-m fading model. The performance of this network is assessed in terms of spectral efficiency as well as outage probability, whose formulations and detailed discussion are presented in [67,68].

Figure 8.27 shows that, in terms of average spectral efficiency (ASE), FD schemes become more attractive and outperform the HD scheme even though suffering additional interference. For instance, for self-interference cancellation

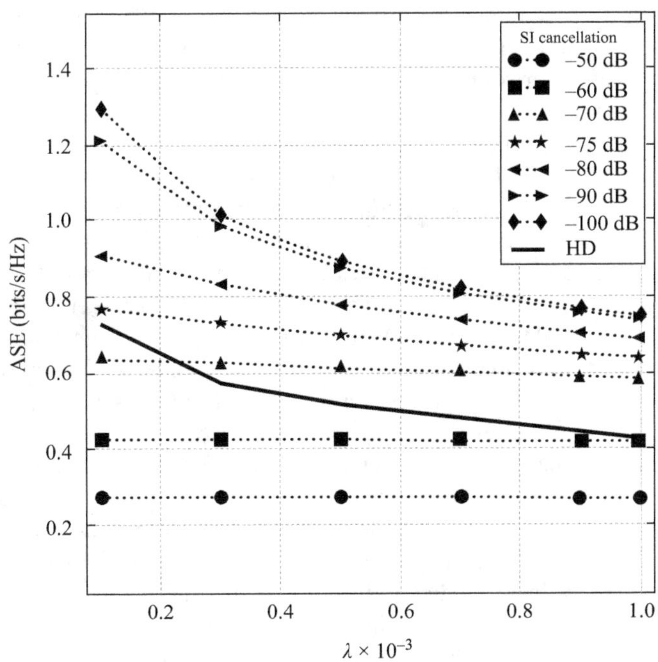

Figure 8.27 Spectral efficiency as a function of the density of the nodes. Different SI cancellation values are considered for the FD case. © 2014 IEEE. Reprinted with permission from [68]

levels of 75 dB (marked as −75 dB in Figure 8.27), the FD scheme presents 50% higher spectral efficiency than HD ones. Even larger gains can be obtained if self-interference cancellation is improved further.

8.5 Summary

FD technology has gained a lot of attention over the last couple of years. The attention has been accelerated because novel and interesting techniques are being developed which attempt to solve one of the main problems of FD communications: the self-interference in the radio nodes. The most promising techniques come from research groups which consider a system approach where the self-interference is rejected and cancelled at different locations in the radio, i.e., at the antenna and in the RF and the digital domain. The interest in applying FD to future evolution of wireless systems has risen significantly. FD wireless transmission is considered as a promising air interface technique for 5G wireless systems as it tackles key areas where system improvements are needed, such as spectral efficiency, link level capacity and network latency.

To achieve an overall self-interference rejection of more than 100 dB, a combination of self-interference isolation and cancellation techniques is required. Although this target is challenging, a recent transceiver design reported in [6] shows already that it is achievable even with single port antenna solution. To achieve this in densely integrated radio consists a further challenge. This target is important as it would open the FD technology applications to compact form factor devices. The transceiver solutions discussed in this chapter are potential candidates for compact form factor FD devices, although their first prototypes still fall behind targeted 100 dB value in performance. Thus, future opportunities and development needs are related to performance improvements (self-interference cancellation capability and bandwidth) and design improvements (improved linearity, architectural solutions to reduce phase noise, improved baseband algorithms).

Smart phones and other radio devices support multiple radio protocols and frequency bands. Therefore, from practical wireless transceiver design point of view it is important that in the FD transceiver design also the integration with other radio protocols in the same radio device is considered. Transceiver design supporting flexible duplexing is also a potential future research direction. The opportunity for FD in larger form-factors should also be considered in relation to future 5G, for instance, with connected cars. Furthermore, development of FD techniques for multi-antenna systems is another potential research direction.

Potential use cases for FD in 5G networks are FD BS serving HD UEs, FD UE, device-to-device connections, relaying, transmission in mesh networks and (self-) backhauling. System level studies indicate, that in small area systems, FD transmission can provide performance gains over half-duplex even with moderate self-interference cancellation levels in the FD transceiver. Thus, it is not necessary to design FD operation based on the worst case link budget in order to benefit at the wireless system level. The system level solution could be adaptive to change of the operation mode between half-duplex and FD transmission. Naturally, high self-interference cancellation capability in the FD transceiver is beneficial in maximising the system level gains of FD transmission.

Radio resource management is a crucial point to consider when aiming at utilising FD transmission efficiently in wireless networks. With the introduction of 5G networks, there will be significant increase in the number of small cells. Thus, coordination, self-organisation, flexible and efficient use of spectral resources, scheduling and radio resource management in general are aspects that will be fundamental for the next generation of wireless communications. For example, the impact of signalling limitations and imperfect channel state information (e.g., in scheduling algorithms) on the system performance, solutions to combat with inter-cell interference, and the co-existence with half-duplex systems require further studies to get more comprehensive view on how FD could be best deployed to future 5G systems. Also MAC protocol and routing solutions can be optimised further to improve their performance in practical mobile ad hoc networks.

Overall, the results discussed in this chapter are encouraging, and pave path for introducing FD transmission as an air interface technology for future 5G systems.

References

[1] Next Generation Mobile Networks (NGMN) Alliance, "5G White Paper," Feb. 2015.

[2] Duarte M., and Sabharwal A. "Full-duplex wireless communications using off-the-shelf radios: feasibility and first results," *Asilomar Conference on Signals, Systems and Computers*, Pacific Grove, CA, USA, Nov. 2010, pp. 1558–1562.

[3] Choi J., Jain M., Srinivasan K., Levis P., and Katti S. "Achieving single channel, full duplex wireless communication," *International Conference on Mobile computing and Networking (MobiCom)*, New York, NY, USA, Sep. 2010, pp. 1–12.

[4] Radunovic B., Gunawardena D., Key P., *et al.* "Rethinking indoor wireless mesh design: low power, low frequency, full duplex," *Wireless Mesh Networks (WIMESH 2010)*, Boston, MA, USA, Jun. 2010, pp. 1–6.

[5] Khandani A. "Two-way (true full-duplex) wireless,"*13th Canadian Workshop on Information Theory (CWIT)*, Toronto, Canada, Jun. 2013, pp. 33–38.

[6] Bharadia D., McMilin E., and Katti S. "Full duplex radios," *ACM Proceedings of the SIGCOMM Conference*, Aug. 2013, pp. 375–386.

[7] Everett E., Sahai A., and Sabharwal A. "Passive self-interference suppression for full-duplex infrastructure nodes," *IEEE Transactions on Wireless Communications*, 2014, vol. 13 (2), pp. 680–694.

[8] Debaillie B., van den Broek D.J., Lavín C., *et al.* "RF self-interference reduction techniques for compact full-duplex radios," *IEEE Vehicular Technology Conference (VTC-Spring)*, Glasgow, Scotland, May 2015, pp. 1–6.

[9] Sabharwal A., Schniter P., Guo D., Bliss D., Rangarajan S., and Wichman R. "In-band full-duplex wireless: challenges and opportunities," *IEEE Journal on Selected Areas in Communications (JSAC)*, 2014, vol. 32 (9), pp. 1637–1652.

[10] Zheng G., Krikidis I., Li J., Petropulu A.P., and Ottersten B. "Improving physical layer secrecy using full-duplex jamming receivers," *IEEE Transactions on Signal Processing*, 2013, vol. 61 (20), pp. 4962–4974.

[11] O'Hara J.F., and Moore G.M., 'A high performance CW receiver using feedthru nulling', *Microwave Journal*, 1963, vol. 6 (9), pp. 63–71.

[12] Slingsby W.T., and McGeehan J. "A high-gain cell enhancer," *IEEE 42nd Vehicular Technology Conference*, Denver, CO, 1992, vol. 2, pp. 756–758.

[13] 3GPP RWS-150074, "Draft Report of 3GPP RAN workshop on 5G," Sep. 2015.

[14] Hong S., Brand J., and Choi J. "Applications of self-interference cancellation in 5G and beyond," *IEEE Communications Magazine*, 2014, vol 52 (2), pp. 114–121.

[15] Kim D., Lee H., and Hong D. "A survey of in-band full-duplex transmission: from the perspective of PHY and MAC layers," *IEEE Communications Surveys & Tutorials*, 2015, vol. 17, no. 4, pp. 2017–2046.

[16] Khojastepour M., Aryafar E., Sundaresan K., Mahindra R., and Rangarajan S. "Exploring the potential for full-duplex in legacy LTE systems," in *Proceedings of the SECON 2014*, Singapore, Jun. 2014, pp. 10–18.

[17] Heo J., Ju H., Park S., Kim E., and Hong D. "Simultaneous sensing and transmission in cognitive radio," *IEEE Transactions on Wireless Communications*, 2014, vol. 13 (4), pp. 1948–1959.

[18] Chang K., Ju H., Kim S., *et al.* "In-band full duplex radios and system performance," IEEE 802.11-15-0043-01-00ax, Jan. 2015.

[19] Meng C., Qiao L., Hongli J., Ping C., and Cheng H. "A 15 GHz full duplex system for microwave backhauling," *IEEE Vehicular Technology Conference (VTC-Spring)*, Glasgow, Scotland, May 2015, pp. 1–5.

[20] Rikkinen K., Tapio V., Alves H., *et al.* "Full-duplex transmission in small area radio communication systems," in *International Workshop on Computer-Aided Modelling Analysis and Design of Communication Links and Networks (CAMAD 2015)*, 2015, Guildford, UK, pp. 1–5.

[21] Rikkinen K. (ed.) "Full-duplex radios for local access, final report," INFSO-ICT-316369 DUPLO – Report D6.3, 2015, available at http://www.fp7-duplo.eu/index.php/deliverables

[22] Debaillie B., van den Broek D.J., Lavin C., *et al.* "Analogue/RF solutions enabling compact full-duplex radios," *IEEE Journal on Selected Areas in Communications (JSAC)*, 2014, vol. 32 (9), pp. 1662–1673.

[23] Khandani A. "Methods for spatial multiplexing of wireless two-way channels," US Patent no. 7817641, issue date: Oct 19, 2010.

[24] Duarte M., Dick C., and Sabharwal A. "Experiment-driven characterization of full-duplex wireless systems," *IEEE Transactions on Wireless Communications*, 2012, vol. 11 (12), pp. 4296–4306.

[25] van den Broek D.J., Klumperink E.A.M., and Nauta B. "A self-interference-cancelling receiver for in-band full-duplex wireless with low distortion under cancellation of strong LX leakage," in *International Solid-State Circuits Conference (ISSCC)*, San Francisco, CA, USA, Feb. 2015, pp. 1–3.

[26] van den Broek D.J., Klumperink E., and Nauta B. "An in-band full-duplex radio receiver with a passive vector modulator downmixer for self-interference cancellation", *IEEE Journal of Solid-State Circuits (JSSC)*, 2015, vol. 50, no. 12, pp. 3003–3014.

[27] van Liempd B., Lavin C., Malotaux S., *et al.* "RF self-interference cancellation for full-duplex," in *International Conference on Cognitive Radio Oriented Wireless Networks (CrownCom)*, Oulu, Finland, Jun. 2014, pp. 526–531.

[28] Mikhael M., van Liempd B., Craninckx J., Guindi R., and Debaillie B. "A full-duplex transceiver prototype with in-system automated tuning of the RF self-interference cancellation," in *International Conference on 5G for Ubiquitous Connectivity (5GU)*, Levi, Finland, Nov. 2014, pp. 110–115.

[29] van Liempd B., Hershberg B., Raczkowski K., *et al.* "A >70 dBm IIP3 single-ended electrical-balance duplexer in 0.18um SOI CMOS," in *International Solid-State Circuits Conference (ISSCC)*, San Francisco, CA, USA, Feb. 2015, pp. 1–3.

[30] Starr T., Cioffi J., and Silverman P. *Understanding Digital Subscriber Line*, Prentice Hall, Upper Saddle River, NJ, 1998.

[31] Nasr K., Cosmas J., Bard M., and Gledhill J. "Performance of an echo canceller and channel estimator for on-channel repeaters in DVB-T/H Networks," *IEEE Transactions on Broadcasting*, 2007, vol. 53 (3), pp. 609–618.

[32] Riihonen T., Werner S., Wichman R., and Zacarias E. "On the feasibility of full-duplex relaying in the presence of loop interference," *10th IEEE Workshop on Signal Processing Advances in Wireless Communications (SPAWC)*, Perugia, Italy, Jun. 2009, pp. 275–279.

[33] Schreurs D., O'Droma M., Goacher A.A., and Gadringer M. (eds.) *RF Power Amplifier Behavioral Modelling*, Cambridge University Press, Cambridge, UK, 2009.

[34] Anttila L., Korpi D., Syrjala V., and Valkama M. "Cancellation of power amplifier induced nonlinear self-interference in full-duplex transceivers," in *Asilomar Conference on Signals, Systems and Computers*, Pacific Grove, CA, USA, Nov. 2013, pp. 1193–1198.

[35] Balatsoukas-Stimming A., Austin A.C.M., Belanovic P., and Burg A. "Baseband and RF hardware impairments in full-duplex wireless systems: experimental characterization and suppression," *EURASIP Journal on Wireless Communications and Networking (2015)*, 2015:142, pp. 1–11.

[36] Korpi D., Anttila L., Syrjala V., and Valkama M. "Widely linear digital self-interference cancellation in direct-conversion full-duplex transceiver," *IEEE Journal on Selected Areas in Communications*, 2014, vol. 32 (9), pp. 1674–1687.

[37] Mango Communications, "WARP v3 User Guide." Available at http://warpproject.org/trac/wiki/HardwareUsersGuides/WARP3

[38] Mango Communications, "WARPLab Reference Design." Available at http://warpproject.org/trac/wiki/WARPLab

[39] Lavin C. (ed.) "Final proof-of-concept validation, results and analysis," INFSO-ICT-316369 DUPLO – Report D5.2, 2015, available at http://www.fp7-duplo.eu/index.php/deliverables

[40] Mietzner J., Schober R., Lampe L., Gerstacker W., and Hoeher P. "Multiple-antenna techniques for wireless communications–a comprehensive literature survey," *IEEE Communications Surveys Tutorials*, 2009, vol. 11 (2), pp. 87–105.

[41] Day B., Margetts A., Bliss D., and Schniter P. "Full-duplex bidirectional MIMO: achievable rates under limited dynamic range," *IEEE Transactions on Signal Processing*, 2012, vol. 60 (7), pp. 3702–3713.

[42] Day B., Margetts A., Bliss D., and Schniter P. "Full-duplex MIMO relaying: achievable rates under limited dynamic range," *IEEE Journal on Selected Areas in Communications*, 2012, vol. 30 (8), pp. 1541–1553.

[43] Foroozanfard E., Franek O., Tatomirescu A., Tsakalaki E., de Carvalho E., and Pedersen G. "Full-duplex MIMO system based on antenna cancellation technique," *Electronics Letters*, 2014, vol. 50 (16), pp. 1116–1117.

[44] Aryafar E., Khojastepour M.A., Sundaresan K., Rangarajan S., and Chiang M., "MIDU: enabling MIMO full-duplex," in *18th Annual International Conference on Mobile Computing and Networking*, Mobicom'12, Istanbul, Turkey, 2012.

[45] Riihonen T., Werner S., and Wichman R. "Mitigation of loopback self-interference in full-duplex MIMO relays," *IEEE Transactions on Signal Processing*, 2011, vol. 59 (12), pp. 5983–5993.

[46] Liang D., Xiao P., Chen G., Ghoraishi M., and Tafazolli R. "Digital self-interference cancellation for full-duplex MIMO systems," in *International Wireless Communications and Mobile Computing Conference (IWCMC)*, Dubrovnik, Croatia, Aug. 2015, pp. 403–407.

[47] Huberman S., and LeNgoc T. "MIMO full-duplex precoding: a joint beamforming and self-interference cancellation structure', *IEEE Transactions on Wireless Communications*, 2015, vol. 14 (4), pp. 2205–2217.

[48] Anttila L, Korpi D., Antonio-Rodriguez E., Wichman R., and Valkama M. "Modelling and efficient cancellation of nonlinear self-interference in MIMO full-duplex transceivers," in *Globecom Workshops*, Austin, TX, 2014, pp. 777–783.

[49] Korpi D., Anttila L., and Valkama M. "Reference receiver based digital self-interference cancellation in MIMO full-duplex transceivers," in *Globecom Workshops*, 2014, Austin, Texas, USA, pp. 1001–1007.

[50] Bharadia D., and Katti S. "Fullduplex MIMO radios," SIGCOM, 2013, pp. 1–13.

[51] Di B., Bayat S., Song L., and Li Y. "Radio resource allocation for full-duplex ofdma networks using matching theory," in *IEEE Conference on Computer Communications Workshops (INFOCOM)*, Toronto, Canada, May 2014, pp. 197–198.

[52] Shao S., Liu D., Deng K., Pan Z., and Tang Y. "Analysis of carrier utilization in full-duplex cellular networks by dividing the co-channel interference region," *IEEE Communications Letters*, 2014, vol. 18 (6), pp. 1043–1046.

[53] Shen X., Cheng X., Yang L., Ma M., and Jiao B. "On the design of the scheduling algorithm for the full duplexing wireless cellular network," in *IEEE GLOBECOM*, 2013, Atlanta, GA, USA, Dec. 2013, pp. 4970–4975.

[54] Goyal S., Liu P., Panwar S., *et al.* "Improving small cell capacity with common-carrier full duplex radios," in *IEEE International Conference on Communications (ICC 2014)*, Sydney, Australia, Jun. 2014, pp. 4987–4993.

[55] Ari R., Kenta T., Yusuke S., Masaki B., Shunsuke S., and Takashi W. "Full duplex media access control by monitoring traffic on adjacent nodes for wireless multi-hop networks," in *IEICE*, Toyama, Japan, 2012, p. 419.

[56] Choi H.-H. "On the design of user pairing algorithms in full duplexing wireless cellular networks," in *International Conference on Information and Communication Technology Convergence (ICTC)*, Busan, Korea, 2014, pp. 490–495.

[57] Cadambe V.R., and Jafar S.A. "Degrees of freedom of wireless networks with relays, feedback, cooperation, and full duplex operation," *IEEE Transactions on Information Theory*, 2009, vol. 55 (5), pp. 2334–2344.

[58] Miura K., and Bandai M. "Node architecture and MAC protocol for full duplex wireless and directional antennas," in *IEEE 23rd International Symposium on Personal, Indoor and Mobile Radio Communications (PIMRC)*, Sydney Australia, Sep. 2012, pp. 369–374.

[59] Singh N., Gunawardena D., Poutiere A., Radunovic B., Balan H., and Key P. "Efficient and fair MAC for wireless networks with self-interference cancellation," in *Modelling and Optimization in Mobile, Ad Hoc and Wireless Networks (WiOpt)*, Princeton, NJ, 2011, pp. 94–101.

[60] Sahai A., Patel G., and Sabharwal A. "Pushing the limits of full-duplex: design and real-time implementation," Rice tech report, Jul. 2011.

[61] Kim J., Mashayekhi O., Qu H., Kazandjieva M., and Levis P. "Janus: a novel MAC protocol for full duplex radio," *CTSR*, 2013, pp. 1–12.

[62] Tamaki K., Raptino H., Sugiyama Y., *et al.* "Full duplex media access control for wireless multi-hop networks," in *Vehicular Technology Conference (VTC Spring)*, Dresden, Germany, Jun. 2013, pp. 1–5.

[63] Li W., and Lilleberg J. "Full-duplex link performance under consideration of error vector magnitude," in *IEEE Wireless Communications and Networking Conference WCNC 2014*, Istanbul, Turkey, Apr. 2014, pp. 654–659.

[64] Li W, Lilleberg J., and Rikkinen K. "On rate region analysis of half- and full-duplex OFDM communication links', *IEEE Journal on Selected Areas in Communications (JSAC)*, 2014, vol. 32 (9), pp. 1688–1698.

[65] Seddar J. (ed.) "Radio resource management and protocol solutions for full-duplex systems," INFSO-ICT-316369 DUPLO – Report D4.2, 2015, available at http://www.fp7-duplo.eu/index.php/deliverables

[66] Alves H., da Costa D., Souza R., and Latva-aho M. "Performance of block-Markov full duplex relaying with self-interference in Nakagami-m

fading," *IEEE Wireless Communication Letters*, 2013, vol. 2(3), pp. 311–314.

[67] de Lima C.H.M., Nardelli P.H.J., Alves H., and Latva-aho M. "Full-duplex communications in interference networks under composite fading channel," in *2014 European Conference on Networks and Communications (EuCNC)*, Bologna, Italy, Jun. 2014, pp. 1–5.

[68] Alves H., de Lima C.H.M., Nardelli P.H.J., Souza R.D., and Latva-aho M. "On the average spectral efficiency of interference-limited full-duplex networks," in *International Conference on Cognitive Radio Oriented Wireless Networks (CrownCom)*, Oulu, Finland, Jun. 2014, pp. 550–554.

[69] Alves H., Souza R.D., and Latva-aho M. "Full-duplex relaying systems subject to co-channel interference and noise in Nakagami-m fading," in *IEEE Vehicular Technology Conference (VTC-Spring)*, Glasgow, Scotland, May 2015, pp. 1–5.

[70] Cirik A.C., Rikkinen K., and Rong Y. "A subcarrier and power allocation algorithm for OFDMA full-duplex systems," in *2015 European Conference on Networks and Communications (EuCNC)*, Paris, France, Jul. 2015, pp. 11–15.

[71] Cirik A.C., Rikkinen K., and Latva-aho M. "Joint subcarrier and power allocation for sum-rate maximization in OFDMA full-duplex systems," in *IEEE Vehicular Technology Conference (VTC-Spring)*, Glasgow, Scotland, May 2015, pp. 1–5.

[72] Cirik A.C., Rikkinen K., Wang R., and Hua Y. "Resource allocation in full-duplex OFDMA systems with partial channel state information," in *IEEE China Summit and International Conference on Signal and Information Processing (ChinaSIP)*, Chengdu, China, Jul. 2015, pp. 711–715.

[73] Cirik A.C., Wang R., Hua Y., and Latva-aho M. "Weighted sum-rate maximization for full-duplex MIMO interference channels," *IEEE Transactions on Communications*, 2015, vol. 63 (3), pp. 801–815.

[74] Cirik A.C. "Fairness considerations for full duplex multi-user MIMO systems,", *IEEE Wireless Communications Letters*, 2015, vol. 4 (4), pp. 361–364.

[75] Cirik A.C., Wang R., Rong Y., and Hua Y. "MSE based transceiver designs for full-duplex MIMO cognitive radios," *IEEE Transactions on Communications*, 2015, vol. 63 (6), pp. 2056–2070.

[76] Nguyen D., Tran L., Pirinen P., and Latva-aho M. "On the spectral efficiency of full-duplex small cell wireless systems," *IEEE Transactions on Wireless Communications*, 2014, vol. 13 (9), pp. 4896–4910.

[77] Nguyen D., Tran L.-N., Pirinen P., and Latva-aho M. "Precoding for full duplex multiuser MIMO systems: spectral and energy efficiency maximization," *IEEE Transactions on Signal Processing*, 2013, vol. 61 (16) pp. 4038–4050.

[78] Malik H., Ghoraishi M., and Tafazolli R. "Cross-layer approach for asymmetric traffic accommodation in full-duplex wireless network," in *European*

Conference on Networks and Communications (EUCNC) 2015, Paris, France, Jul. 2015, pp. 265–269.
[79] Seddar J., Khalife H., Al Safwi W., and Conan V. "A full duplex MAC protocol for wireless networks," in *IEEE International Wireless Communications and Mobile Computing Conference (IWCMC)*, Dubrovnik, Croatia, Aug. 2015, pp. 244–249.
[80] Al-Imari M., Ghoraishi M., Xiao P., and Tafazolli R. "Game theory based radio resource allocation for full-duplex systems," in *IEEE 81st Vehicular Technology Conference (VTC Spring)*, Glasgow, Scotland, May 2015, pp. 1–5.
[81] Al-Imari M., Ghoraishi M., and Xiao P. "Radio resource allocation and system-level evaluation for full-duplex systems," in *International Workshop on Computer-Aided Modelling Analysis and Design of Communication Links and Networks (CAMAD 2015)*, Guildford, UK, Sep. 2015, pp. 1–5.
[82] Pirinen P. (ed.) "Performance of full-duplex systems," INFSO-ICT-316369 DUPLO – Report D4.1.1, Jan 2014, available at http://www.fp7-duplo.eu/index.php/deliverables
[83] 3GPP TS 36.104 V10.2.0, "LTE; evolved universal terrestrial radio access (E-UTRA); base station (BS) radio transmission and reception (Release 10)," May 2011.
[84] 3GPP TR 36.828 V11.0.0, "Evolved universal terrestrial radio access (E-UTRA); further enhancements to LTE time division duplex (TDD) for downlink–uplink (DL–UL) interference management and traffic adaptation," (Release 11), Jun. 2012.
[85] 3GPP TR 25.996, "Spatial channel model for multiple input multiple output (MIMO) simulations," Release 11, Sep. 2012.
[86] 3GPP TR 36.828 V11.0.0 (2012-06), "Evolved universal terrestrial radio access (E-UTRA): further enhancements to LTE time division duplex (TDD) for downlink–uplink (DL–UL) interference management and traffic adaptation (Release 11)," Jun. 2012.

Chapter 9

Decoupled uplink and downlink access in heterogeneous networks

Hisham Elshaer[1,2], Maria A. Lema[1], Toktam Mahmoodi[1] and Mischa Dohler[1]

9.1 Introduction

Data centric UEs are currently not only demanding more capacity from wireless networks, but also service oriented Quality of Service (QoS) and comparable Quality of Experience (QoE) in both Downlink (DL) and Uplink (UL). With the introduction of Internet of Things (IoT), machine-to-machine communications, cloud services and the widely used social media, UEs and devices in general are increasingly more content generators than they were before. As a consequence, the new generations of mobile communications must devise strategies that improve the device's overall experience. Adding intelligence into the Media Access Control (MAC) and Radio Resource Control (RRC) layers has shown significant improvements in terms of data rate and fairness. However, the non-constant QoE of UEs along the cell area motivates to increase the number of serving eNBs by adding cells of shorter coverage, bringing the network closer to the UE. This change in the deployment allows to further improve the network capacity by enabling better load balancing among the cells and eliminating coverage holes. However, this new paradigm of system design brings with it significant open challenges to ensure a correct operation such as: backhaul improvement, mobility and interference management, cell association and UL/DL relationship.

One of the key issues when moving towards a user or service centric network is to provide the mobile network with sufficient flexibility to select the serving cell that better suits the device or service requirements. To this direction, Downlink and Uplink Decoupling (DUDe) goes one step further, and allows the user to independently transmit and receive to and from different base stations. Essentially, DUDe breaks the hard and classical constraint of cell selection based on downlink received power, and provides the network with the flexibility to associate users to different eNBs in the DL and UL.

[1]King's College London, Strand, London WC2R 2LS, United Kingdom
[2]Vodafone Group R&D, Vodafone House, The connection, RG14 2FN, Newbury, UK

This chapter describes in detail the DUDe technique. First, Section 9.2 goes through the main challenges HetNets must face in terms of radio planning and interference management, and reviews the potential solutions to address these problems. Thereupon, decoupling UL and DL is recognised as a new technique that can effectively improve the HetNet performance, as shown in Section 9.4, and also can impact positively the performance of other radio access technologies, such as Carrier Aggregation (CA), Coordinated Multipoint transmission and reception (CoMP) or millimetre wave (Section 9.5). The enabling radio access network architectures are discussed in Section 9.6 and higher layer opportunities and challenges are addressed in Section 9.7.

9.2 HetNets challenges in 4G networks

LTE-A has to face a variable QoE within cells because of the difference in spectral efficiency depending on the UE position. Adding spectrum or improving the link adaptation provides faster connectivity, however, no homogeneous performance is actually met. One of the challenges 4G and future technologies, such as the forthcoming 5G, must meet is the new requirements of area spectral efficiency and user rate distribution. Further improvements to achieve a fair per-user performance along the cell are possible by increasing the eNB deployment density. Nowadays, mobile networks are shifting from a single-tier homogeneous network approach to a multi-tier heterogeneous one, the so-called HetNets. It has become a popular approach in the past few years as an efficient and scalable solution to improve the network capacity in hot-spots; it is also a viable solution to improve fairness, since the network gets closer to the UE.

A HetNet is a network deployment consisting of Macro Cells or MCells and overlaid low-power nodes such as pico-cells, femto-cells, RRHs or relays, referred to more generally as Small cells or SCells. The MCells are high power eNBs typically located along the geographical area with purposes like coverage maximization or interference reduction. SCells are in charge of eliminating coverage holes or improving capacity by off-loading the MCells at hot-spot areas. Figure 9.1 shows a typical HetNet deployment. Based on the frequency deployment, two solutions arise: co-channel deployment, where both MCells and SCells share the same frequency band; and dedicated deployment, where each type of cells transmits at different frequencies. While the former allows to maximize the spectrum utilization and capacity, interference can become a major concern in the system design. On the other hand, dedicated deployments may solve the interference problem, at the expense of over provisioning SCells with frequency resources. In this sense, this new paradigm of network deployment potentially increases the spectral efficiency; however, significant challenges are introduced to ensure the correct operation of HetNets.

9.2.1 Radio planning challenges in HetNets

The trend of cell hyperdensification with multiple tiers require different approaches in network planning and design to meet the fundamental objectives of maximizing user rate and empowering fair per-user spectral efficiency, among others.

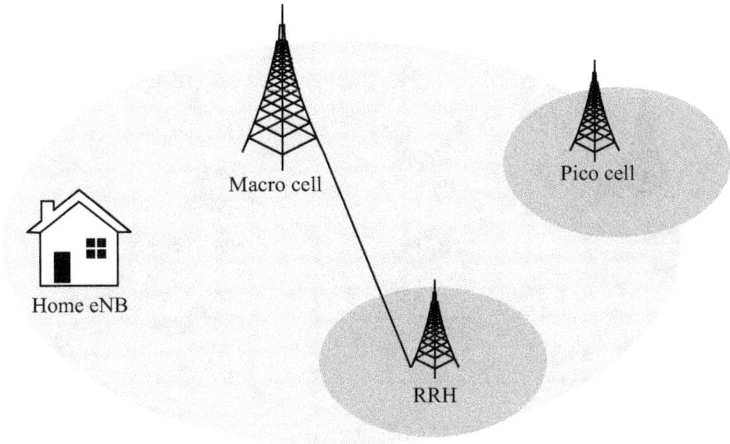

Figure 9.1 Example of HetNet deployment

An obvious challenge in the HetNets network planning is the frequency deployment: if maximizing capacity is the main driver, or access to spectrum is limited, MCell frequencies should be reused by the other tiers. In co-channel deployments the inter-cell interference is an important constraint. In both UL and DL, there is a higher number of interfering nodes compared to classical MCell deployments; a large number of SCells reusing the same carriers may generate high levels of interference. In such cases, sophisticated radio resource management procedures shall be considered to minimize the interference generated. On the other hand, dedicated deployments are attractive in large bandwidth availabilities, which will also allow to better control the interference.

By definition SCells are low power nodes. Very low DL transmit power may result in low SCell coverage, providing very little user migration from the MCell. On the other hand, higher transmit powers increase the SCell size, which may increase the SCell load. The DL transmit power disparities among the different nodes and cell selection based on the DL received power result in imbalance problems between both UL and DL connections. The DL/UL imbalance problem has been recognised by 3GPP in [4,15]. A UE is said to be in this situation if the best serving cell from an UL received power perspective and the best DL serving cell are different. The UE optimal connection in the UL is to the SCell as it is closer, but in the DL it would be connected to the MCell. Classical cell association rules have a huge impact in the load imbalance and the UL performance. Figure 9.2 shows a graphical example of the HetNet imbalance problem.

SCells need to be provided with energy efficient and low cost backhaul, and this has quite often been proven tricky in SCells deployments. The lack of high capacity backhaul can limit the SCell availability, which may impair the potential improvements brought by cell densification. Hence, cell association procedures should consider the availability of such resources which provide the user with sufficient quality of experience.

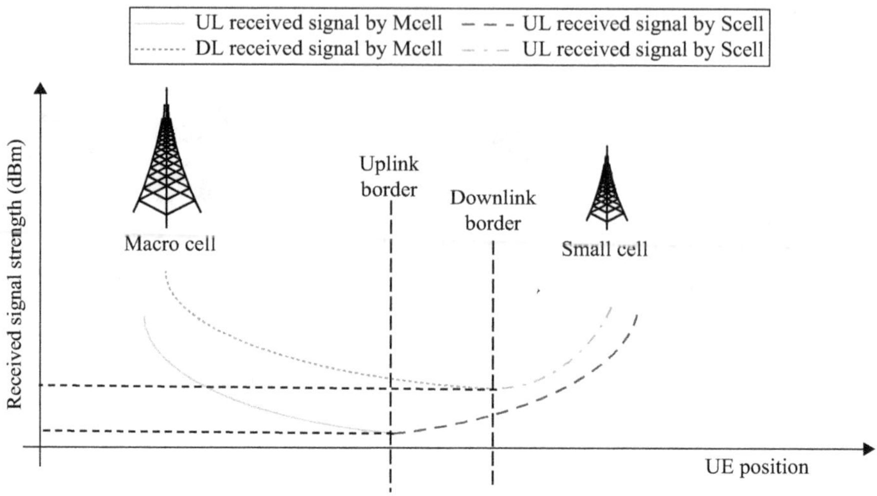

Figure 9.2 Example of HetNet imbalance problem

9.2.2 Strategies for improvements in HetNets

Cell selection based on the reference signal received power (RSRP) result in imbalance problems, since the DL coverage of the MCell is much larger than that of the SCell. One strategy that can address this problem and bring some fairness to the UL is the cell Range Extension (RE), which allows the UEs to associate to cells that do not provide the highest DL RSRP, increasing the SCell coverage region. The SCell range is expanded by adding a cell selection offset to the RSRP measured from the SCell, and the UEs in the range expanded area are associated to the SCell in both UL and DL. In this situation, the UL of UEs is fairly improved, while the DL is associated in a suboptimal way, since the maximum received power is still provided by the MCell. Although expanding the range of low power nodes partially compensates UL interference issues, DL interference on RE UEs is substantially increased. Studies in [10,20] have shown that using high offsets (greater than 3–6 dB) increases the DL interference levels. For this reason, eICIC mechanisms have also been investigated in this context. These techniques are mostly variations of the same idea: frame muting and coordinated scheduling. Indeed, 3GPP introduced the possibility of using almost blank subframes (ABS) since LTE-A Release 10.

Eventually, the range extension technique is limited to moderate offset values and the adjustment is not trivial for very heterogeneous coverage footprints as it will occur in ultra-dense deployments. In addition, current research work assumes slow updates (scale of seconds) for interference coordination [19,32]. The reasons are an increased RTT (due to X2 signalling delays) but more importantly, the need to guarantee the system stability.

One of the major design goals for HetNets is the use of radio resources across MCells and SCells to achieve figures of per-user throughput and system capacity

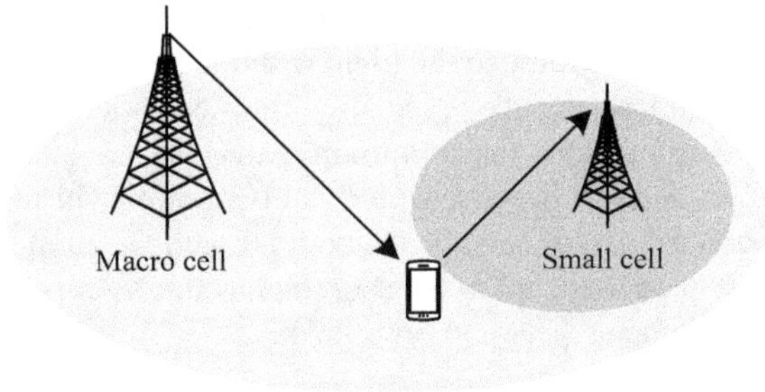

Figure 9.3 Example of HetNet with DL and UL decoupling

similar to ideal backhaul deployments. The 3GPP has introduced the *Dual Connectivity* concept in Release 12, where the user consumes radio resources provided by at least two different network points. Dual connectivity is one of the 3GPP potential solutions to improve user performance by combining the benefits of the MCell coverage and the SCell capacity [4]. Work in [37] addresses this topic in a DL scenario where MCells share resources with other cells, where a CA window is proposed to determine if CA-capable UEs should be included in inter-site CA. A dedicated frequency deployment is considered. The benefit of aggregating resources from both cells is verified for different traffic patterns, as well as for different load situations. This topic has been addressed as well in the UL in [38] where results show improvement of UL throughput with the use of inter-site CA in low load situations due to larger bandwidth accessibility. The improvements in UE performance with the use of shared resources provide a strong indication that cooperative techniques are becoming mandatory to maximize resource utilization and meet the requirements for innovative and more demanding applications.

One step further in the optimization of HetNet is the relationship between UL and DL and how the association policies affect the performance on both links. Both UL/DL power and MCell/SCell load and power imbalance motivates the decoupling of both links, which is particularly beneficial in co-channel heterogeneous deployments. The basic decoupling scheme between a SCell and a MCell is shown in Figure 9.3. In Release 12, 3GPP provided an initial evaluation of the HetNet performance when including UL and DL split, results show improvements particularly at the cell edge for both low and medium load scenarios [4,14]. The literature has tackled the power and load imbalance problem recently and some relevant references can be identified. Authors in [12] present the path-loss cell association solution to the power imbalance problem. Results in terms of gain that can be achieved in the UL capacity are very promising. A detailed analysis of the decoupled access in terms of association probability, coverage and capacity are presented in [35], where prior work is extended by adding the analytical evaluation using stochastic geometry and architectural considerations. Results show same trend between the stochastic geometry analysis and the real-world experimental data. Work in [13] introduces cell load and the backhaul limitation into the cell association process. SINR variance is reduced with the enhanced DUDe solution presented; also, the interference-aware

UL power control applied allows a further improvement in the UL throughput. Finally, Singh *et al.* [34] contribute to the topic with the analysis of the UL SINR and rate distributions as a function of the association rules considering UL power control design parameters. Results show that minimum path-loss association leads to identical load distribution across all cells, which is also optimal in terms of rate, irrespective of power control parameters. When both UL and DL joint coverage must be maximized, the decoupled association is the optimal solution. It is beneficial because it reduces the QoS imbalance between both links.

9.2.3 Decoupling as a solution

The current technologies that address the main HetNet challenges in the context of radio planning and interference management can be classified into three main groups:

- Dual connectivity. The literature has verified the UL improvements in dedicated deployments. However, there is yet no study that includes dual connectivity as a solution to the UL/DL imbalance problem, the association rules proposed in the prior art are based on the DL RSRP or RSRQ.
- Cell RE with eICIC. The literature has verified the UL improvement in co-channel deployments. This strategy helps to reduce the UL/DL imbalance, however, while the UL improvement is maximized the DL inter-cell interference is also increased. Therefore RE techniques must always be accompanied by eICIC solutions for the DL such as ABS. Nevertheless, the RE technique is limited to moderate offset values due to the harsh interference in the DL. Cell selection rules are based on DL RSRP and RSRQ with a RE offset added.
- UL/DL decoupling. There are verified improvements in the UL in co-channel deployments with the use of DUDe. The decoupled association policies in the UL and DL can cater for the different requirements of both links and, in turn, can successfully solve the UL/DL imbalance problem in terms of coverage, load and interference.

Based on this comparison, it is clear that DUDe brings the benefits of having very high RE offsets in the UL without the interference effects in the DL, since both links are separated, and connected to the best serving cell.

9.3 Simulation setup

In this section, the simulation setup used in the performance evaluation in Section 9.4 is presented. The simulation setup follows largely from [12] where the simulation scenario is based on the Vodafone LTE SCell test network in the London area shown in Figure 9.4. The test network covers an area of approximately one square kilometre and consists of two sectored Macro sites and sixty-four SCells illustrated by the black shapes and small circles respectively. The users distribution is based on traffic data extracted from the live test network. Additionally, a high-resolution 3D ray tracing pathloss prediction model is used which takes into account clutter, terrain and building data. This guarantees a realistic and accurate

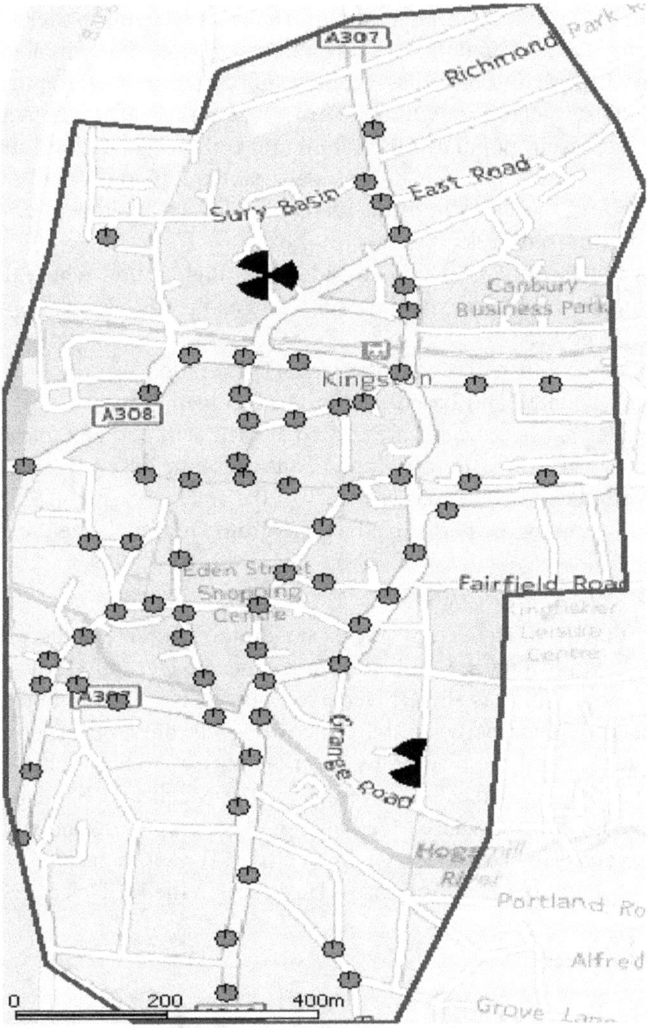

Figure 9.4 Vodafone SCell LTE test network in London. © 2014 IEEE. Reprinted, with permission, from [12]

propagation model. The UL power control is based on the open-loop fractional pathloss compensation algorithm as specified by 3GPP [3].

9.4 Performance evaluation

As explained in Section 9.1, DUDe is considered to be a major paradigm shift from the cell-centric approach used up to now in designing cellular networks into a more flexible device-centric approach, which is envisioned to be one of the main building blocks of future 5G networks. DUDe can offer substantial gains in the UL in terms of coverage, throughput, reliability, load balancing and interference behaviour. Improving the UL performance has become more and more important due to the emergence and exponential growth of the IoT where the traffic is often UL centric and also the increasing popularity of symmetric traffic applications such as social networking, video calls and real-time video gaming. In this section, simulation results that confirm the gains in terms of the previously mentioned criteria will be presented and discussed in details.

The simplest form of DUDe is considered in this section where the UL association is based on the UL received power whereas the DL association is based on the DL received power.

DUDe is compared with conventional LTE operation where both UL and DL associations are coupled and based on the DL received power. Two LTE baseline cases are considered where SCells are treated as Pico cells and Femto cells and these two cases are termed as Pico-Baseline and Femto-Baseline, respectively. The transmit powers of Macro, Pico, Femto cells are 46, 30, 20 dBm, respectively. Subsequently, a set of results illustrating the performance gains from DUDe will be presented.

9.4.1 Coverage and capacity

In a HetNet, the UL and DL coverages are quite different as discussed in Section 9.2 and shown in Figure 9.2, therefore basing the UL and DL associations on the same criterion, which is the DL received power, is highly suboptimal. In this part, the great difference between the UL coverage in the LTE baseline cases and DUDe is illustrated and the resulting gains in capacity are highlighted.

Figure 9.5 illustrates the UL coverage of the Mcell and Scell layers for the three cases in comparison, where Mcell and Scell coverages are shown in black and grey, respectively. The UL coverage of Scells is shown to be very small in the Femto-Baseline and Pico-Baseline cases. However, in the DUDe case Scells have a

Figure 9.5 Uplink coverage of the Femto-Baseline (left), Pico-Baseline (middle) and DUDe (right) cases where black and grey represent the MCells and SCells coverage respectively

much larger UL coverage, which is shown to be much less dominated by Mcells than in the LTE baseline cases. This effect results in a more homogeneous distribution of UEs between the nodes which, in turn, leads to a much more efficient use of resources as will be demonstrated in the following results.

The homogeneous coverage has a large effect on the throughput and specially the cell edge UEs throughput, which is represented by the fifth percentile throughput. Figure 9.6 shows the effect of increasing the number of Scells in the simulated scenario on the fifth percentile throughput. In the DUDe case, it can be noticed that the fifth percentile throughput is improving quickly with the number of Scells, this is partly due to the increased UL coverage of Scells in DUDe as shown in Figure 9.5. As the number of Scells increase, the fifth percentile UEs throughput starts to saturate as these UEs become limited by the channel quality and transmit power and the gains start to be more pronounced in the 50th and 90th percentiles. Looking at the Femto-Baseline and Pico-Baseline cases, it can be noticed that adding Scells has little effect on the fifth percentile throughput due to the very limited coverage of Scells in both cases, which makes them more effective for the 50th and 90th percentile UEs. In addition, in these two cases the throughput is fluctuating as the number of Scells is increased. This is due to the high interference that the Scells UEs create to the Mcell cell edge UEs, as these UEs are closer to the Scells so they suffer from a high level of interference. This effect is emphasised more in the Femto-Baseline case as the fifth percentile throughput starts to decrease after a certain point whereas in the DUDe case the throughput increases in a more stable way since UEs always connect to the node to which they have the best UL

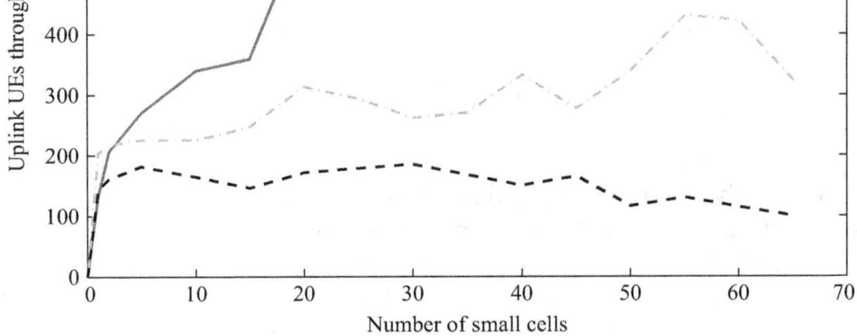

Figure 9.6 Uplink cell edge (fifth percentile) throughput evolution with the number of SCells

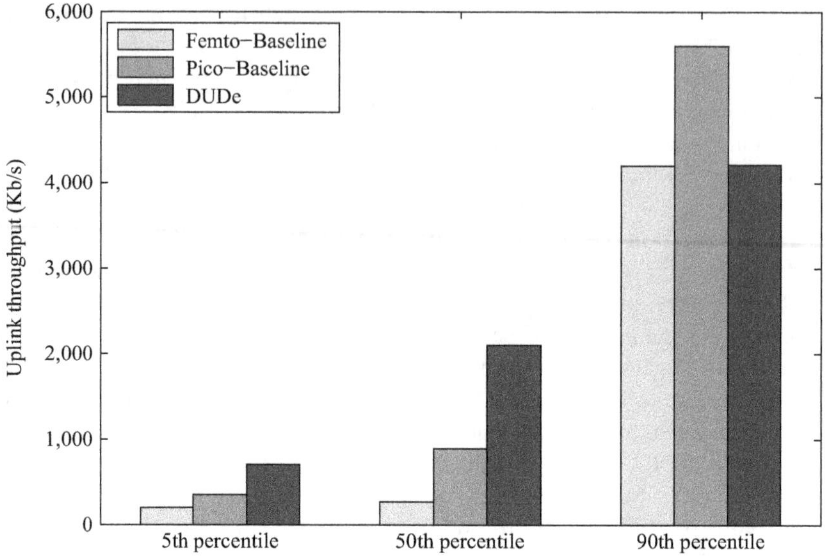

Figure 9.7 Uplink 5th, 50th and 90th percentile throughput

channel which guarantees a lower interference level. This effect on interference is further addressed later on in this section.

Figure 9.7 shows the 5th, 50th and 90th percentile throughput results for the three cases in comparison. The fifth percentile UL throughput in DUDe is increased by more than 200% and 100% compared to Femto-Baseline and Pico-Baseline, respectively. Similarly, the 50th percentile throughput is improved in the DUDe case by 600% and 100% compared to Femto-Baseline and Pico-Baseline. The gains result from the improved coverage of Scells in the DUDe case which results in a better distribution of UEs between the nodes giving way to a more efficient resources utilization. In addition, the fact that UEs connect to the node to which they have the best UL channel results in an improved Signal to Interference and Noise Ratio (SINR), which allows UEs to use a higher modulation and coding scheme and in turn achieve a better utilization of resources and a higher throughput. As for the 90th percentile throughput, DUDe does not achieve the highest throughput, which makes sense since DUDe aims at improving the load balancing effect that will naturally result in a reduction of the peak data rate.

Additionally, it can be seen that Pico-Baseline achieves the highest 90th percentile throughput, which seems counter intuitive since it would be expected that the Femto-Baseline would be the one achieving the highest peak data rate. However, looking at the 98th percentile throughput, the throughput reaches 15 Mb/s and 10 Mb/s in the Femto-Baseline and Pico-Baseline cases, respectively. This shows that the effect of Scells in the Femto-Baseline case is limited to a small number of UEs.

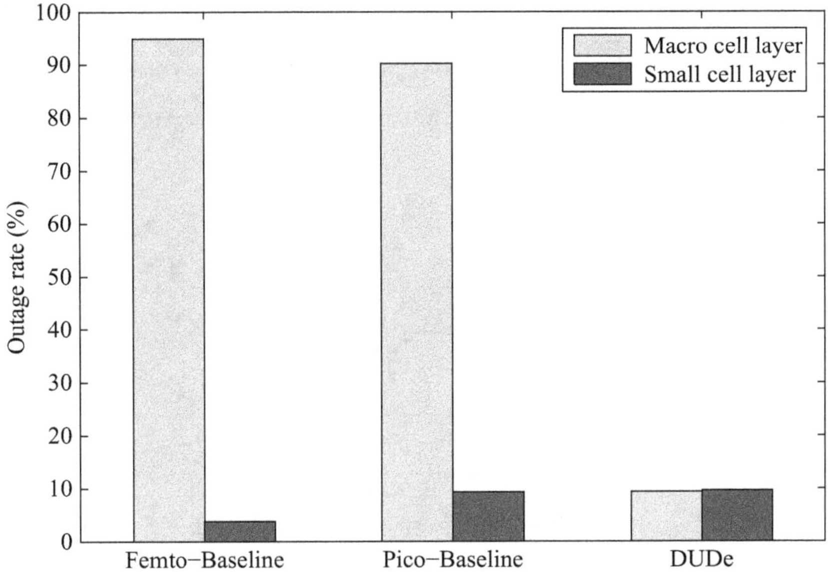

Figure 9.8 Outage probability

9.4.2 Reliability and load balancing

Reliability is becoming one of the most important requirements of future cellular systems due to the proliferation of the IoT which is an umbrella for several applications such as sensor networks, factory automation and many more. The reliability of wireless systems can be affected by several factors including congestion and coverage loss, where a device could have its connection dropped due to lack of available resources or due to very low link quality resulting from the device being out of coverage or being subject to high level of interference. DUDe plays a crucial role in improving the UL reliability by reducing the outage rate as will be shown next.

Figure 9.8 illustrates the average outage rate for the Mcell and Scell layers for the three cases where a high traffic scenario is simulated by setting the minimum requested throughput for each UE to 1 Mb/s. The outage is defined as the percentage of UEs that cannot achieve the 1 Mb/s minimum throughput. Since this scenario is considered to be a high traffic congestion scenario it requires a very efficient use of resources in order to satisfy the high throughput requirements of the UEs. As can be noticed from the figure, the Macro layer has a very high outage rate in the LTE baseline cases, which is explained by the fact that the Macro layer is very congested in the UL and MCells cannot serve all the UEs with the required throughput resulting in a high number of dropped UEs. However, in the DUDe case, UEs are distributed more evenly among the nodes so the outage rate can be drastically reduced to less than 10% on the MCell and SCell layers.

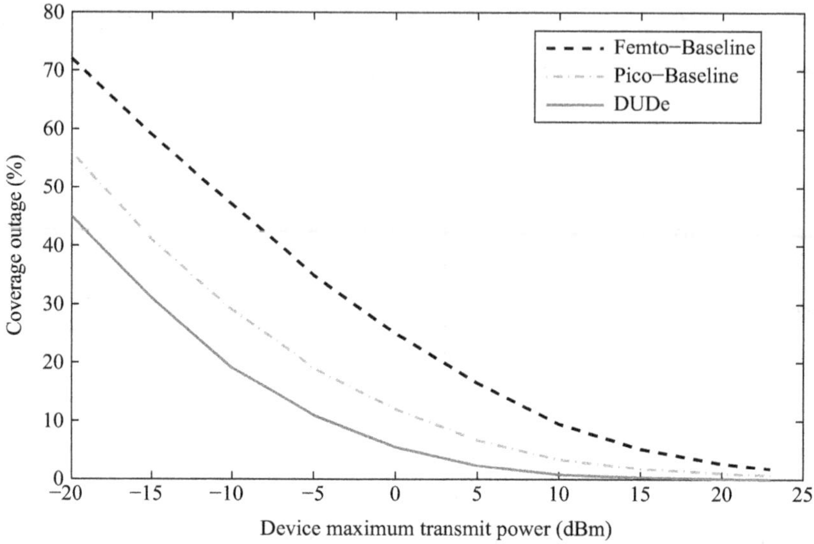

Figure 9.9 Outage probability versus the maximum uplink transmit power

Another trend that affects reliability is outage caused by the loss of coverage or poor channel conditions. Figure 9.9 shows the outage rate against the maximum transmit power of the UE. Outage here is defined by the fraction of UEs whose UL signal quality is lower than the one needed to access the lowest LTE modulation and coding scheme (MCS) and therefore are dropped. This result shows that for the same maximum UE transmit power DUDe can reduce the UL outage rate by up to 25% compared to baseline LTE. Alternatively, for the same outage rate the maximum transmit power can be reduced by up to 10 dB using DUDe. These results are very crucial for IoT applications, where either link reliability or battery life or both are of paramount importance and, it can be seen from the figure, that both can be drastically improved using DUDe.

The load at a given BS in the UL maybe different than the load in the DL at the same BS. This implies that it is not optimal, in terms of load balancing, to have the same set of UEs connected to the same BS in the UL and DL.

In addition, DUDe has been shown in the previous results to improve the UL coverage of SCells, which results in a much better distribution of the UEs among the different tiers of the cellular network. This load balancing effect is illustrated in Figure 9.10 where the average number of UEs per cell in each tier is shown for the three cases in comparison. DUDe results in a much more homogeneous distribution of UEs between the different node types, which is directly translated into a drastically improved spectral efficiency and outage reduction as shown in the previous results.

9.4.3 Interference behaviour

Interference is one of the main limiting factors in cellular networks. DUDe has the ability to decrease the UL interference due to multiple complementary effects.

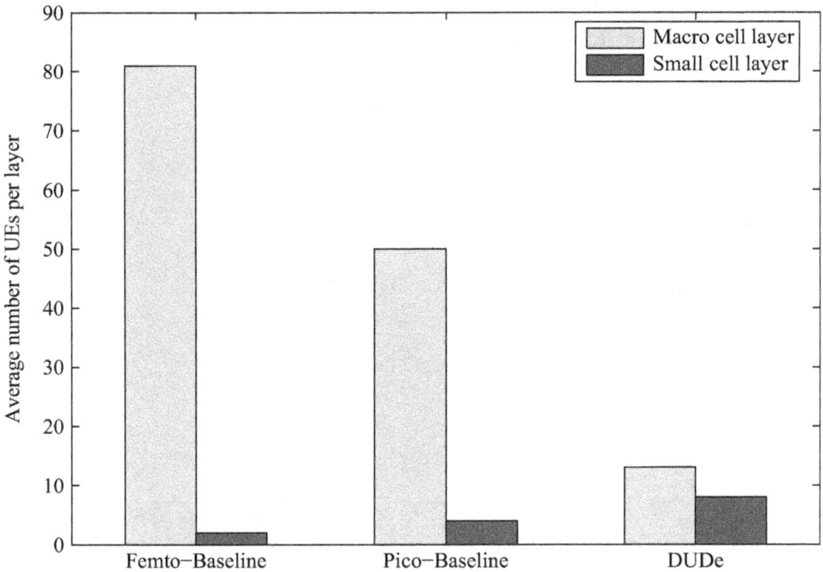

Figure 9.10 Average number of UEs per base station for MCell and Scell layers

First, and as a result of the UL association that tends to connect the UEs in the UL to their closest node or the node to which they have the best UL received power, this can be translated into a reduction in UEs transmit power as shown in Figure 9.9 and emphasized in [8]. This has the effect of reducing the UL interference to other base stations, which is quite significant especially for UEs with low UL SINR.

Secondly, DUDe provides the ability to independently select the association that minimizes the interference at the UE as well as the BS. The UL interference depends on the UEs position relative to the interfered BS, the UL power control algorithm used and the UL precoding weights. In contrast, the DL interference toward a certain UE depends on the transmit power of the BS, the DL beamforming weights and the distance from the different BSs. On top of this, the UL and DL have different traffic and scheduling behaviours that can be considered nearly independent. For all the above reasons, a decoupled UL and DL association that allows the UE or BS to seek the best UL or DL connection that reduces the interference on the respective link makes a lot of sense and is expected to outperform the conventional coupled association.

In addition to reducing the average amount of interference, DUDe allows a reduction in the standard deviation of the UEs UL SINR over time as shown in Figure 9.11. Reducing the variance of the SINR means that the channel is more stable and predictable, which has a substantial impact on reducing the complexity of radio resource management (RRM) and self-organizing network (SON) functions.

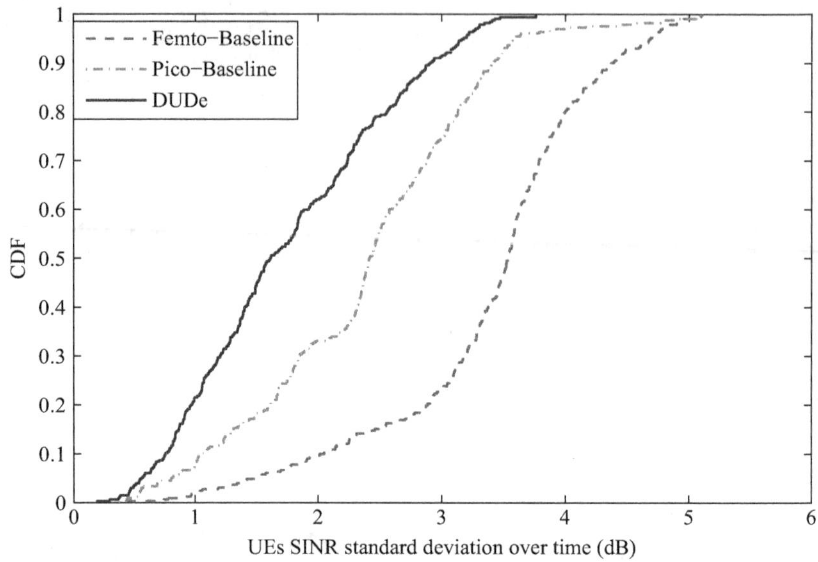

Figure 9.11 Uplink UEs SINR standard deviation over time

9.5 Interoperability of DUDe with 4G and 5G features

The deployment of a reliable high speed spectral efficient network needs the inclusion of a variety of innovative features, provided that link level solutions have evolved to near Shannon limit capacity with advanced Modulation and Coding Schemes (MCS). Given this, 4G and the forthcoming 5G must offer pioneering solutions or improved versions of earlier releases in spectrum management and cooperative communications.

In particular, the interoperability of DUDe with other radio access technologies can lead to an overall improvement of these features. The reduced interference variability, the enhanced network flexibility or the reduced transmit power are some of the advantages that help to make the most out of the radio management techniques.

9.5.1 Inter-band carrier aggregation

CA provides great advantages when carried out in HetNets. In particular, the inclusion of CA in such context has been recognised as a feasible way of providing multi-site radio resource allocation feature that was first introduced in Release 11 [3]. Also, CA in general (multi or same site aggregation) allows to improve capacity by extending the available bandwidth, and supports mobility and interference management techniques.

Several studies have focused specifically on the implementation of CA in the UL, where power capabilities of the UEs constitute the most limiting constraint [23,24]. It is observed that the potential gains of CA transmissions are strongly related to the power demanded, which essentially depends on the bandwidth allocations and UE pathloss. Cell association determines which eNB is serving the UE. The eNB being closer to the UE means that less power is needed to transmit data in an allocated piece of bandwidth. To assure a correct performance of the aggregated bandwidth transmissions in the UL, it is crucial to account for the UE maximum transmit power. In MCell-only deployments, cell edge UEs are less likely to transmit in CA; however, in heterogeneous deployments, the distance to the eNB is shorter given the higher cell density. In such a context, if UEs are associated based on the DL RSRP, the UL CA transmission is going to be highly restricted. Decoupling strategies are more lax in adopting aggregated transmissions, mainly, due to the improvement in UL power availability brought by decoupled associations. This is important, since CA is intended to be applied in both UL and DL, and with traditional DL received power association rules, the applicability of CA in the UL is seriously conditioned by the lack of power availability.

9.5.2 Cooperative multi-point

Base station cooperation, in the form of Coordinated Multi-Point (CoMP) transmission or reception, has gained popularity in the context of HetNets as a means to increase the UE achievable throughput. eNBs within the same cluster communicate via backhaul links (i.e., via the X2 interface) with the objective of minimizing the inter-cell interference and capitalize on the benefits of distributed antenna systems. In fact, interference within a cooperation cluster can be effectively cancelled [1,18]. This level of coordination and cooperation can be carried out in both UL and DL, and the realization of such coordination relies strongly on the availability of sufficient backhaul capacity, first to serve the UE in the cell cluster, and second to communicate with other cells in the cooperation cluster. This backhaul dependency can be very limiting in situations of high load, and in capacity limited links.

The increased flexibility provided by decoupled UL and DL associations provides advantages when selecting UL and DL coordinated transmissions or receptions. In particular, there is no need to have both UL and DL simultaneous connection to the entire cooperating set of base stations and the UE could have unequal UL and DL active links (as in the case of CA). This flexible association inside the cluster, and the interoperability of DUDe with CoMP goes one step further in the device-centric network, since the UE can select independently the number and position of DL and UL serving cells, according to several input parameters, as backhaul capacity, power limitation, throughput maximization, among others.

9.5.3 Millimetre wave

The ever increasing network traffic demands have led to several trends in cellular networks including the densification of the network and the emergence of heterogeneous networks where SCells are playing a key role in providing capacity in hotspots and dense urban areas. Another approach was to explore new unused frequency bands that would help satisfy the projected future traffic needs. Popular sub 6 GHz frequencies are becoming scarce and would no longer be able to cope with the increasing network traffic.

A solution to this is to resort to higher frequencies in the millimetre-wave bands (20–100 GHz) where a significant amount of spectrum is underutilized or completely unused. The several GHz of available spectrum promise a spectacular increase in capacity which qualifies millimetre-wave technologies as one of the main enablers of future 5G networks.

In reality, millimetre-wave bands will not replace sub 6 GHz bands, at least initially, where sub 6 GHz will still be needed to provide coverage and ubiquitous service since millimetre-wave frequencies have poor propagation properties and are quite sensitive to blockages. The existence of millimetre-wave networks as an overlay to the conventional sub 6 GHz networks would require a change in the conventional cell association techniques. DUDe is expected to play a key role in millimetre-wave networks for several reasons.

Recent studies on electromagnetic field exposure [9] have shown that to be compliant with the mandatory exposure limits at frequencies above 6 GHz, the maximum transmit power of devices would have to be reduced by several dBs below the conventional power levels in current cellular networks. This has significant implications on the UL; since the UL coverage depends on the transmit power of devices and since mmWaves have unfavourable propagation properties, a reduction in the transmit power of devices could result in serious degradations in the link quality in the UL. A possible solution to this problem would be to decouple the UL and DL where for some UEs the UL will be served by UHF MCells with a better link budget, whereas the DL would be served from the mmWave SCells. That is, while previously DUDe was discussed in the context of associating UEs to MCells in the DL and Scells in the UL, for mmWaves the opposite strategy might prove useful.

In addition, mmWave SCells are expected to have a very limited coverage area and consequently the conventional association techniques based on the DL received power will leave the mmWave SCells very much under-utilized considering the vast amounts of spectrum available in the millimetre-wave bands. SCell biasing or range extension is a technique that was introduced in LTE release 10, where effectively a cell selection bias is applied to SCells in order to expand the coverage of SCells and improve their offloading effect. A similar approach could be employed for millimetre-wave SCells to solve the aforementioned problem by using aggressive range extension values to attract as many UEs as possible in order to make use of the large spectrum chunks available at millimeter-wave frequencies.

DUDe would play an important role in this setup as it would allow to set different values of range extension for UL and DL in order to meet the requirements of both links. For instance, if the UL is required to be highly reliable whereas the DL can be less reliable but is more demanding in terms of throughput, a high value of DL SCell range extension can be used while not applying any range extension in the UL. Therefore DUDe would offer the flexibility to cater for the different needs for the UL and DL in millimetre-wave scenarios.

9.5.4 Different duplexing techniques

DUDe can function with both FDD and TDD, with different implications from a system level and spectrum point of view.

TDD allows much more flexibility in trading DL and UL resources as compared to FDD. With DUDe, it was demonstrated that fewer UL resources are needed to achieve the same UL rate as compared to coupled operation. This could lead the way to DUDe having a positive effect on the DL rate by allowing the DL to use more resources via dynamic TDD.

Another benefit of TDD is the possibility of estimating the DL channel via UL reference signals. This is quite important especially for channels with large dimensionality like in the case of massive MIMO. However, when DUDe is used this reciprocity is broken as DL and UL transmissions originate and terminate at different BSs. Much of the existing spectrum is paired FDD spectrum, therefore for these two reasons massive MIMO may need to be supported without channel reciprocity.

In the medium-to-long term, DUDe along with different emerging technology trends could lead to a rethinking of the traditional FDD/TDD paradigms. DUDe, hyper-densification, the use of millimetre-wave frequencies and highly directional antennas, could allow for duplexing approaches over the spatial domain. For instance, the same band could be used for two different devices located in different locations, one receiving in the DL from a certain BS and the other transmitting in the UL to another BS. A spatial UL/DL coordinated scheduling mechanism would effectively allow full-duplex like gains without the complicated interference cancellation mechanisms of full-duplex. In addition, once analogue/digital interference-cancellation mechanisms are realised to support full temporal duplex, DUDe can be beneficial as it enables a generalized decoupled access that would allow the support of a DL and not necessarily the same UL user in the same frequency band.

9.6 Enabling radio access network architectures for decoupling

The access network configuration presents several challenges to enable the use of decoupled connections, while maintaining interference and energy efficiency at reasonable levels. Holding more than one UL connection is less power efficient for users that are placed near the cell edge, which are the ones more likely to decouple [22,35]; also, the use of one carrier to exclusively transmit control signals will lead to poor spectral efficiency. In particular, the most challenging part of devising architectures that support DUDe is that an increased amount of control information needs to be signalled back to the corresponding serving cell.

The 3GPP has proposed several architectural alternatives for DL dual connectivity and the architecture needed to support the U-Plane aggregation from different eNBs is expected to be very similar to those proposed for dual connectivity, based on the bearer split concept [4]. Alternatively, those architectures that enable a full UL and DL decoupling should support a feasible cooperation among both serving cells while not jeopardising the improvements in the UL in terms of reliability and capacity. To achieve this it is necessary to assure the delivery of layer 1 and layer 2 control signals while maintaining the RAN latency requirements.

9.6.1 Network procedures

When a UE accesses the system and associates with eNB, a radio resource control connection is set up. The RRC is the protocol that handles all the control plane signalling of Layer 3 between the UE and the EUTRAN. Among the different tasks, the RRC is in charge of the connection configuration, radio bearer establishment and release, mobility procedures and configuration of power control mechanisms. The amount of resources the RRC consumes dynamically changes, depending on the cell traffic type, number of UEs being served and the connection timers. Short timer values generate high RRC signalling overhead whereas high timer values result in poor use of idle mode [2].

Layer 1 and layer 2 control channels are crucial to support the user plane data transmission. PDCCH and PUCCH physical control channels are in charge of delivering information related to the active transmission, as the scheduling decisions or the acknowledgements of the transmitted information, and the channel state information to perform accurate link adaptation. In particular, RAN control signals that are of paramount importance to handle scheduling and other MAC layer procedures are:

- control information carried in the PUCCH: DL CQI, buffer status reports, scheduling requests and power headroom reports;
- downlink control information (DCI) carried in the PDCCH in charge of indicating, among others, both UL and DL physical resource blocks for transmission (UL-SCH and DL-SCH), as well as link adaptation forms and transmit power for the UL;
- downlink HARQ acknowledgment messages carried in the Physical Hybrid-ARQ Indicator Channel (PHICH); and
- RRC messages that configure the UE connection and release, as well as the PUCCH position and resources and SRSs for UL channel state information configuration.

Among all these control information, the most stringent one in terms of latency requirement is the HARQ RTT, which is approximately 8 ms, considering information processing at both sides and frame transmission. The HARQ in the UL follows a synchronous process: with a periodicity of 8 sub-frames the same HARQ id process is used, and no explicit information is exchanged about the process id. This means that there is a strict relation among the sub-frame number and the HARQ process id, which results in strong delays if one of the acknowledgements cannot be sent in the corresponding sub-frame. The architecture that enables the use of DUDe shall be the one that minimizes delay in the acknowledgment synchronization process and assures that the RAN RTT is maintained.

9.6.2 DUDe with assisting connections

The use of supporting UL and DL connections in order to transmit the control signals is an immediate solution to feasibly implement DUDe. The terminal is

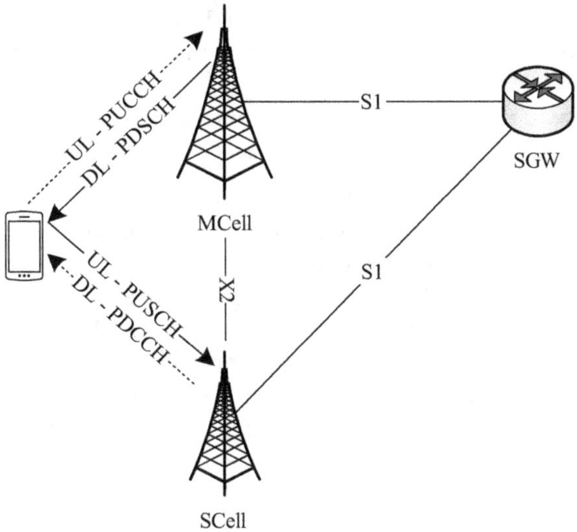

Figure 9.12 Example of DUDe configuration with assisting connections

connected to both cells and aggregates the data flows; in such a way, all signalling layer 1/layer 2 and RRC can be handled locally. Architecture alternatives along these lines were presented in [14].

A simple way to support this is with the use of CA, where each carrier component is configured separately to carry a shared and a control channel. This configuration is shown in Figure 9.12. This solution allows to keep the RAN latency at desired levels since no further delay is sensibly introduced. However, potential disadvantages of this configuration are: first, having two simultaneous UL connections can seriously jeopardise the UL performance in terms of UE energy efficiency. Second, this configuration does not maximize capacity over the available spectrum, since one component carrier is exclusively used to handle control information.

9.6.3 User plane bearer split for DUDe

Latest releases of LTE-A (Releases 12 and 13) consider new architecture alternatives for dual connectivity, with the introduction of the novel bearer split concept, an alternative that eases having two simultaneous transmissions in different eNBs, a MCell and a SCell for instance [4]. On the other hand, the UL feasibility of adopting the bearer split can be argued in terms of power consumption, and UL data should be either transmitted directly to the MCell, or forwarded to the MCell by the SCell [22].

Based on this, in a DUDe context, assisting PUSCH or PUCCH connections may not be carried out in the MCell; and PDSCH and PDCCH connections may not be carried out in the SCell. Having no control information signalled back to the corresponding serving cell through the user interface (i.e., using supporting

UL and DL connections), the delivery of layer 1 and layer 2 signalling and RRC relies on the non-ideal backhaul connection, the X2 interface, between both serving cells.

Following the LTE architecture of distributed cells and using the bearer split concept two possible solutions arise for the decoupling:

1. The radio bearer is managed at the serving gateway (SGW), and the DL flows though the S1 from the SGW to the MCell and the UL flows through the S1 to the SGW from SCell. This option has reduced flow control among both serving eNBs, since only part of the control signals, for example, HARQ acknowledgements, needs to be forwarded through the X2. The architecture largely follows the one in Figure 9.12, where information flow for DL and UL goes through the S1 interface.
2. The Radio bearer is managed at the MCell, so the configuration is the master/slave MAC for UL and DL control feedback. This means that real-time MAC PDUs need to be forwarded to the corresponding eNB via the X2 backhaul interface, while respecting the 8 ms HARQ round trip time requirement. Master/slave configuration is for UL and DL, and the processing of each MAC PDU is done in the corresponding cell. Figure 9.13 shows a diagram of the information flow for DL and UL.

Both configurations allow to maximize the spectrum usage since all carriers can be configured for U-Plane information transmission. Current heterogeneous networks pose a big challenge to meet the RAN RTT fixed by the HARQ process, since delays that range from 5 ms to 30 ms are expected in the X2 interface [4].

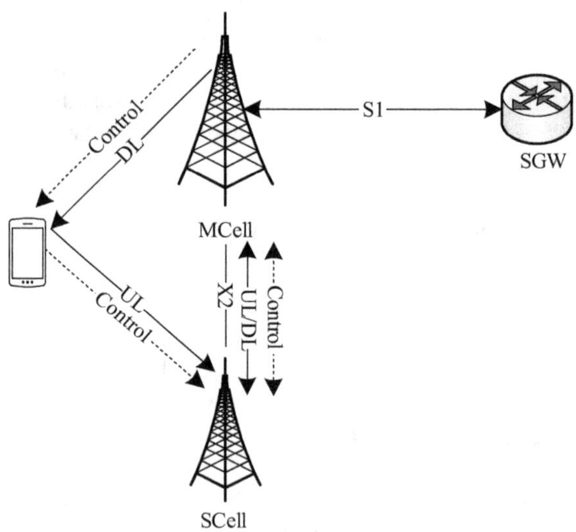

Figure 9.13 Protocol stack information flow for radio bearer managed at MCell

Similar to what has been done in the user plane, the control plane that enables the full DUDe is analysed based on the RRC protocol architecture presented in [4] for dual connectivity. Two alternatives have been presented:

- Based on prior negotiation of parameters between SCell and MCell, the MCell generates the final RRC message. This alternative is good, as only one RRC connection is active, but requires an assisting UL connection with the MCell.
- In this case, the RRC message is built by the SCell based on input parameters provided by the MCell. This solution adds complexity in the UE side as it needs to handle parallel RRC procedures [40].

In light of this, to completely decouple UL and DL poses a huge effort in the RRC connection.

9.6.4 Centralized solution

Previous architectural solutions present limitations in terms of spectrum usage, power efficiency and meeting latency requirements. If both eNBs, MCell and SCell, are able to share the same BBU, then complete DUDe can be handled and mitigate most of the drawbacks presented hitherto. The concept under the BBU sharing is the centralized based architecture, also known as C-RAN, which breaks the static relationship between BBU and RRH, and each RRH does not necessarily needs to be mapped with one specific BBU. In particular, the radio signals from/to a RRH can be processed by a virtual eNB, which facilitates enhanced real-time cooperation among them. Given this, virtualization technology (network function virtualization, NFV) will maximize the flexibility in the RAN network allowing different levels of coordinated transmissions, or separated associations. Figure 9.14 shows the C-RAN architecture for DUDe.

The C-RAN architecture needs to use a new fronthaul interface that allows the communication among the RRH and the BBU, which can potentially increase the RAN latency. Different protocol functional split options can be recognised which variate the delays and capacity requirements over the fronthaul interface [11,25,30]. There is a benefit in locating part of the signal processing functions near the RRH, the bandwidth and latency requirements are brought to a level that can be fulfilled by cost-effective transport networks (i.e., dark fibre, wireless or other technologies) and at the same time enable the possibility of having enhanced cooperative radio resource management procedures. Realistic fronthaul delays can range from 250 μs to 4 ms [30].

When the DL serving RRH and the UL serving RRH share the same BBU, one RRC connection is held. To assume this situation, the BBU has one single physical cell identifier (PCI).

9.6.5 On the road to 5G-based architectures

The next few years will witness a surge in research and development in the different areas of 5G. The ITU and 3GPP are ramping up their activities on requirements and standardization of 5G which is expected to be rolled out commercially

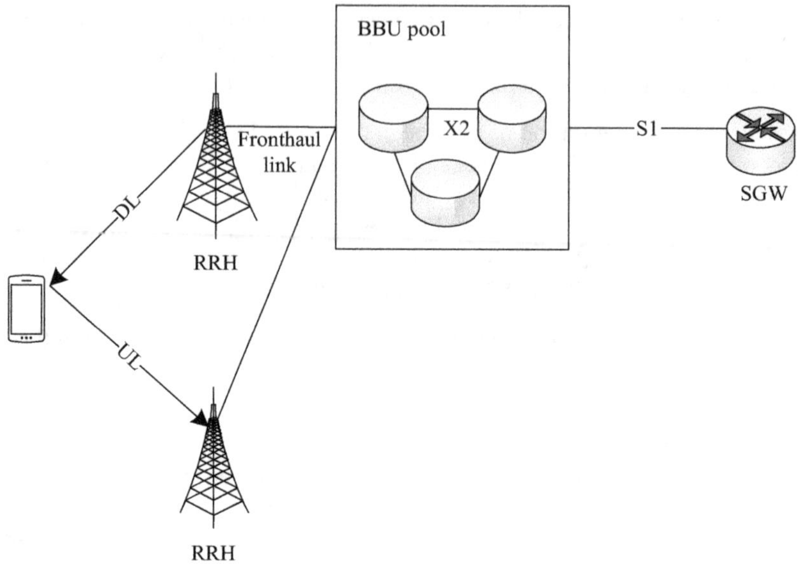

Figure 9.14 C-RAN Architecture with DUDe

by the early 2020s. In addition, there is an increasing consensus about the main requirements of 5G regarding capacity and latency as well as the key technical features of 5G, including ultra base station densification, massive MIMO, millimetre-wave frequencies and possibly a "cell-less" architecture [5,7].

With this view of 5G, does the new 5G standard need to include extra features to natively support DUDe?

An important question to answer is whether a simple evolution of current 3GPP architecture design would be able to efficiently support DUDe in future heterogeneous 5G deployments. In the previous subsections, a discussion was undertaken on how the LTE-A architecture already supports an implementation of DUDe when different BSs are linked via fibre to the same baseband unit. It was also discussed how the support of DUDe in 4G could be limited to different frequencies in the case of carrier aggregation or dual connectivity. Intuitively, any future 5G releases in 3GPP should allow for same-frequency dual connectivity, which, despite having implications on resource and interference management, is not considered to be a major upgrade.

Another issue to address is to ensure proper encryption of all data and control channels, especially when communication via the X2 interface is used between BSs. In LTE, each BS can support tens of IP Security (IPSec) tunnels. However, the management of security via IPsec is so problematic that operators tend to deploy only a few IPsec gateways per country, where most of the gateways are deployed near the SGW. This means that traffic that logically goes via the X2 is actually routed via the SGW, which adds a lot of delay that renders the DUDe operation rather inefficient. Whilst LTE-A has more IPSec gateways deployed closer to the mobile edge, 5G architecture designs need to consider efficient encryption of the X2 traffic to reduce the extra latency in 4G.

In addition, some integration work is needed for some emerging techniques that have proved efficient for coupled systems. The integration of DUDe with decoupled Control/Data planes and License Assisted Access (LAA) will require some architecture changes. Self-organizing networking (SON) will also be instrumental in managing and coordinating the increased degrees of freedom introduced by DUDe.

Given, the above discussion, it can be concluded that a native support of DUDe does not require major design changes in 5G from an architectural perspective.

9.7 DL/UL challenges and opportunities for higher layers

In this section, the main focus is on the transport layer and in particular on the TCP, given that this is the prominently used transport layer protocol to achieve reliable end-to-end data transfer in IP based networks. Despite the fact that TCP has been initially designed for elastic applications, it is currently commonly used in various popular streaming applications. It is worth noting that Real Media and Windows Media, the two dominant streaming media applications, both are based on TCP streaming. Here, we discuss challenges in the higher layer and the TCP in twofold: First, the effect of UL channel on the TCP performance and second, is the support available at the TCP to enable DL/UL decoupling. In this section, both sides of the argument are widely elaborated and the following questions are mainly answered:

1. How DL/UL asymmetry can impact performance of TCP?
2. What are the effects of DL/UL joint resources allocation on the TCP?
3. What are the effects of DUDe on the TCP?
4. How DUDe can be enabled by the available mechanisms in TCP?

9.7.1 Effect of UL on the performance of higher layers

TCP connections, as being inherently bi-directional, require ACKs from the receiver in order to achieve reliable communication. On the one hand, the common assumption in data centric networks is that the DL carries heavier data load, thus less bandwidth is specified for the UL path. On the other hand, channel measurements are carried out based on the DL channel. Moreover, in many of today's applications such as MTC, the sender can be the mobile terminal, which is different from traditional applications, and therefore UL capacity plays an important role.

The effect of link asymmetry on the performance of TCP is widely studied in wired networks. Limited available bandwidth and congestion on the reverse path breaks down the principle of ACK clocking, and may cause an increase in the RTT, which can degrade the TCP throughput on the forward path [6]. Several research works explore these issues and a range of solutions have been proposed. Some of these proposals require explicit support from routers or middle boxes, whereas others are end-to-end schemes. For example, ACK congestion control [16] attempts to reduce the sending rate of ACK traffic, with the assumption that a reduction in ACKs rate may help reduce the congestion itself.

Within mobile and wireless communications, a wealth of work has gone into optimizing the performance on the DL channel of wireless networks such as for example, honing its congestion awareness mechanism so that it is minimally affected by random wireless losses, and optimizing achieved fairness of the end-to-end TCP rate. Limited volume of research in fact addresses the effect of such imbalance on the TCP performance. The work in [26], however, brings the requirements of TCP on the UL path, into the actual radio resource allocation mechanisms for DL. Therefore, the limited capacity or lower channel quality on path will not adversely affect the end-to-end performance. This solution can be implemented by adding one bit in the optional part of TCP header to explain the state of congestion control TCP is in, i.e., *Congestion Avoidance* or *Slow Start.*

9.7.1.1 Modifications to the TCP congestion control

The core of TCP congestion control algorithm are the slow start and the congestion avoidance phases. In the slow start, the achievable throughput by the TCP flow simply depends on the actual value of CW and RTT. On the other hand, throughput in the congestion avoidance phase can be expressed by the TCP steady state throughput [31]. Introducing a new flag in the TCP header, SS/CA flag, can illustrates the actual state of TCP, i.e., SS/CA equal to zero represents being in *Slow Start*, and one being in *Congestion Avoidance.*[1]

9.7.1.2 TCP congestion control and DL/UL resource allocation

While resource allocation mechanisms commonly consider DL only, it is often useful to consider the joint DL/UL allocation problem. Such allocation techniques can provide a dynamic border between DL capacity and UL capacity, in order to allocate the total amount of resources more efficiently. This proportion can be defined dynamically based on various system constraints to guarantee the requested QoS. To fully illustrate the joint resource allocation, here an OFDMA resource allocation problem is detailed that is also constrained with the requirements of the bi-directional connection of the TCP so that the delivery of TCP ACK packets for the specific allocated bandwidth to the DL is guaranteed.

The DL resource allocation problem per se, could aim to maximize sum rate [39] subject to power consumption, minimize the overall power consumption while satisfying the minimum rate requirements [21], consider certain degree of QoS or fairness [27,28]. Here, a joint allocation of DL/UL resources, to provide the maximum throughput on the DL, and also to guarantee the delivery of the DL packets with the appropriate data rate on the UL, is discussed. The objective of DL allocation scheme is presented in two variants. The first defined problem, i.e., (P1), aims to maximize sum rate, constrained to the proportional data rate on each individual TCP flow. The proportional rate is weighted with the theoretical TCP throughput, which is the throughput that a TCP flow can achieve, depending on the

[1]The SS/CA flag can be implemented in the TCP header, option section with kind = 30, and length = 1 [36].

end-to-end RTT and the PER of the corresponding flow-this throughput can be considered as the actual capacity of the end-to-end path. In the second problem, i.e., (P2), the difference between the allocated wireless link rate and the theoretical achievable TCP throughput is studied. The (P2) aims to maximize the sum rate while minimizing the gap between the allocated data rate to each flow and the theoretical throughput that can be achieved by that TCP flow.

9.7.1.3 Formulation of the OFDMA resource allocation – (P1)

The formulated optimization problem (P1) aims to maximize DL sum rate, while TCP fairness is assured by imposing a set of nonlinear constraints into the problem. In other words, the proportional DL rate among users with respect to the TCP theoretical throughput, are subject to constraints. In the UL, assuming TCP receiver acknowledges every single packet, the minimum required data rate would be a proportion of the DL data rate for each specific TCP flow depending on the size of the ACK packet, which can be increased for example using the SACK option, $R_{u_i} \geq \rho R_{d_i}$, i.e., $0 < \rho < 1$.

$$\text{(P1)}: \text{Maximize} \sum_{i=1}^{n}\sum_{j=1}^{m} c_j a_{ij} w \log_2\left(1 + \frac{p_{ij}G_{ij}}{\sigma^2 c_3}\right),$$

$$\text{subject to}: \sum_{i=1}^{n} a_{ij} \leq 1, \qquad \forall j \in \{1, \ldots, m\} \tag{9.1}$$

$$\sum_{i=1}^{n}\sum_{j=1}^{m} c_j a_{ij} p_{ij} \leq P_T, \tag{9.2}$$

$$\sum_{j=1}^{m} (1 - c_j) a_{ij} p_{ij} \leq P_t, \qquad \forall i \in \{1, \ldots, n\} \tag{9.3}$$

$$\frac{R_{d_i}}{B_i} = \frac{R_{d_1}}{B_1}, \qquad \forall i \in \{2, \ldots, n\} \tag{9.4}$$

$$R_{u_i} \geq \rho R d_i, \qquad \forall i \in \{1, \ldots, n\} \tag{9.5}$$

$$\sum_{j=1}^{m} c_j \leq m_d, \qquad m_d \in \{1, \ldots, m\}, \tag{9.6}$$

$$p_{ij} \geq 0, \qquad \forall i \in \{1, \ldots, n\}, j \in \{1, \ldots, m\} \tag{9.7}$$

$$a_{ij} \in \{0, 1\}, \qquad \forall i \in \{1, \ldots, n\}, j \in \{1, \ldots, m\} \tag{9.8}$$

$$c_j \in \{0, 1\}. \qquad \forall j \in \{1, \ldots, m\} \tag{9.9}$$

In (P1), despite the classic approaches in solely maximizing throughput, subcarriers (a_{ij}) and transmission power over each subcarrier (p_{ij}) are allocated so that certain performance metrics of TCP are guaranteed. These performance metrics are provided via the constraints that are detailed below. Constraints (9.1) ensure that every subcarrier is assigned to only one user. We assume c_j represents the allocation of subcarrier j to DL $(c_j = 1)$ or UL $(c_j = 0)$. Thereby, constraints (9.2) and (9.3) restrict the total available power at the base station, P_T, and at each mobile user, P_t. Moreover, constraint (9.4) provide fairness among TCP flows with maintaining proportional rate with respect to the TCP throughput for each user. Constraint (9.5) provide the required data rate for UL, in order to guarantee delivery of the DL allocated resources, and finally constraint (9.6) bounds the number of DL subcarriers to m_d. Note that the optimal value of m_d can be found solving the problem (P1) iteratively for different values of m_d.

9.7.1.4 Formulation of OFDMA resource allocation – (P2)

Similar to (P1), (P2) aims to maximize DL sum rate, but the instantaneous rate allocation is constrained by the TCP theoretical throughput. The difference between the allocated DL rate to the ith user and the theoretical TCP throughput of flow i is denoted by D_i:

$$D_i = |\alpha B_i - R_{d_i}|. \tag{9.10}$$

In (9.10), α represents the overhead of the TCP/IP header. The resource allocation problem in this case attempts to minimize D_i while maximizing the sum rate. Therefore, (P2) can be defined as a multi objective optimization problem. There are various approaches to formulate such a multi-objective problem; we use a well-studied approach that combines the multiple objectives into a single objective function whose solution is Pareto optimal:

$$\text{(P2)}: \text{Maximize} \ \sum_{i=1}^{n}\sum_{j=1}^{m} c_j a_{ij} w \log_2\left(1 + \frac{p_{ij}G_{ij}}{\sigma^2 c_3}\right) - \mu \sum_{i=1}^{n} D_i.$$

$$\text{subject to:} \sum_{i=1}^{n} a_{ij} \le 1, \qquad \forall j \in \{1, \ldots, m\} \tag{9.11}$$

$$\sum_{i=1}^{n}\sum_{j=1}^{m} c_j a_{ij} p_{ij} \le P_T, \tag{9.12}$$

$$\sum_{j=1}^{m} (1 - c_j) a_{ij} p_{ij} \le P_t, \qquad \forall i \in \{1, \ldots, n\} \tag{9.13}$$

$$R_{u_i} \ge \rho R d_i, \qquad \forall i \in \{1, \ldots, n\} \tag{9.14}$$

$$\sum_{j=1}^{m} c_j \le m_d, \qquad m_d \in \{1, \ldots, m\}, \tag{9.15}$$

$$p_{ij} \ge 0, \qquad \forall i \in \{1, \ldots, n\}, j \in \{1, \ldots, m\} \tag{9.16}$$

$$a_{ij} \in \{0, 1\}, \qquad \forall i \in \{1, \ldots, n\}, j \in \{1, \ldots, m\} \tag{9.17}$$

$$c_j \in \{0, 1\}. \qquad \forall j \in \{1, \ldots, m\} \tag{9.18}$$

Constraints (9.11)–(9.18) are the same as (9.1)–(9.3) and (9.5)–(9.9). As mentioned above, problem (P2) has a Pareto optimal solution; thus the solution is not unique and it depends on the value of μ that balances the two objectives. In the above problem, increasing the value of μ shift the allocation balance towards TCP throughput, while decreasing the value of μ shift the balance towards a data rate maximization problem.

9.7.1.5 Finding Solutions for (P1) and (P2)

The well-used method for solving the multiple variable optimization problems is to decouple the problem [33], i.e., to allocate a_{ij} and p_{ij} separately. In the subcarrier allocation, it is assumed that power is equally distributed in all the subcarriers, therefore the solution is suboptimal. Afterwards, to a certain subcarrier allocation, an optimization problem can be reformulated over the continues variable p_{ij}. Thus, using the water filling approach, power will be distributed optimally.

The principle of the DL algorithm is to allocate the subcarrier with the highest channel gain available for each user. In addition to that, in the first round of allocation, users with the highest theoretical TCP throughput (largest value of B_i) are first to select subcarrier. Thereafter, in each round of the allocation, in solving problem (P1) users with the lowest proportional rate have priority to select the best available subcarrier. This step performs differently solving problem (P2), i.e., user with the smallest objective function selects the next subcarrier.

Subcarrier allocation in the UL in both problems is in order to satisfy the UL minimum rate requirements. The initial value of m_d maintains the proportion of ρ between number of DL and UL subcarriers – $m_d = m \cdot \frac{1}{1+\rho}$. Afterwards, in few iterations, the largest value of m_d which satisfies constraint (9.4) will be found; clearly this value maximizes the objective function. The above procedure for problem (P1) is detailed in Algorithm 1, in which Ω_i is the set of allocated subcarriers to the user i in the DL and Ψ_i is the set of allocated subcarriers to this user in the UL. Algorithm 2 details the subcarrier allocation solving problem (P2) in which steps (a), (b), (d), and (e) are similar to Algorithm 1, and only step (c) is restated.

The problem of power allocation with pre-defined subcarrier allocation, is based on the reformulation of (P1) into a maximization problem over continuous variable p_{ij} (similarly for (P2)).

Algorithm 1 Subcarrier allocation algorithm for the optimization problem (P1)

(a) Initialization

1. $m_{d_1} = \frac{1}{1+\rho} \cdot m.$
2. Set $R_{d_i} = 0$ and $\Omega_i = \phi$ for i = 1 to n and $C_d = \{1, \ldots m_d\}$.
3. Set $R_{u_i} = 0$ and $\Psi_i = \phi$ for i = 1 to n and $C_u = \{m_d + 1, \ldots m\}$.
4. Sort the users' index in the descending order of B_i.

(b) for $i = 1$ to n
1. Find the subcarrier k satisfying $|G_{ik}| > |G_{ij}|$ for all $j \in C_d$.
2. Let $\Omega_i = \Omega_i \cup \{k\}$ and $C_d = C_d - \{k\}$.
3. Update R_{d_i}

(c) while $C_d \neq \phi$
1. Find user l satisfying $R_{d_l}/B_l < R_{d_i}/B_i$ for all $i \in \{1, \ldots, n\}$.
2. For user l, find the subcarrier k satisfying $|G_{ik}| > |G_{ij}|$ for all $j \in C_d$.
3. Let $\Omega_l = \Omega_l \cup \{k\}$ and $C_d = C_d - \{k\}$.
4. Update R_{d_i}

(d) for $i = 1$ to n
1. Find the subcarrier k satisfying $|G_{ik}| > |G_{ij}|$ for all $j \in C_u$.
2. Let $\Psi_i = \Psi_i \cup \{k\}$ and $C_u = C_u - \{k\}$.
3. Update R_{u_i}

(e) while $C_u \neq \phi$
1. Find user l satisfying $(R_{u_l} - \rho R_{d_l}) < (R_{u_i} - \rho R_{d_i})$ for all $i \in \{1, \ldots, n\}$.
2. For user l, find the subcarrier k satisfying $|G_{ik}| > |G_{ij}|$ for all $j \in C_u$.
3. Let $\Psi_l = \Psi_l \cup \{k\}$ and $C_u = C_u - \{k\}$.
4. Update R_{u_i}

(f) If $\sum_{i=1}^{n} R_{u_i} \geq \rho \cdot \sum_{i=1}^{n} R_{d_i}$.
1. Find the largest $m_d \in \{m_{d_1}, \ldots, m\}$ such that constraint (9.5) are satisfied.
2. Else, find the largest $m_d \in \{1, \ldots, m_{d_1}\}$ such that constraints (9.5) are satisfied.

Algorithm 2 Subcarrier allocation algorithm for the optimization problem (P2)

(c) while $C_d \neq \phi$
1. Find user l satisfying $R_{d_l} - \mu D_l < R_{d_i} - \mu D_i$ for all $i \in \{1, \ldots, n\}$.
2. For user l, find the subcarrier k satisfying $|G_{ik}| > |G_{ij}|$ for all $j \in C_d$.
3. Let $\Omega_l = \Omega_l \cup \{k\}$ and $C_d = C_d - \{k\}$.
4. Update R_{d_i}

In Algorithms 1 and 2, Ω_{i_1} and Ω_{i_2} are mutually exclusive, if $i_1 \neq i_2$; the same assumption is also valid for Ψ_i. Problem (P2) also can be rewritten as (P2′) over the continuous variable p_{ij}.

9.7.1.6 Reducing the UL effect on TCP performance

Numerical evaluations are carried out using parameters listed in Table 9.1, to demonstrate reduced effect of UL imbalance on the overall system KPIs (based on [26]). The benchmarks are the sum rate maximization problem, denoted by (BM1),

Table 9.1 Parameters used in the numerical study

Bandwidth	5 MHz
Total number of OFDMA subcarriers	52
Maximum power at the base station	43 dBm
Maximum power at the mobile device	23 dBm
Thermal noise power (Johnson–Nyquist noise)	107 dBm
Target BER	10E4
Average SNR of wireless channel	15 dB
Wireless channel model	ITU pedestrian model
TCP maximum segment size	1,460 bytes
TCP flow type	Long-lived with SS/CA = 1
Number of mobile users	1–15 users
End-to-end RTT	Uniform in [10 ms, 200 ms]
Initial number of DL subcarriers (m_d)	32

and also the sum rate maximization with an equal rate constraint, denoted by (BM2). The m_d is initialized with 32 subcarriers for the DL, while $\rho = 0.2$. Afterwards and in step (f) of Algorithm 1, the optimal value of m_d is calculated in few iterations. The two benchmark problems (BM1) and (BM2), operate at $m_d = 32$; thus the results comparison here mainly show how the end-to-end performance is benefitted from setting the border between DL and UL adaptive to the current load of the system.

The two examined KPIs here are DL TCP throughput that also reflects the UL TCP throughput, and fairness among TCP flows. Fairness is quantified based on Jain's fairness index, which is a well-used index in computer and communication networks. The presented numbers in Figures 9.15–9.17 are average values taken from 150 Monte Carlo simulations. From Figure 9.15, it can be observed that the achieved fairness index by resource allocation (P1) is increased approximately 30% in average compared with the results of allocation scheme (BM1). In addition, Figure 9.15 shows that, as the number of mobile users competing over the wireless link is increased, distribution of the resources in a fair manner is more challenging. The highest increase in fairness index is in the 15-user case, where (P1) performs 70% better than the benchmarks. Moreover, Figures 9.16 and 9.17 show that joint allocation of the DL and UL improve the total aggregated throughput as well as the aggregated DL throughput by approximately 15%.

9.7.2 DL and UL, joint and decoupled

Enabling DUDe greatly contributes to the success of DL/UL TCP-ware resources allocation, while the joint resource allocation improves the efficiency of both UL and DL communications. First, the joint resource allocation will allow UL to have sufficient capacity for carrying TCP ACKs for the DL and vice versa. Therefore, allocated resource at the RAN will be efficiently utilized by the end-to-end connection. Second, the decoupled DL/UL will provide additional flexibility to the joint resource allocation. Hence, decoupled DL/UL while resources to the two are

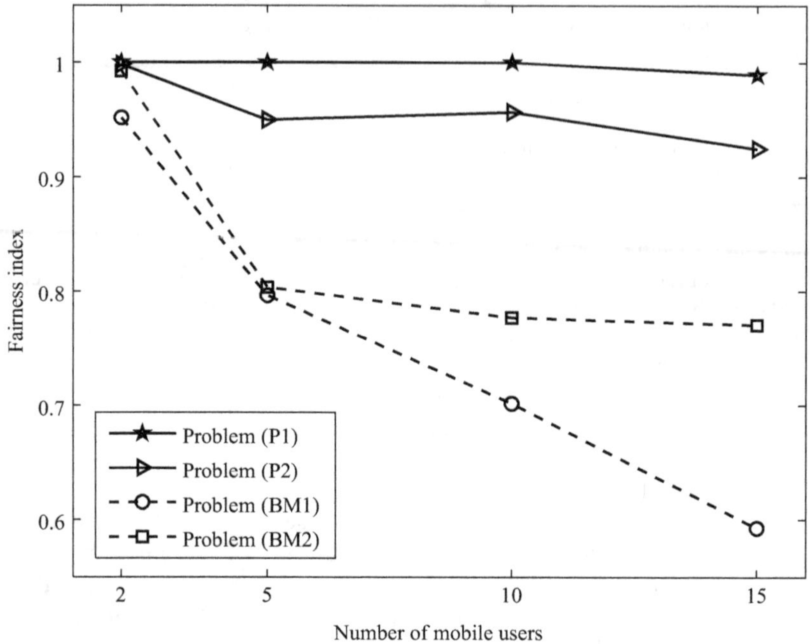

Figure 9.15 Jain's Fairness Index as achieved by solving resource allocation problems (P1), (P2), (BM1), and (BM2) vs. the number of mobile users

allocated in coordination with each other and in awareness from the higher layers can provide significant extra efficiency to the end-to-end communications both in terms of latency and data rate.

9.7.2.1 Enabling DUDe through multiple TCP flows

One of the challenges facing DUDe architecture is the possibility of having multiple TCP flows, that can handle the separate DL and UL connections. One of the major existing efforts in the Internet is the MultiPath TCP (MPTCP) [17]. MPTCP enables TCP to present a regular TCP interface to applications, while several IP interfaces are used simultaneously. In other words, data communication of a flow is spread over multiple *subflows*. The MPTCP connections begin regularly, and if there are extra paths available, additional TCP sessions, termed *subflows*, are created on these paths, and are combined with the existing session, which continues to appear as a single connection to the applications at both ends. Therefore MPTCP will allow for switching between coupled and decoupled DL/UL for different traffic flows. While MPTCP could be a great enabler for DUDe architecture, the challenge here is the availability of two different IP addresses. In other words, MPTCP can only be used to handle the DL and UL, if the UE is multihomed. Example of other efforts in running multiple and parallel TCP flows includes MulTCP [29], which behaves as if it was a collection of multiple virtual TCP

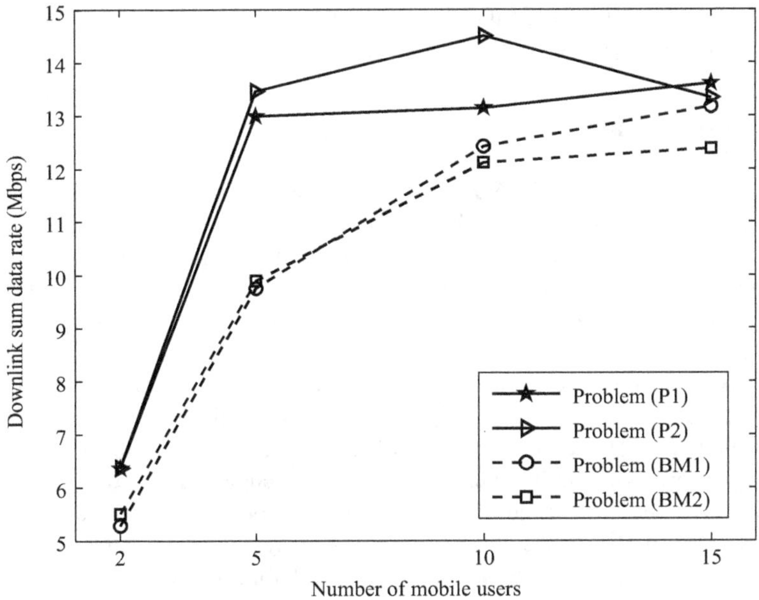

Figure 9.16 DL sum data rate (Mbps) as achieved by solving resource allocation problems (P1), (P2), (BM1), and (BM2) vs. the number of mobile users

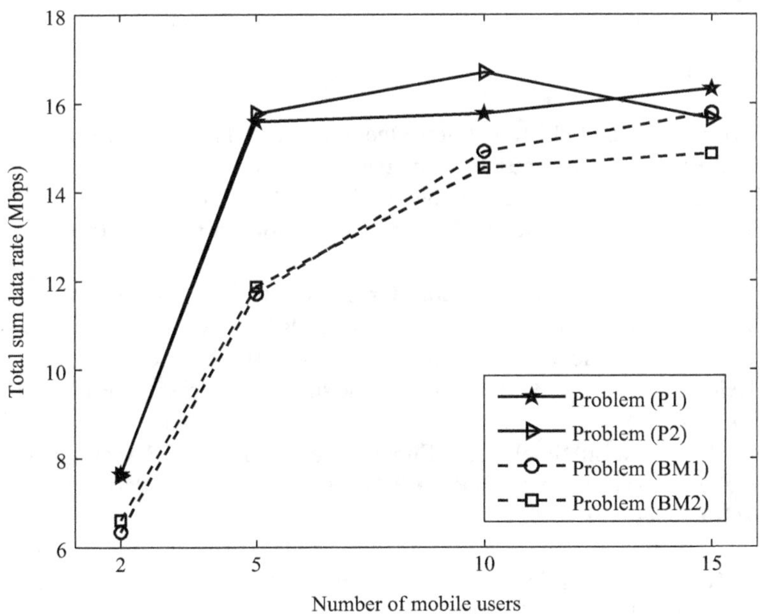

Figure 9.17 Total (UL+DL) sum data rate (Mbps) as achieved by solving resource allocation problems (P1), (P2), (BM1), and (BM2) vs. the number of mobile users

connections. Such initiative in the Internet domain is often taken for increasing data rate and better utilization of the end-to-end bandwidth.

9.8 Conclusions

Since the inception of cellular networks, the UL connection was always associated with the same base station that has been chosen for the DL reception. In this chapter, this postulate has been revisited and the feature of Downlink/Uplink Decoupling (DUDe) was introduced. DUDe is a new architectural paradigm where UL and DL are not constrained to associate to the same base station and are effectively treated as separate networks. This concept is becoming very relevant with the expected densification in future cellular networks where each terminal will have multiple access point in proximity and allowing the terminal to choose the best UL and DL connection(s) will result in a much more efficient operation of cellular networks. In this chapter, the main challenges faced by heterogeneous networks have been identified and how DUDe can be a solution to some of these challenges. The performance gains from DUDe were also demonstrated in terms of UL capacity, coverage, load balancing and reliability. The interoperability of DUDe with the different new trends in cellular networks has been highlighted. Finally, the realization of DUDe in terms of architecture has been studied thoroughly, where the different options of implementing DUDe in LTE have been illustrated, with a glimpse of what is needed in 5G networks for DUDe to be natively supported as well as a study on the effects of DUDe on higher layers.

References

[1] 3GPP. Coordinated Multi-Point Operation for LTE Physical Layer Aspects. TS 36.819, 3rd Generation Partnership Project (3GPP), Sep. 2012.

[2] 3GPP. LTE Radio Access Network (RAN) Enhancements for Diverse Data Applications. TR 36.822, 3rd Generation Partnership Project (3GPP), Sep. 2012.

[3] 3GPP. Technical Specification Group Radio Access Network; Evolved Universal Terrestrial Radio Access (E-UTRA) and Evolved Universal Terrestrial Radio Access Network (E-UTRAN); Overall description; Stage 2 (Release 11). TS 36.300, 3rd Generation Partnership Project (3GPP), Sep. 2012.

[4] 3GPP. Study on Small Cell Enhancements for E-UTRA and E-UTRAN; Higher Layer Aspects. TR 36.842, 3rd Generation Partnership Project (3GPP), Sep. 2014.

[5] J.G. Andrews, S. Buzzi, W. Choi, *et al.* "What will 5G be?" *IEEE Journal on Selected Areas in Communications*, 32(6):1065–1082, Jun. 2014. doi: 10.1109/JSAC.2014.2328098.

[6] H. Balakrishnan, V.N. Padmanabhan, G. Fairhurst, and M. Sooriyabandara. "TCP performance implications of network path asymmetry". *IETF RFC 3449*, Dec. 2002.

[7] F. Boccardi, R.W. Heath, A. Lozano, T.L. Marzetta, and P. Popovski. "Five disruptive technology directions for 5G". *IEEE Communications Magazine*, 52(2):74–80, Feb. 2014. doi: 10.1109/MCOM.2014.6736746.

[8] F. Boccardi, J. Andrews, H. Elshaer, *et al.* "Why to decouple the uplink and downlink in cellular networks and how to do it". 2015. Available at: http://arxiv.org/abs/1503.06746.

[9] D. Colombi, B. Thors, and C. Tornevik. "Implications of EMF exposure limits on output power levels for 5G devices above 6 GHz". *IEEE Antennas and Wireless Propagation Letters*, 14:1247–1249, 2015. doi: 10.1109/LAWP.2015.2400331.

[10] A. Daeinabi, K. Sandrasegaran, and X. Zhu. "Performance evaluation of cell selection techniques for picocells in LTE-advanced networks". In *2013 10th International Conference on Electrical Engineering/Electronics, Computer, Telecommunications and Information Technology (ECTI-CON)*, pp. 1–6, May 2013.

[11] U. Dtsch, M. Doll, H.-P. Mayer, F. Schaich, J. Segel, and P. Sehier. "Quantitative analysis of split base station processing and determination of advantageous architectures for LTE". *Bell Labs Technical Journal*, 18(1):105–128, 2013.

[12] H. Elshaer, F. Boccardi, M. Dohler, and R. Irmer. "Downlink and uplink decoupling: a disruptive architectural design for 5G Networks". In *2014 IEEE Global Communications Conference (GLOBECOM)*, pp. 1798–1803, Dec. 2014.

[13] H. Elshaer, F. Boccardi, M. Dohler, and R. Irmer. "Load & backhaul aware decoupled downlink/uplink access in 5G systems". In *2015 IEEE International Conference on Communications (ICC)*, pp. 5380–5385, Jun. 2015.

[14] ST. Ericsson. Further Discussions on UL/DL Split. Technical Report R2-131678, 3GPP TSG-RAN, May 2013.

[15] ST. Ericsson. Physical Layer Aspects of Dual Connectivity. Technical Report R1-130566, 3GPP TSG-RAN, Feb. 2013.

[16] S. Floys, A. Arcia, and J. Iyengar. Adding Acknowledgement Congestion Control to TCP. *IETF Internet Draft*, Jan. 2009.

[17] A. Ford, M. Handley, and O. Bonaventure. TCP Extensions for Multipath Operation with Multiple Addresses. *IETF RFC 6824*, Jan. 2013.

[18] D. Gesbert, S. Hanly, H. Huang, S. Shamai Shitz, O. Simeone, and W. Yu. "Multi-cell MIMO cooperative networks: a new look at interference". *IEEE Journal on Selected Areas in Communications*, 28(9):1380–1408, Dec. 2010.

[19] X. Gu, X. Deng, Q. Li, L. Zhang, and W. Li. "Capacity analysis and optimization in heterogeneous network with adaptive cell range control". *International Journal of Antennas and Propagation*, 2014:10, Apr. 2014. doi:10.1155/2014/215803.

[20] I. Guvenc, M.-R. Jeong, I. Demirdogen, B. Kecicioglu, and F. Watanabe. "Range expansion and inter-cell interference coordination (ICIC) for picocell networks". In *2011 IEEE Vehicular Technology Conference (VTC Fall)*, pp. 1–6, Sep. 2011.

[21] Z. Han, Z. Ji, and K.J. Ray Liu. "Power minimization for multi-cell OFDM networks using distributed non-cooperative game approach". In *Global Communications Conference (GLOBECOM)*. IEEE, Nov. 2004.

[22] Huawei. Handling of UL Traffic of a DL Split Bearer. Technical Report R2-140054, 3GPP TSG-RAN, Feb. 2014.

[23] M. Lema, M. Garcia-Lozano, and S. Ruiz. "MPR-aware scheduler for carrier aggregation transmissions in LTE Uplink". *Wireless Personal Communications*, 84(2):1417–1438, 2015.

[24] M.A. Lema, M. Garcia-Lozano, S. Ruiz, and D.G. Gonzalez. "Improved component carrier selection considering MPR information for LTE-A uplink systems". In *2013 IEEE 24th International Symposium on Personal Indoor and Mobile Radio Communications (PIMRC)*, pp. 2191–2196, Sep. 2013.

[25] A. Maeder, M. Lalam, A. De Domenico, *et al.* "Towards a flexible functional split for cloud-RAN networks". In *2014 European Conference on Networks and Communications (EuCNC)*, pp. 1–5, Jun. 2014.

[26] T. Mahmoodi, V. Friderikos, and H. Aghvami. "Using traffic asymmetry to enhance TCP performance". *Computer Networks*, 57(1):317–329, Jan. 2013.

[27] T. Mahmoodi, V. Friderikos, O. Holland, and H. Aghvami. "Cross-layer optimization to maximize fairness among TCP flows of different TCP flavors". In *Global Communications Conference (GLOBECOM)*. IEEE, Nov. 2009.

[28] T. Mahmoodi, V. Friderikos, O. Holland, and H. Aghvami. "TCP-aware resource allocation problem in OFDMA based wireless networks. In *International Workshop on Cross layer Design (IWCLD)*. IEEE, Jun. 2009.

[29] P. Mulroy, S. Appleby, M. Nilsson, and B. Crabtree. "The use of MulTCP for the delivery of equitable quality video". In *International Packet Video Workshop*. IEEE, May 2009.

[30] NGMN. Further Study on Critical C-RAN Technologies. Deliverable, NGMN, 2015.

[31] J. Padhye, V. Firoiu, D. Towsley, and J. Kurose. "Modeling TCP throughput: a simple model and its empirical validation". *SIGCOMM Computer Communications Review*, 28(4):303–314, 1999.

[32] K.I. Pedersen, Y. Wang, B. Soret, and F. Frederiksen. "eICIC functionality and performance for LTE HetNet co-channel deployments". In *2012 IEEE Vehicular Technology Conference (VTC Fall)*, pp. 1–5, Sep. 2012.

[33] Z. Shen, J.G. Andrews, and B.L. Evans. "Adaptive resource allocation in multiuser OFDM systems with proportional rate constraints". *Transactions on Wireless Communications*, 4(6):2726–2737, 2005.

[34] S. Singh, X. Zhang, and J.G. Andrews. "Joint rate and SINR coverage analysis for decoupled uplink–downlink biased cell associations in HetNets". *IEEE Transactions on Wireless Communications*, 14(10):5360–5373, Oct. 2015.

[35] K. Smiljkovikj, P. Popovski, and L. Gavrilovska. "Analysis of the decoupled access for downlink and uplink in wireless heterogeneous networks". *IEEE Wireless Communications Letters*, 4(2):173–176, Apr. 2015.

[36] W. Richard Stevens. *TCP/IP Illustrated, Volume I: The Protocols*. Addison-Wesley Longman Publishing Co., Inc., Boston, MA, 1993.

[37] H. Wang, C. Rosa, and K.I. Pedersen. "Dedicated carrier deployment in heterogeneous networks with inter-site carrier aggregation". In *2013 IEEE Wireless Communications and Networking Conference (WCNC)*, pp. 756–760, Apr. 2013.

[38] H. Wang, C. Rosa, and K.I. Pedersen. "Uplink inter-site carrier aggregation between macro and small cells in heterogeneous networks". In *2014 IEEE 80th Vehicular Technology Conference (VTC Fall)*, pp. 1–5, Sep. 2014.

[39] I.C. Wong and B.L. Evans. "Optimal downlink OFDMA resource allocation with linear complexity to maximize ergodic rates". *Transactions on Wireless Communications*, 7(2):962–971, 2009.

[40] ZTE. Comparison between CP solution C1 and C2. Technical Report R2-132383, 3GPP TSG-RAN, Aug. 2013.

Chapter 10
Wireless networks virtualization

Chih-Lin I[1], *Jinri Huang*[1], *Changming Bai*[1], *Ran Duan*[1] *and Rongwei Ren*[1]

10.1 Introduction

With the impressive development of smart terminals and data cards, mobile Internet service has experienced an exponential increase in the past few years [1]. However, what people cannot see is that mobile operators always are paying a lot of effort to respond to the ever-increasing traffic. In fact, while the traffic has an exponential increase, the growth rate of income and profit of most operators is no more than 10% or even negative. The reason is that the rapidly developing mobile Internet has a huge pressure on mobile networks. In order to handle the daily increasing traffic, operators have to pay more resources on the network deployment, reconstruction or upgrade, leading to huge amount of investment and Operating Expense (OPEX). It can be easily imagined that if mobile Internet keeps increasing in the coming years, mobile networks will face more and more pressure. Mobile operators are facing unprecedented challenges.

How can operators efficiently reduce OPEX, increase network utilization or even accelerate the activation of service development to face the challenges of mobile Internet era? To address this issue, let us start from the knowledge on traditional telecom equipment, include its form, development and debugging mode, network maintenance and update.

The telecom network equipment has two main features, namely "proprietary platform" and "dedicated chipsets". "Proprietary platform" means that the platform on which the equipment is developed is proprietary, different from vendor to vendor. "Dedicated chipsets" means that chipsets such as Digital Signal Processing (DSP), Field-Programmable Gate Array (FPGA) and Application-Specific Integrated Circuit (ASIC) are mainly adopted for equipment development. As a result, proprietary platforms cause poor interoperability and compatibility issues while dedicated chipset usage cause issues of poor programmability, poor portability and slow secondary

[1]Green Communication Technology Research Center, China Mobile Research Institute, No. 32, Xuanwumenxi Avenue, Xicheng District, Beijing 100053, China

development. In addition, this industry requires engineers with high-level skills who cannot be found easily. Thus, the competitiveness and service innovation of the whole industry is restricted. Besides, equipment usually takes a long development cycle, complex testing and verification process. Beyond the above, the telecom network consumes a huge amount of power with low utilization. It can be foreseen that the ever-increasing demands of mobile Internet will increase the requirements on the next generation wireless networks in terms of network upgradability, innovation and timely introduction of services, power efficiency and so on, which are what traditional equipment development modes are unable to satisfy.

Then let us turn our eyes to IT industry. The landscape there is different. IT industry is famous for its openness and softness. Various standardized interface makes interoperability much easier so that the developers can focus on development and optimization of certain components without having to consider interconnection issues. Good examples include video cards, CPU, memories and so on. In addition, general-purpose platform (GPP), e.g. x86 architecture, is widely adopted which facilitates software development, maintenance and portability. This, in some sense, encourages service innovation and accelerates time-to-market.

The development of IT industry, in particular the prosperity of data centres, is greatly attributed to a key technology, namely the virtualization technology.

10.2 Virtualization basics

As for the origin of virtualization technology, the term was used for the first time by Dr Christopher Strachey from Oxford University. In his book *Time Sharing in Large Fast Computers* the virtualization concept was mentioned in the 1960s. Then back in 1964 IBM initiated a project which was called CP/CMS system and evolved to the Virtual Machine Facility/370 [2]. From then on, the term of virtual machine (VM) has been used to differentiate from the physical machine.

In virtualization technology area IBM and the research community stated the basic concepts for virtualization and Goldberg published works [3] that addressed the concepts and benefits of this technology. But for some years the interest decreased and nothing related to virtualization technology happened until 1999 when the company VMware Inc. presented its product VMware Virtual Platform for the x86-32 architecture [4].

But what is virtualization in short? One opinion is that virtualization is the act of ABSTRACTING the physical boundaries of a technology [5]. To achieve physical abstraction there are several ways. One of them is that applications or servers no longer need dedicated physical hardware in order to run as independent entities and a virtual machine could come into play. With virtual machines the traditional hardware is emulated and presented to an operating system as if the hardware truly existed. Beyond that, multiple VMs have the capability to run over different operating systems on the same hardware in parallel.

Another opinion prefers a simple analogy of virtualization to the picture-in-picture feature of some televisions and set-top boxes because it displays a small

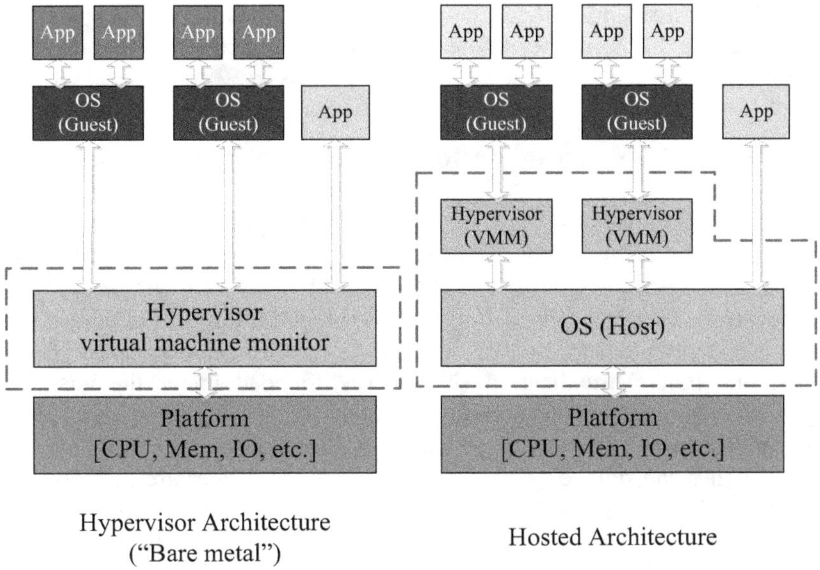

Figure 10.1 Two types of virtual machine monitors

virtual television image on top of another television image, thereby allowing both programs to play simultaneously [6]. Based on the virtualized platform, which is the combination of computer, network and storage virtualization, the traditional applications or multiple sets of software can execute at the same time.

In virtualization technology evolution path, there are two types of virtualization machine monitors (VMM). One runs directly on the hardware with the same functions as in traditional operating systems (OS) to control or manage the hardware. It is called "Hypervisors" architecture. The other runs over the existed OS, and manages and supports the OS running over itself. In this scenario the OS below VMM is called Host and the one over is called guest OS. The mode is called "Hosted" architecture.

Figure 10.1 shows an example of those two types of VMMs and hypervisors.

For the system level virtualization, there are three approaches: full virtualization, paravirtualization and OS-virtualization. We can simply introduce the differences between those three methods [6]:

- Full virtualization has one dedicated VMMs or hypervisors to manage the underlying hardware. With hardware assistant, Intel and AMD processors could support such virtualization. It could support various guest OS running over it.
- Paravirtualization is slightly different from full virtualization. It has a better performance than full virtualization. But it also requests guest OS modification.
- OS virtualization asks one layer to do the virtualization and has the best performance scalabilities. But it just supports a single OS.

For the above three approaches, there is no need to make any changes to the applications. The difference lies in the complexity, the capabilities and the number of supported OS.

10.3 Cloud RAN virtualization and its challenges

On the one hand, operators are facing the challenges imposed by the ever-increasing mobile traffic. On the other hand, it is witnessed that the virtualization technology has passed beyond the IT industry and extended its impact on several other industries. People begin to think about the possibility of adopting virtualization into telecom industry. In 2012, an Industry Specification Group (ISG) named Network Functions Virtualization (NFV) was founded under the auspices of European Telecommunications Standards Institute (ETSI). The basic idea of NFV is to "consolidate many network equipment types onto industry-standard high-volume servers, switches and storage, which could be located in data centres, network nodes and in the end-user premises" [7]. People expect that virtualization could bring operators such benefits as quicker development process, software-like system upgrade, less complex management, quicker time-to-market and improved power efficiency and so on.

Wireless networks consist mainly of the core network and the radio access network (RAN). It is clear that RAN is much more complicated than the core network from the computation and real-time requirements perspective. Therefore there is a wide consensus in the industry that realization of the RAN virtualization is much more challenging than that of the core network.

In this chapter, we will mainly focus on the RAN virtualization. In particular, we will focus on the Cloud RAN (C-RAN) which is an essential architecture towards 5G.

10.3.1 C-RAN basic

The concept of C-RAN was first proposed by China Mobile in the white paper in 2009 [8,9]. The "C" here indeed stands for "Centralized, Collaborative, Cloud and Clean". The basic idea of C-RAN starts from centralization, which is to aggregate different baseband units (BBUs) that in a traditional deployment are geographically separated into the same location. Once centralized, it would be possible to make different BBUs to communicate with each other in a more timely way by connecting them with high-speed switch networks, therefore allowing implementation of the cooperative algorithms to improve system performance.

Simple centralization does not support resource sharing among different physical BBUs. With virtualization, this is about to change. The ultimate goal of C-RAN is to realize the so-called feature of "resource cloudification", which is to pool the baseband processing resources so that the resources could be managed and dynamically allocated on demand.

Virtualization is the key to realize resource cloudification. Virtualization of RAN network aims to architect RAN networks with evolving standard IT

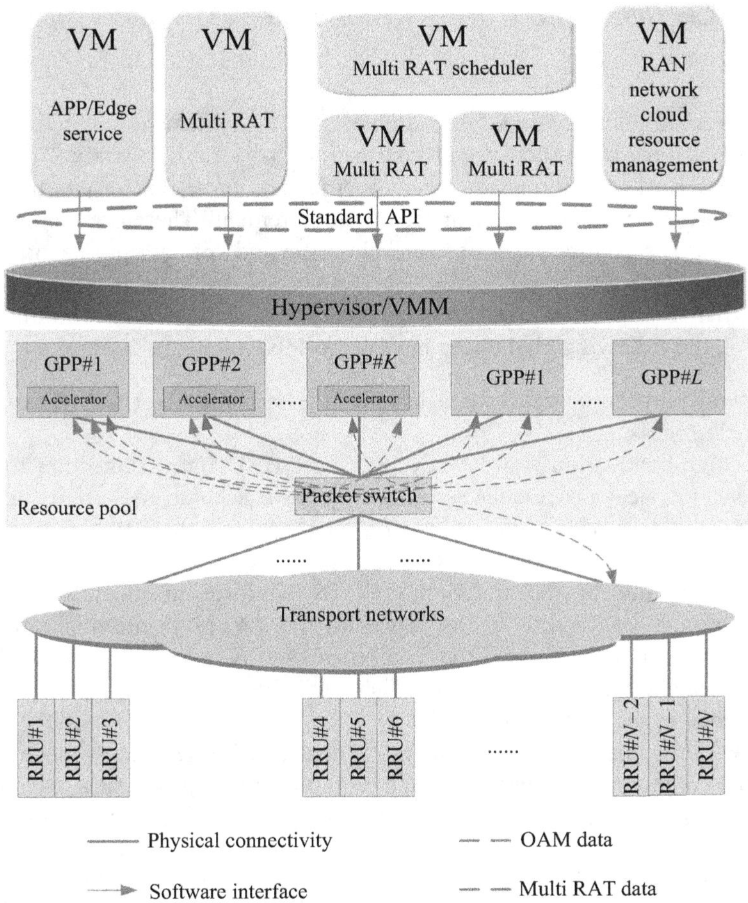

Figure 10.2 Virtualized C-RAN system

virtualization technology to consolidate traditional network equipment types onto industry-standard high-volume servers, switches and storage, which would be located in wireless sites or central offices. It involves the implementation of parts of RAN functions in software that can run on a range of industry-standard server hardware, and that can be moved to, or instantiated in, various locations in the network as required, without the need for installation of new equipment.

Figure 10.2 illustrates a virtualized C-RAN system based on the Commercial-off-the-shelf (COTS) platform. With virtualization, standard IT servers are used as the general platform with computation and storage as the common resources. As shown in Figure 10.2, on top of the servers run different applications in the form of virtual machines (VM). The indispensable applications in C-RAN are those to realize different radio access technologies (RAT) including 2G, 3G, 4G and future 5G. Additional user applications, such as content delivery network, can also be deployed on the same platform.

10.3.2 Challenges on C-RAN virtualization

Wireless communication, especially the baseband processing in RAN, is distinct from IT data centres in that wireless communication has extremely strict requirements on real-time processing. For example, for LTE systems it is required that an ACK/NACK must be produced and sent back to the UE/eNB within 3 ms after a frame is received. Traditional data centre virtualization technology with General Purpose Processor (GPP) cannot meet this requirement. Therefore, applying virtualization to base stations requires careful design and special optimization of key functional blocks.

In order to implement real-time virtualized base stations in a centralized base band pool, the following challenges have to be solved:

- General purpose processor and advanced processing algorithm for real time signal processing.
- The high-bandwidth, low latency, low cost BBU inter-connection topology among physical processing resources in the baseband pool. It includes the interconnection among the chips in a BBU, among the BBUs in a physical rack, and among multiple racks.
- The capability of the traditional COTS platform to meet the real-time requirements of RAN functions. It will involve the optimization of various key modules including the OS, the hypervisor, the I/O, the switch networks, the management systems and so on. There will be detailed introduction on the solutions in the next section.
- Study of the virtualization granularity is also a challenge. Due to the difference of the communication protocols, the requirements of processing resources for GSM, TD-SCDMA and TD-LTE are quite different in terms of computing capability and real-time performance. In practice, different virtualization granularity should be considered accordingly. For example, in practical deployments, it is possible for a virtual machine to handle only one carrier within the BBU or even smaller units. There is another layered approach in which a virtual machine handles just one layer such as L1, L2 or L3, or even a certain part of a layer of a carrier.

10.4 Key technologies and solutions to RAN virtualization

To accomplish RAN virtualization, a wide range of technology challenges need to be addressed. From hardware and software perspective, the virtualization platform should have specified enhancements and modifications. Research and experience has clearly demonstrated that standard desktop and enterprise class IT solutions do not provide the determinism, performance and reliability required for a successful RAN implementation—a purpose built, optimized solution is required.

The following key technologies are important in achieving effective RAN virtualization.

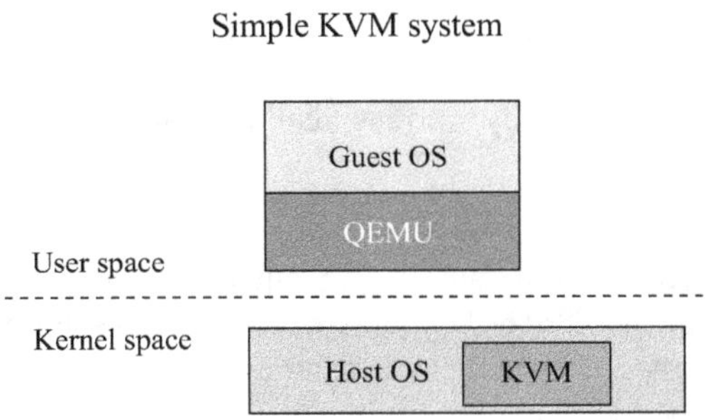

Figure 10.3 Simple KVM systems

10.4.1 Enhancement on host Linux

We firstly will introduce the Kernel based Virtual Machine (KVM). KVM is designed for x86 and x86_64 to take advantage of native support for hardware virtualization, while KVM also supports other processor architectures such as ARM and Power. KVM builds upon Quick Emulator (QEMU) emulation to provide a comprehensive virtualization solution.

The simple usage for KVM is as Figure 10.3 shows.

RAN virtualization is a more complex use case scenario than some core network applications, since there are several virtual machines or guest OS running and the performance is very sensitive to stringent requirements (hard real-time requiring processing every 500 μs without jitter). Similarly, performance without determinism is useless in this context. The complete virtualized solution must deliver the same level of performance with guaranteed predictability to every workload associated with the virtualized RAN application.

The following limitations need to be addressed:

- Support of real-time KVM and QEMU.
- Improved guest isolation with CPU interference minimization techniques.
- Improved system configuration, partitioning and isolation of virtual machines and environments using flexible scripting capabilities.
- Well-defined security contexts for virtual machines and environments.
- Support of Full NO_HZ (the tickles kernel). A set of kernel patches to improve the isolation of a particular CPU from sources of system interference such as Inter Processor Interrupts (IPI), timer ticks, memory reclaim and others. Modifications to global state machines (RCU) and services (timekeeping) are made to work without periodic access to a core.

In addition, there are still a lot of other key system parameters, which are called options in Linux and may affect the real-time performance. System tuning work will be required for different hardware and software. The options are passed to the kernel by the grub boot loader, and appear in the /proc/cmdline file once the target completes the boot process.[1] In the case of a guest kernel, they are passed to the qemu-kvm binary at launch time.

The following lists the key options that may impact the system real-time performance.

1. clocksource
   It is used to override the default clock source, with jiffies being the fall back clock source. For the x86_64 architecture, the options are hpet and tsc. While QEMU can emulate a HPET clock source for a guest, OVP uses the tsc as clocksource to avoid emulation overhead.
2. idle
   It is used to define the sleeping behaviour of a CPU when it is idle. Valid parameters are pool, mwait, halt, and nomwait. When using poll, the idle task keeps the CPU awake by periodic polling preventing it from going into power saving mode. This reduces latencies associated with context switching at the expense of higher power consumption. When using MWAIT, the idle task uses the special MONITOR/MWAIT instructions that achieve nearly the same performance of the poll option but with less power consumption. All newer Intel processors support these instructions. The nomwait option allows the CPU to sleep in C power saving states but makes no use of the MONITOR/MWAIT instructions. Note that since Linux 3.0, there has been a boot-time warning when idle=mwait is invoked, stating that this option will be removed in future releases. This is in fact the case since kernel 3.9. Kernels configured with CONFIG_ACPI=n and CONFIG_INTEL_IDLE=n when run on hardware that supports MWAIT will simply use HLT. If MWAIT is desired on those systems, these two configuration options should be enabled.
3. irqaffinity
   It is used to set the default IRQ affinity mask, that is, the set of CPUs that should process interrupts. Valid parameters are a list, or a range, or a combination of CPU identifiers. For example, if on a 16 core machine users want to prevent interrupts from going to CPUs 1, 2 and 3. They could use irqaffinity=0,4–15 on the host boot line. The irqaffinity option is particularly convenient when used in conjunction with the isolcpus option to more fully isolate one or more CPUs from the Linux kernel's scheduler and interrupt management subsystems.
4. isolcpus
   This option is used to specify one or more CPUs in order to isolate from the general SMP balancing and scheduling algorithms. Users can move a process onto or off an isolated CPU with the CPU affinity system calls or the taskset and cpuset commands. The cpu number begins at 0 and the maximum value is number of CPUs in system −1.

[1]All the Linux parameters mentioned in this section could be found at https://www.kernel.org/doc/Documentation/kernel-parameters.txt

This option is also used in the guest QEMU-KVM kernel's command line to isolate virtual CPUs which are intended to be used for specific tasks. For example, for a two virtual CPU guest instance isolcpus=1 would leave virtual CPU zero available for housekeeping tasks while virtual CPU 1 will be dedicated to only those tasks explicitly set to run on it. In general, achieving full CPU isolation is not entirely possible. In particular, some kernel tasks such as soft IRQ processing through the ksoftirqd thread will likely be always present. The ksoftirqd allowed CPU list in the kernel is essentially read only.

5. rcu-nocbs
   It is used to prevent RCU callback routines from being executed in the targeted CPUs. Valid parameters are a list, or a range, or a combination of CPU identifiers. RCU callbacks are functions that perform cleanup work after an RCU grace period passes. With some workloads, the number of callbacks can get quite large. As an example, to request that CPUs 1, 2, 3 and 6 are not be used to execute RCU callbacks, users can use rcu-nocbs=1–3,6 on the host boot line. It is not possible to assign every CPU in the system to the no callback list; at least one processor, CPU 0, must remain in the traditional mode or RCU grace period processing will not function properly. Typically the rcu-nocbs range will match the isolcpus parameters to further improve the isolation status of the targeted CPUs.
6. rcu_nocb_poll
   It is used to force polling mode on RCU threads that handle the execution of RCU callbacks. This option has no parameters. When a CPU runs without RCU callbacks, there will be a separate process charged with callback handling. When that process wakes up, it will grab the list of outstanding callbacks for its assigned CPU, wait for the RCU grace period to expire, and then run through all the callbacks. The rcu_nocp_poll boot option instructs this process to poll occasionally for new work instead. Polling mode contributes to alleviate the load on offloaded CPUs, those still executing the RCU callbacks, by sparing them the effort to wake up the cleanup processes regularly.
7. tsc
   It is used to make assumptions about the availability and behaviour of the Time Stamp Counter. Using tsc=perfect avoids some of the runtime checks performed on the TSC reliability. It is used in host and guest.

10.4.2 VM communication improvement

Inter-VM communication is very important for the virtualization platform. One of the outstanding properties of virtualization is its ability to isolate co-resident virtual machines on the same physical platform. These virtual machines often need to communicate and exchange a considerable amount of data. Many RAN applications have different modes of operation within the node that need to exchange messages and data during the span of a voice or data call. It is important for the RAN vendors to be able to minimize the changes to the software so that same software can run on dedicated hardware or in VMs, leading to the requirement for high performance inter-VM messaging. The performance will be bottleneck if the throughput or latency cannot meet the requirements of the virtualization RAN. Since the RAN is not the same as the legacy IT system, the requirements are exceedingly strict.

There are many types of inter-VM communication methods, including

- Shared memory-based technique
- Network-based technique
  - Host virtual switch
  - Hardware assistant switch

Here we will focus on the network communication. The major reason is that using the network communication will require very few modifications of the virtual machine.

The standard KVM and QEMU has some features supported such as

- Virtio devices. VirtIO is a specification for para-virtualized I/O in a virtual machine. Traditionally, the hypervisor emulates real devices such as an Ethernet interface or disk controller to provide the VM with I/O. This emulation is often inefficient.
- Vhost. The vhost drivers in Linux provide in-kernel virtio device emulation. Normally the QEMU userspace process emulates I/O accesses from the guest. Vhost puts virtio emulation code into the kernel, taking QEMU userspace out of the picture. This allows device emulation code to directly call into kernel subsystems instead of performing system calls from userspace.

There are several virtual switches available currently. The basic, default choice is stable, but poorly performing Open vSwitch, which is open source. There are also some commercial virtual switches such as the high performance Wind River Accelerated vSwitch, integrated into the Titanium Server NFV infrastructure platform.

To provide the high throughput inter-VM network, virtual switches may have the following features:

- User space packet handling
- Dedicated CPU core assigned

To provide the low latency inter-VM network, virtual switches may have the following features:

- Polling mode driver
- No lock system design

10.4.3 Real-time enhancement

To achieve high performance and predictability, the operating system distribution must be built and tested with a specific set of configuration and tuning parameters:

- Preempt-RT should support both host and guest kernels: A set of kernel patches for the Linux kernel with the goal of improving determinism and real-time behaviour through lock semantic changes, algorithm modifications and driver changes. All preemption levels are supported up to PREEMPT_RT_FULL depending on the use case.

- Real-time drivers and features: Adaptations of common tools and operations to work within the preempt-rt locking semantics.
- Full NO_HZ: A set of kernel patches to improve the isolation of a particular CPU from sources of system interference, such as Inter Processor Interrupts (IPI), timer ticks, memory reclaim and others. Modifications to global state machines (RCU) and services (timekeeping) are made to work without periodic access to a core.
- Real-Time Enhanced KVM and KVM Guest: In addition to the base preempt-rt kernel functionality, KVM guest and host support has been optimized for interrupt latency and process isolation. This allows virtual machines to run with a performance that nears that of a native BSP, with additional security, isolation, abstraction, emulation and acceleration while leveraging KVM feature support such as live migration.

10.4.4 Live migration

Live migration refers to the process of moving a running virtual machine or application between different physical machines or between different CPUs in the same physical machine without stopping the virtual machine. Memory, storage and network connectivity of the virtual machine are transferred from the original guest machine to the destination.

The migration could be used for the following use cases:

- Load balancing—guests can be moved to hosts with lower usage when a host becomes overloaded.
- Hardware failover—when hardware devices on the host start to fail, guests can be safely relocated so the host can be powered down and repaired.
- Energy saving—guests can be redistributed to other hosts and host systems powered off to save energy and cut costs in low usage periods.
- Geographical migration—guests can be moved to another location for lower latency or in critical circumstances.

To achieve live migration, the state of the guest OS on the source host must be replicated on the destination host. This requires migrating processor state, memory content, local storage and network state.

At the virtualization RAN field, the downtime of the live migration is of essence. Live migration should not affect end user quality of experience and no calls or data should be dropped. To achieve the best live migration performance, the special memory migration technique can be adopted.

An increasingly important factor to consider when selecting a virtualization platform for vRAN deployment is support for the live migration of Intel® Data Plane Development Kit (DPDK) based applications. As more and more applications are built using DPDK, live migration for these applications will be blocked unless the virtualization solution explicitly supports this functionality.

10.4.5 Network acceleration

For network acceleration, we will focus on Data Plane Development Kit (DPDK). DPDK is a set of libraries and drivers for fast packet processing. It was designed to run on any processor. The first supported CPU was Intel x86 and it is now extended to IBM Power 8, EZchip TILE-Gx and ARM. It runs mostly in Linux userland. A FreeBSD port is available for a subset of DPDK features. DPDK is an Open Source BSD licensed project [10] (Figure 10.4).

The main goal of the DPDK is to provide a simple, complete framework for fast packet processing in data plane applications.

The DPDK implements a run-to-completion model for packet processing, where all resources must be allocated prior to calling Data Plane applications, running as execution units on logical processing cores. The model does not support a scheduler and all devices are accessed by polling. The primary reason for not using interrupts is the performance overhead incurred by interrupt processing.

In addition to the run-to-completion model, a pipeline model may also be used by passing packets or messages between cores via the rings. This allows work to be performed in stages and may allow more efficient use of code on cores.

The Intel DPDK vSwitch is a fork of the open source Open vSwitch multilayer virtual switch. It reimplements the kernel forwarding module (data plane) by building the switching logic on top of the Intel DPDK library. This significantly boosts packet switching throughput and brings Open vSwitch closer to the overall performance of commercial vSwitching solutions. Unfortunately the overall

DPDK architecture overview

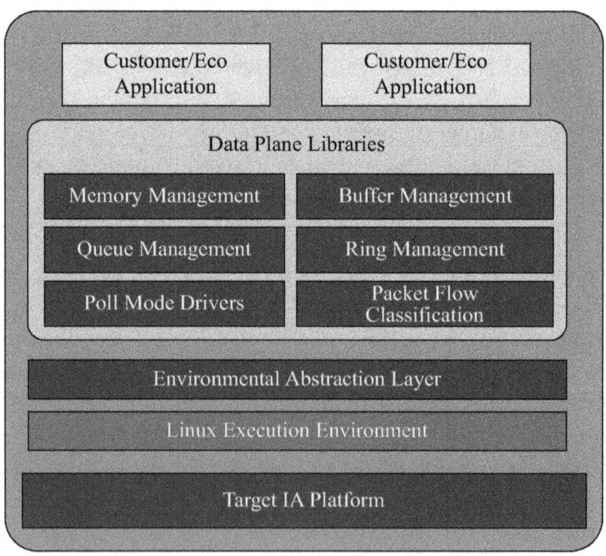

Figure 10.4 DPDK architecture overview

implementation is still immature, suffering from usability and reliability limitations. While not yet ready for commercial use, it looks promising.

The forwarding engine incorporates Intel DPDK Huge Page Tables, and the Open vSwitch control Daemon is modified to connect to these tables. The forwarding module runs in Linux user space with BSD license rights. Intel DPDK vSwitch implements a subset of the switching functionality of Open vSwitch.

10.4.6 Container

Containers are the result of operation system virtualization and separation. They provide a lightweight virtual environment that groups and isolates a set of processes and resources such as memory, CPU, disk, etc., from the host and any other containers.

Core container technologies include:

- cgroups
- namespaces
- chroot
- Linux Security Module (LSM), Mandatory Access Control (MAC)
- Capabilities/SECCOMP
- Traditional virtualization
- Runtime/management

Understanding the difference between hypervisor and container virtualization can better prepare users to design and develop virtualizations based on the end-system requirements. Two virtual nodes are illustrated in Figure 10.5, one running hypervisor virtualization with two virtual machines and another running container virtualization with two virtual environments (VE).

A common, shared Linux kernel provides:

- User space management tools for Linux containment features.
- Isolation of hardware resources (CPU, memory, block I/O, network, etc.) without the need for full VMs.
- Isolation of applications' view of O/S resources (process trees, network, user IDs and mounted file systems).

Container technology is light weight than VM level virtualization. Since no hypervisor is used all containers share the same host kernel.

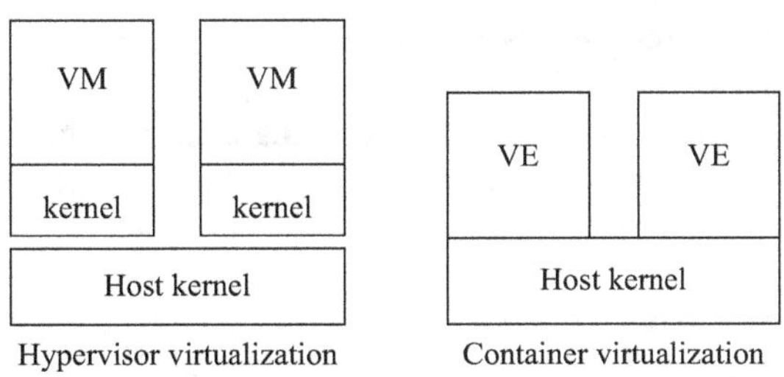

Figure 10.5 Hypervisor virtualization vs. container virtualization

10.4.7 CAT/CMT

Cache Allocation Technology (CAT) is a new feature that allows an OS, hypervisor or VMM to control allocation of a CPU's shared last-level cache. Once CAT is configured, the processor allows access to portions of the cache according to the established class of service (COS). The processor obeys the COS rules when it runs an application thread or application process. This can be accomplished by performing the following steps:

- Determine if the CPU supports the CAT feature.
- Configure the COS to define the amount of resources (cache space) available. This configuration is at the processor level and is common to all logical processors.
- Associate each logical processor with an available COS.
- Run the application on the logical processor that uses the desired COS.

Cache Monitoring Technology (CMT) is a new feature that allows an operating system or hypervisor or virtual machine monitor (VMM) to determine the usage of cache by applications running on the platform. CMT could be used to do the following:

- To detect if the platform supports this monitoring capability;
- For an OS or VMM to assign an ID for each of the applications or VMs that are scheduled to run on a core. This ID is called the Resource Monitoring ID (RMID);
- To monitor cache occupancy on a per-RMID basis, and
- For an OS or VMM to read last level cache occupancy for a given RMID at any time.

10.4.8 Support of OpenStack

OpenStack [11] is a free and open-source software platform for cloud computing, mostly deployed as an infrastructure-as-a-service (IaaS). The software platform consists of interrelated components that control hardware pools of processing, storage and networking resources throughout a data centre. OpenStack includes lots of core services such as Nova, Swift, Neutron, etc. which are critical for RAN virtualization.

10.5 Proof-of-Concept development of virtualized RAN

In order to evaluate the feasibility of RAN virtualization, a prototype was set up with various optimization functionality implemented on the virtualization platform, which is described in this section.

10.5.1 Virtualized RAN demo system

Figure 10.6 shows the topology of the C-RAN Proof-of-Concept (PoC) setup. Two controller nodes run on the Titanium Server services to manage the cloud

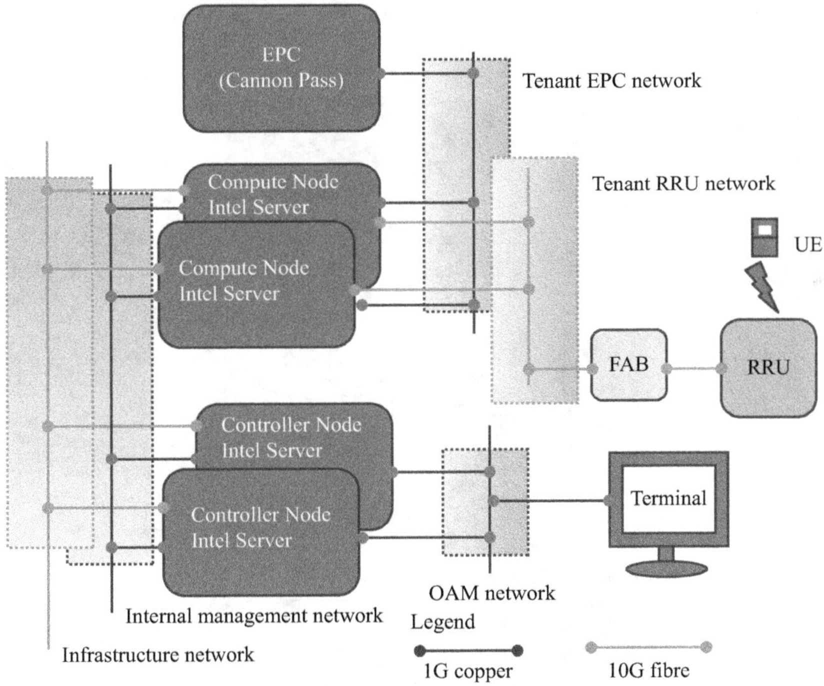

Figure 10.6 Topology of virtualized C-RAN demo

infrastructure. Two compute nodes run on the Titanium Server to provide computing services and host the virtual machines.

The system uses the Wind River Titanium Server as the NFV infrastructure with the following features:

- Automatic VM recovery on failure of a host compute node (node failure detection in seconds rather than minutes which are typical of standard IT grade solutions)
- Automatic VM recovery on failure of a VM (VM failure detection 60× faster than standard IT grade)
- Live migration of VMs using Intel DPDK (not available in IT-grade OpenStack)
- Controller node redundancy and automatic failover (not available in IT-based OpenStack)
- VM monitoring tied into application health checks within the guest (provides fine grained levels of application specific fault detection)
- VM protection groups (ensuring VMs of the same group are created on different compute nodes)

The data flow is indicated in Figure 10.7.

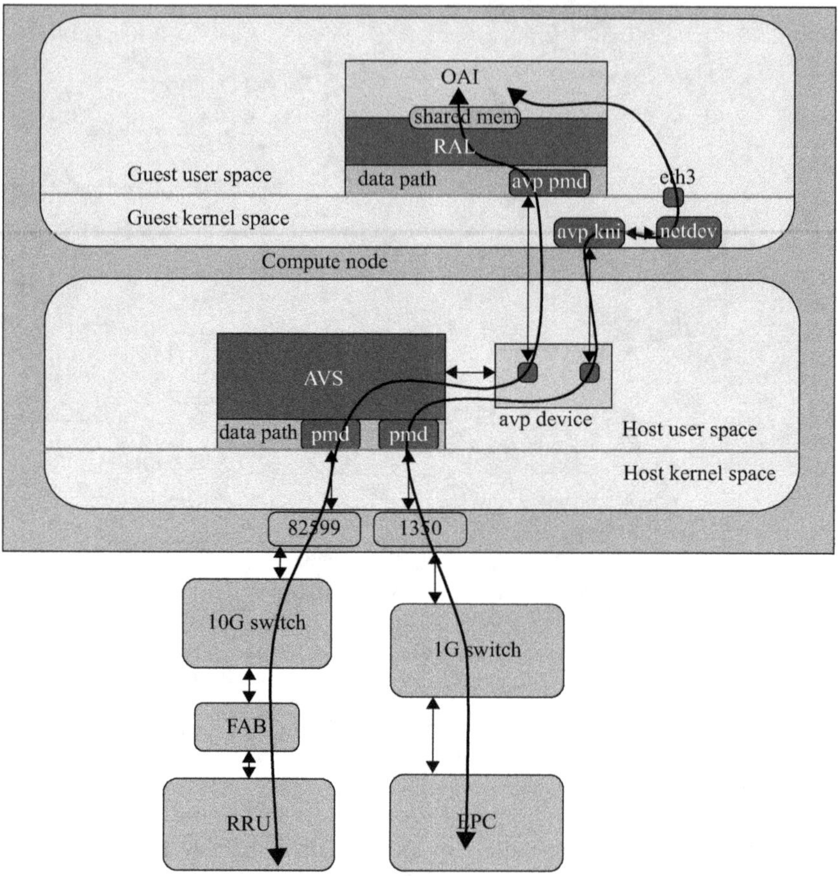

Figure 10.7 Data flow of the virtualized RAN demo

On the compute node, the Titanium Server Accelerated vSwitch (AVS) is responsible for all the Ethernet traffic. A high performance user space vSwitch based on the DPDK enables high performance VM-to-VM communication without the need to use the Linux kernel, as well as high performance packet processing from the NIC to applications in VMs. Ethernet with support for Intel DPDK, SRIOV, and 1G, 10G and 40G ensures ultra-fast packet processing.

Key modules in the figure include:

- Accelerated Virtual Port (AVP) is a TiS-specific accelerated Ethernet interface for increased throughput.
- avp pmd is the device which is the poll-mode driver compatible with the Intel DPDK.
- avp_kni is the device which is compatible with Linux kernel standard network device.

Every virtual Ethernet device in virtual machine will have a corresponding device in AVS (such as "avp devices").

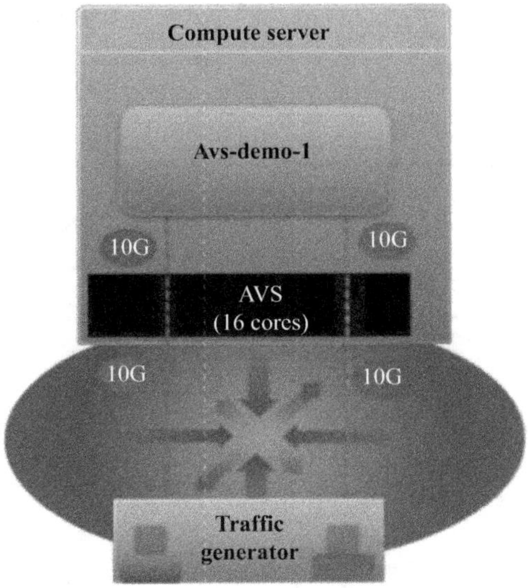

Figure 10.8 vloop scenario illustration

All the traffic is bidirectional. Let us take uplink as an example. The signal from remote radio unit is transformed into Ethernet packets. Then it goes into 10G switch. When compute node 82599 NIC receives the packets, it will be polled by AVS. AVS will make a decision based on MAC/IP, then forward it to dedicated virtual NIC of virtual machine. After the VM process, the packets will be forwarded to the core network. The packets go to specified virtual NIC, then enter the AVS again. AVS will make the decision based on MAC/IP, then forward the packet to I350, another NIC.

In this experimental environment, HP DL380 is used as a server, and other equipment elements are listed below:

- Server: HP DL380 Gen8 with 2 sockets, 8 core per socket,
- CPU: Intel(R) Xeon(R), E-2690 v2, at 3.00 GHz
- Network Interface Card (NIC): Intel 82599
- Memory: 64 GB (compute node)

Software running in the system is listed below:

- WindRiver Linux OVP 6 (Open Virtualization Platform)
- Enhanced OpenStack kilo
- WindRiver AVS (Advance Vswitch)

10.5.2 Test results

The tests are running in two scenarios by using an Ethernet tester. In each scenario, time delay, jitter and throughput are measured, respectively.

I. vloop scenario

In the scenario shown in Figure 10.8, the packets are sent from the Ethernet tester to the host OS. After that, the host OS passes packets to AVS. When the packets arrive

Table 10.1 Throughput in percentage of different frame sizes

Frame size(byte)	**64**	**128**	**256**	**512**	**1024**	**1518**
Throughput (%)	33.203	58.516	99.298	99.298	99.298	99.298

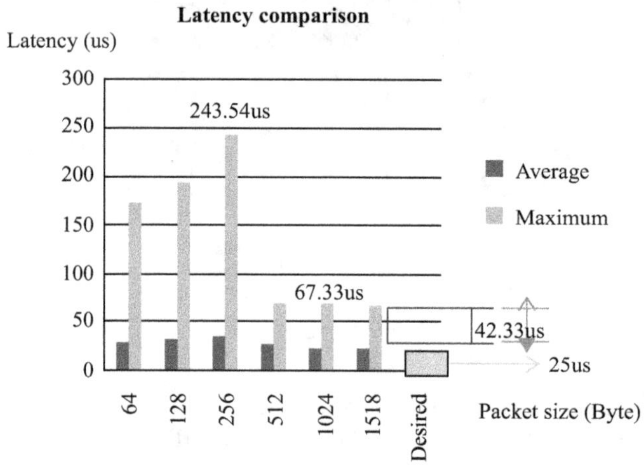

Figure 10.9 Latency comparison

at Avs-demo1 which is a VM, L2 forwarding process is done within demo1. Finally, after packets are sent out from vNIC of demo1, they will travel back to Ethernet tester and do packet analysis.

The throughput is measured for different frame size in the environment mentioned above, as shown in Table 10.1.

According to Table 10.1, when the frame size becomes larger, the throughput is increasing. The reason is, for the same amount of data, more packets are needed for smaller frame size. The system will put more efforts on disassembling packages. As a result, the throughput is low when the size is small.

For different frame sizes, the loop delay is shown in Figures 10.9 and 10.10.

The graph of jitter against frame size can be observed in Figure 10.10.

From Figures 10.9 and 10.10, it could be observed that:

1. Under zero packet loss condition, when the packet size is no less than 256 bytes, the throughput could achieve almost full capacity of 10 Gbps.
2. When the packet size is no less than 512 bytes, packet transmission delay almost remains constant, with a maximum latency of 67.33us.
3. All the tested jitters are smaller than 66.67us, which fulfils the requirement of wireless communication.

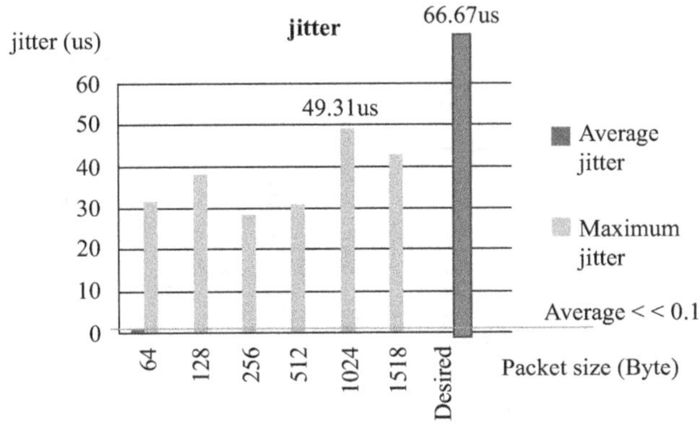

Figure 10.10 Jitter of different packet sizes

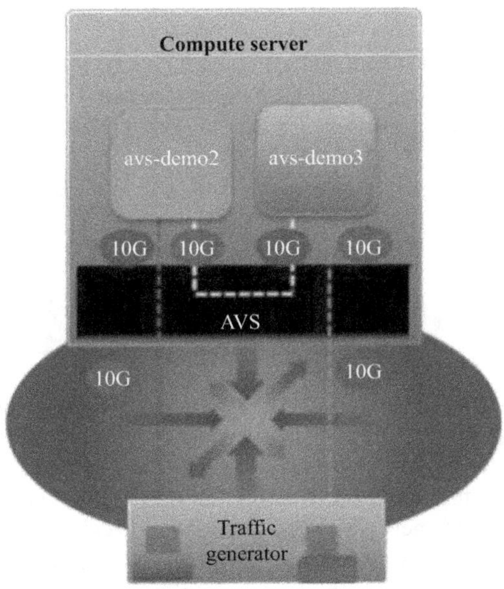

Figure 10.11 M model scenario illustration

II. M model scenario

The test is designed to evaluate the latency for VM communication, which is critical for live migration (Figure 10.11).

In this test, the frame size is fixed at 1024 bytes. Data is generated from traffic generator and sent to demo2. After L2-forward packet process, it is passed to demo3. Different from the vloop scenario, in this scenario, after the test packets arrive at avs-demo2, the destination MAC address is set as another VM (avs-demo3).

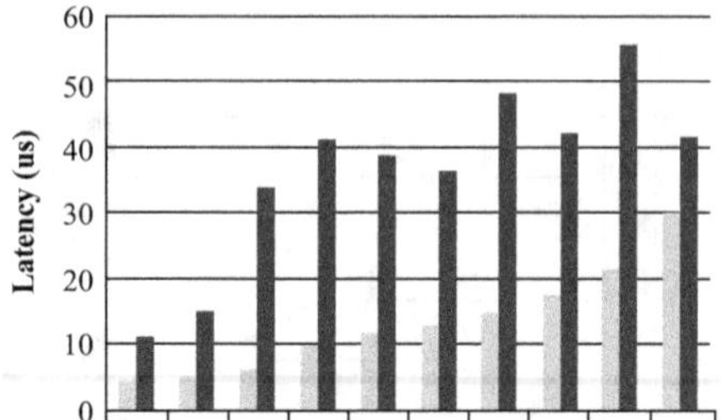

Load (%)	10	20	30	40	50	60	70	80	90	99.2
Average interval	4.2	4.8	6	9.4	11.7	12.6	14.6	17.3	21.3	30
Maximum interval	10.8	15	33.8	40.9	38.6	36.4	48.2	42	55.8	41.5

Figure 10.12 The latency between VMs transmission

Then packets travel from avs-demo3 back to host OS. Finally, packets will be captured by the Ethernet test centre.

Based on the latency of the vloop scenario and the current scenario, the difference between them is measured as the latency of VM communication.

According to Figure 10.12, the latency between VMs is smaller than 60us at any time. This delay is better than the performance in current commercial servers and meets the demand of dynamic migration.

10.6 Conclusion

On the road towards 5G, virtualization is viewed as one of the critical technologies to maximize the network flexibility, scalability and efficiency in order to accommodate diverse applications and scenarios. Despite the simplicity of the idea of adopting virtualization to telecom industry, the implementation is more difficult in practice. Wireless communication is distinct from IT data centres in that wireless communication has extremely strict requirements on real-time processing.

In this chapter, the virtualization on radio access networks, in particular C-RAN, is carefully studied with the analysis on the challenges and potential solutions. Given that RAN functions, such as baseband processing, require intensive computation and real-time processing, the RAN virtualization then requires joint optimization on various system modules, including the operating systems, the hypervisor, etc. It also requires enhancement and optimization on existing traditional IT technologies including the network acceleration, live migration, VM communication, support of OpenStack and so on.

A prototype of optimized virtualization platform with enhanced real-time performance was developed. The tests have shown that the optimized round-trip L2 processing latency of the platform could be reduced to around 30us on average, while it is usually around 300–500us in traditional Linux system. In the meantime, the VM communication latency could be optimized to around 10us, compared to several hundred microseconds for traditional Linux system.

Enhancement on the platform real-time performance is just the first step. In future, much more work should be involved such as real-time management, virtualization granularity, hardware acceleration, design of high-speed low-latency switch networks, etc. to realize completely virtualized C-RAN systems.

Acknowledgements

The authors would like to express their sincere gratitude to their partners, Wind River and Intel for the development of the virtualized RAN prototype. The authors also owe special thanks to all C-RAN team members in China Mobile for helpful discussion and valuable comments.

References

[1] I, C.-L., Rowell, C., Han, S., Xu, Z., Li, G., and Pan, Z., "Toward green and soft: a 5G perspective," *IEEE Communications Magazine* 52(2) (2014) 66–73.

[2] Creasy, R. J., "The origin of the VM/370 time-sharing system," *IBM Journal of Research and Development* 25(5) (1981) 483–490.

[3] Rodriguez-Haro, F., Freitag, F., Navarro, L., *et al.*, "A summary of virtualization techniques," *Procedia Technology* 3 (2012) 267–272.

[4] Goldberg, R. P., "Architecture of virtual machines," in *Proceedings of the Workshop on Virtual Computer Systems*, ACM Press, New York, NY, USA, 1973, pp. 74–112. doi: http://doi.acm.org/10.1145/800122.803950.

[5] Chris, W. and Halter, E. M., *Virtualization from the Desktop to the Enterprise*, Apress Press Inc., New York, NY, USA, 2005.

[6] Bauer, E. and Adams, R., *Reliability and Availability of Cloud Computing*, Wiley-IEEE Press, Hoboken, NJ, USA, 2012.

[7] ETSI NFV ISG (2012). Network Functions Virtualisation. [Online]. Available: http://portal.etsi.org/portal/server.pt/community/NFV/367

[8] C. M. R. Institute (2014). C-RAN: The road towards green ran. [Online]. Available: labs.chinamobile.com/cran

[9] I, C.-L., Huang, J., Duan, R., Cui, C., Jiang, J. X., and Li, L., "Recent progress on C-RAN centralization and cloudification," *IEEE Access*, 2 (2014) 1030–1039.

[10] [Online] Available: http://dpdk.org

[11] [Online] Available: http://en.wikipedia.org/wiki/OpenStack

Chapter 11

Licensed shared access (LSA) and three-tier spectrum sharing models: regulation, business and technology perspectives

Marja Matinmikko[1], *Seppo Yrjölä*[2], *Miia Mustonen*[3], *Petri Ahokangas*[4] *and Kari Horneman*[5]

Abstract

Spectrum sharing is an important enabler for future mobile broadband systems to meet the growing end user data traffic needs. Considering spectrum sharing from regulation, business and technology perspectives is the key for the successful development of sharing models that can be deployed in practical systems. This chapter studies recent spectrum sharing concepts from regulatory, business and technology perspectives. We present the European Licensed Shared Access (LSA) spectrum sharing concept covering its regulatory, business, and technical aspects with a focus on the LSA case example of sharing between LTE and incumbent program making and special events (PMSE) services in the 2.3–2.4 GHz band. The LSA evolution including more dynamic sharing toward the US three-tier spectrum sharing model for Citizen Broadband Radio Service (CBRS) based on Spectrum Access System (SAS) is depicted in terms of its ecosystem, business benefits, and technical approaches. Future outlook is given to envisage the development of sharing models for the mobile broadband.

11.1 Introduction

Wireless data traffic growth has motivated industry and academia to develop new spectrum sharing models for the adoption by spectrum regulators. Spectrum sharing refers to the situation where two or more radio systems operate in the same frequency band. Starting from the cognitive radio research over 10 years ago, technical

[1]Centre for Wireless Communications (CWC), University of Oulu, Oulu, Finland
[2]Nokia, Oulu, Finland
[3]VTT Technical Research Centre of Finland Ltd., Oulu, Finland
[4]Oulu Business School, Martti Ahtisaari Institute, Oulu, Finland
[5]Nokia Bell Labs, Oulu, Finland

approaches for spectrum sharing have become mature to be applied to specific sharing scenarios between real-life wireless systems in specific frequency bands. The development of spectrum sharing models needs a thorough consideration of Regulation – Business – Technology relations as these three dimensions are highly related and their distinct requirements need to be met [1].

After extensive investigations of spectrum sharing in the so-called TV white spaces often based on unlicensed access using the geolocation database approach, the interest in the spectrum regulatory domain has expanded to licensed sharing approaches that can protect the incumbents while introducing additional licensed users on a shared basis. Licensed sharing can result in a controlled interference environment where both the incumbent and the entrant systems operating in the same band could enjoy quality of service (QoS) guarantees. After its introduction in 2011, the Licensed Shared Access (LSA) concept has received considerable attention in European regulation as a means to introduce additional licensed users on bands with currently exclusive use while protecting the existing use [2,3].

As a hybrid of the two approaches for sharing, namely unlicensed type and licensed, the Federal Communication Commission (FCC) in the US has proposed a three-tier sharing model [4] for providing Citizen Broadband Radio Services (CBRS) [5] by introducing two additional sharing layers while protecting incumbents. The model consists of both a licensed sharing layer and an opportunistic sharing layer with different levels of access rights and guarantees of interference protection. This model is being developed for the 3.5 GHz band as the first application case but could be adopted more widely if proven its success.

This chapter will look into the LSA and CBRS sharing concepts from the regulation, business, and technology perspectives. The LSA and CBRS spectrum sharing concepts are presented covering the state of the art from these perspectives, and a future outlook is given.

11.2 Spectrum sharing

To understand the development of new spectrum sharing models, this section first places the spectrum sharing in the larger picture of spectrum access methods and introduces the regulation, business, and technology perspectives. Then the LSA and CBRS concepts are introduced.

11.2.1 Spectrum access models and perspectives to sharing

Traditional spectrum access models are mostly based on exclusive licensing and license-exempt operations, see e.g., [6,7] for discussion. The license-exempt operations enable easy access to specific spectrum bands to devices that operate in accordance with pre-defined operational conditions such as power and duty cycle limits. Since the amount of devices admitted to the band cannot be controlled, this type of operations cannot guarantee a certain level of QoS. Until now spectrum access for mobile network operators (MNOs) has been based solely on exclusive licenses with long license durations and wide coverage areas. However, increasing challenges in finding spectrum resources that are free (or even can realistically be

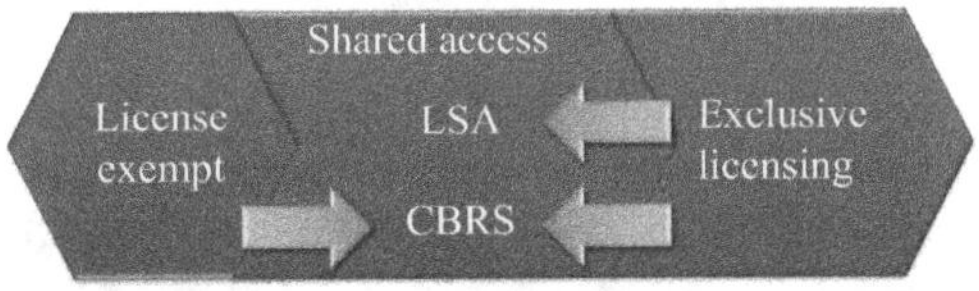

Figure 11.1 Spectrum access models

freed) from existing usage have inspired more flexible ways to appoint spectrum resources to the MNOs. Spectrum sharing allows two or more radio systems to operate on the same frequency band under certain rules and conditions that provide a feasible operational environment for the systems. Figure 11.1 illustrates two of the shared spectrum access models currently discussed in spectrum regulation, one in Europe and one in the US. The general regulatory access models are not limited to mobile use only but can be applied between any two or more radio systems. However, in the scope of this paper they are discussed as tools to enable mobile access to new spectrum bands on a shared basis.

The European LSA spectrum sharing concept [3] resembles exclusive licensing with the differentiation that the spectrum band is not reallocated solely for mobile use, but the existing usage on the band is maintained in a long term. The MNO, as an LSA Licensee, is allowed to use those parts of the spectrum band that are unused by the existing spectrum user at certain location, at certain times according to agreed rules and conditions. The existing user maintains higher usage rights and may reclaim the spectrum band or parts of it in a certain area, and the mobile network needs to implement management tools that enable to respond to changing spectrum usage of the existing system. The CBRS, considered in the US, is a more complex sharing concept that enables additional usage on a band with existing Incumbent Access (IA) usage on both licensed (Priority Access Licenses (PALs)) and license-exempt (General Authorized Access (GAA)) basis, see Figure 11.1. The CBRS introduces additional licensed users based on PALs, which resemble the LSA concept. Two main differences between the two are that the center frequency for the operations is not fixed in the PAL and the spectrum resource is available only when being used, thus, the PAL resource may be used opportunistically by the license-exempt (opportunistic) General Authorized Access (GAA) devices when not in use by the PAL holder. The CBRS concept also introduces an opportunistic GAA layers whose main difference compared to traditional license-exempt use is that GAA users should be registered as CBRS users.

Spectrum sharing is more than a technical matter, as considered in the early days of cognitive radio system research. It needs to be considered carefully also from regulation and business perspectives, see e.g., [1]. Figure 11.2 illustrates the business, regulation, and technology dimensions of spectrum sharing and topics that are of interest from these perspectives. From the regulatory perspective, national regulatory authorities (NRAs) have a set of criteria for a sharing model to be feasible, see [8] for details. For example, the protection of incumbents from harmful interference, fairness, and legal certainty are important matters from

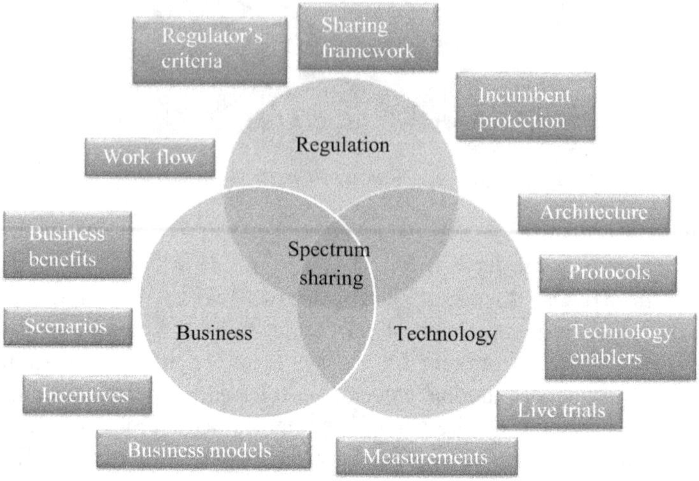

Figure 11.2 Aspects of spectrum sharing from regulation, business, and technology perspectives

regulatory point of you that need to be fulfilled by the technical implementations to deploy the sharing framework. From business point of view, spectrum sharing needs to provide business benefits for the stakeholders involved and business models and incentives for those who share. Finally, from technology perspective and allowance of making business point of view, it is important to develop feasible architecture and protocols for sharing that implement the sharing framework in a harmonized way that allows economies of scale. Live trials and performance measurements are important to show the feasibility of sharing concepts for the regulators for the adoption of the new models.

We will use the three perspectives of regulation, business and technology throughout this chapter and discuss the recent sharing concepts from these perspectives.

11.2.2 Spectrum sharing concepts

Today the most promising sharing models currently are the European LSA model [2] and the US three-tier model [4] for provision of CBRS [5]. The LSA concept introduces additional licensed users on bands that currently have incumbent exclusive use. The CBRS concept introduces two layers of additional users with different access level rights while protecting the incumbents. Figure 11.3 illustrates the tiers of different levels of access rights in the LSA and CBRS concepts in a high level. In the following these two concepts are presented.

11.2.2.1 Licensed shared access

Licensed Shared Access (LSA) concept is a spectrum sharing model proposed by the European Commission (EC) to introduce any radio system to a frequency band with existing usage. Sharing is based on the LSA license issued by the NRA and a

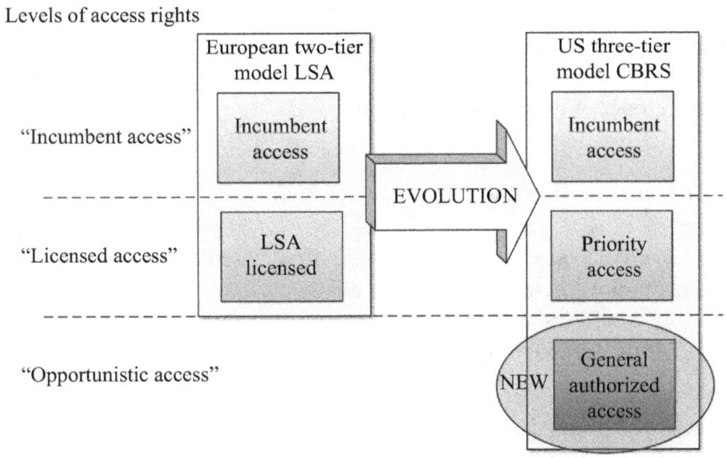

Figure 11.3 High level overview of tiers in LSA and CBRS

sharing framework agreed between the involved stakeholders, see [2,9] for details. LSA is based on voluntariness and the incumbent user can decide which frequency bands to appoint for sharing in which geographical areas. In such a way, bands with permanent or frequent incumbent usage may be left outside the LSA arrangement and more stable sharing conditions for the LSA licensee can be anticipated. LSA license from the regulator guarantees the LSA licensee with operational certainty for the duration, area, and conditions on spectrum bands underutilized by incumbent, as defined in the license and sharing framework. The incumbent spectrum users maintain higher level usage rights and may reclaim the band or parts of it at any location any time. The management mechanisms and techniques which LSA licensee needs to implement in order to respond to the changing availability of the LSA band are discussed later.

11.2.2.2 CBRS

In April 2015 the FCC adopted a three-tier spectrum sharing model called Citizens Broadband Radio Service (CBRS) [5] as proposed in the 2012 President's Council of Advisors on Science and Technology (PCAST) Report [4]. CBRS utilizes 'fast track' band 3,550–3,700 MHz and introduces there two types of sharing layers while protecting the incumbents. The model consists of both a licensed sharing layer and an opportunistic sharing layer with different levels of access rights. Rules are optimized for small cell use, but also accommodate point-to-point and point-to-multipoint particularly for rural services. Management of interference protection is accomplished through a Spectrum Access Systems (SAS) that coordinate the spectrum usage of the entrants to protect the incumbents. In addition to the spectrum database approach that was adopted for the LSA concept, the CBRS introduces the notion of Environmental Sensing Capability (ESC) for monitoring the incumbent activity which would detect the appearance of specific incumbents such as naval radar in coastal areas and near inland military bases. According to the discussions in [5], in case of detecting incumbent activity the ESC communicates that information to an SAS which should reconfigure local access points to avoid harmful interference with incumbent radars.

11.3 Technical approaches

The LSA and CBRS sharing concepts are first discussed from the technology perspective providing their key technical elements.

11.3.1 LSA technical aspects

One of the main advantages of the LSA is that the implementation is foreseen to be based on two additional functional units on top of the existing mobile network architecture, as was discussed in [10]. LSA does not require modifications to the existing mobile network internal procedures or Radio Resource Management functions. On the user equipment (UE) side, there is no impact beyond implementing the support of any new frequency band. Therefore, the additional LSA spectrum can be utilized without lengthy regulatory and standardization process. Two required additional functional units are the LSA Repository and the LSA Controller [11]. The LSA Repository stores and updates the information about the availability and use of LSA spectrum band together with the prevailing policies and conditions (transmission powers, etc.) for the use of it. The LSA Repository also coordinates the information exchange between incumbent users and LSA Controllers. The LSA Controller ensures the protection and interference-free operation of the incumbent user and mobile network, by calculating the protection areas based on the information received from the LSA Repository and the information on the mobile network layout, used transmission powers, etc. Current LTE and LTE-Advanced standard include several features that enable and support the use of LSA spectrum with variable availability by the mobile network. These features are discussed in detail in [7] and are summarized in Figure 11.4.

The cell/sector on LSA band may need to be de-activated due to evacuation requests from the incumbent user and then re-configured when they are switched on again. Using self-configuration, the LSA cell can be brought back into service with minimum manual intervention. In the case of activation of a cell on the LSA band, cell re-selection procedure allows the UE to select the cell on an LSA band based on its measurements of the signal strength. In case of a de-activation of the cell on the LSA band, terminals are able to automatically start a cell re-selection procedure. Additionally, tools such as traffic steering, frequency handover and load balancing help to steer users to the best serving cell based on a predefined criteria. Active antenna systems (AAS) can be used for reducing the size of exclusion zones and interference between the incumbent user and the mobile network by optimizing network coverage and capacity. Using AAS, the size of the cells can be automatically adjusted and capacity directed by changing antenna parameters such as tilt, azimuth, or beam shape. Carrier aggregation (CA) enables utilization of radio

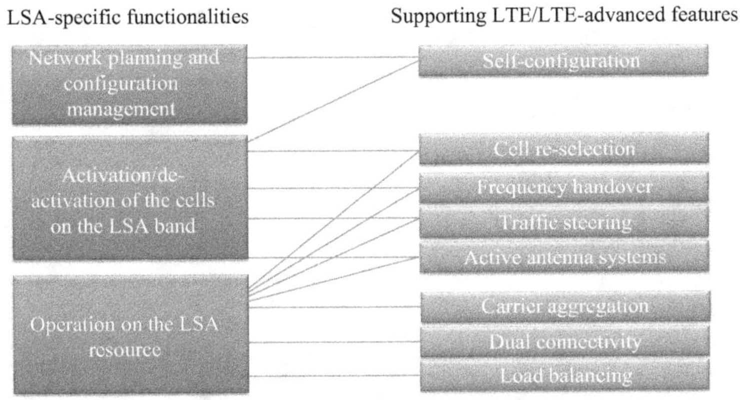

Figure 11.4 LTE/LTE-A features that enable the use of LSA bands

resources across multiple component carriers to provide a wider effective bandwidth to the end user. Using CA, a carrier on the LSA band may be used together with a carrier on a licensed band to provide additional capacity to its users, without the risk of connection break. Dual connectivity allows the UE to be connected to both the existing LTE FDD cell and the TD-LTE LSA cell, at the same time.

The world's first LSA trial for sharing between LTE and incumbents in the 2.3 GHz band was shown in Finland in 2013 [12]. Figure 11.5 illustrates the Finnish LSA trial evolution. In 2015, the Finnish LSA trials demonstrated how a mobile application is able to automatically track an incumbent user and the mobile network is able to adapt to this automatically [13]. Additionally, a new interference estimation algorithm, which is able to take into account aggregate interference from all the base stations, was demonstrated. This trial was presented in [14] showcasing enhanced power control for interference coordination between the LTE network and the incumbent PMSE system.

11.3.2 CBRS technical aspects

The main technical element of the CBRS is the SAS. In addition to the discussed SAS functionalities, the CBRS concept consists of SAS connected spectrum databases, ESCs, CBRS devices (CBSD), End User Devices (EUD) and optionally Domain Proxies and Network Management System (NMS) as shown in Figure 11.6. The CBSDs are fixed or portable base stations or access points, or networks of such stations and can only operate under the authority and management of a centralized the FCC selected SAS which could be multiple. Commission approved CBRS devices (CBSD) and must register with the SAS with information required by the rules, e.g., operator ID, device identification and parameters, and geolocation information. In case of CBSD is a managed network as in the typical case of MNOs, CBSD includes the domain proxy and network management functionality. Proxy could be a pure bidirectional information processing and routing engine or a more intelligent mediation function, e.g., combining the small cells of a mall or sports

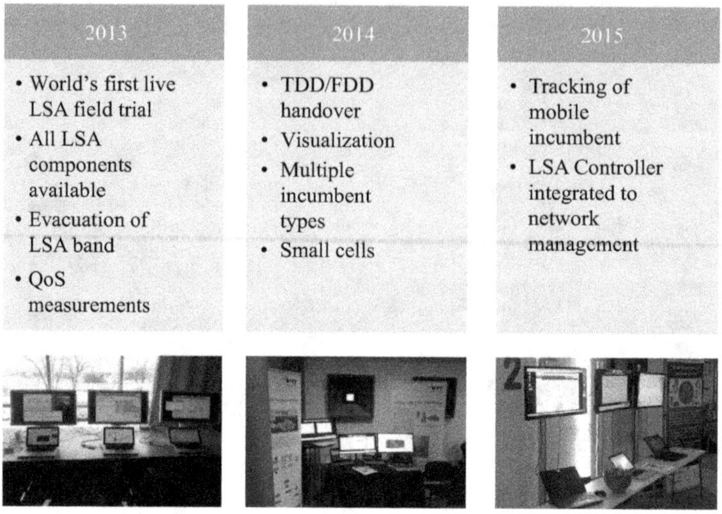

Figure 11.5 Finnish LSA trial evolution

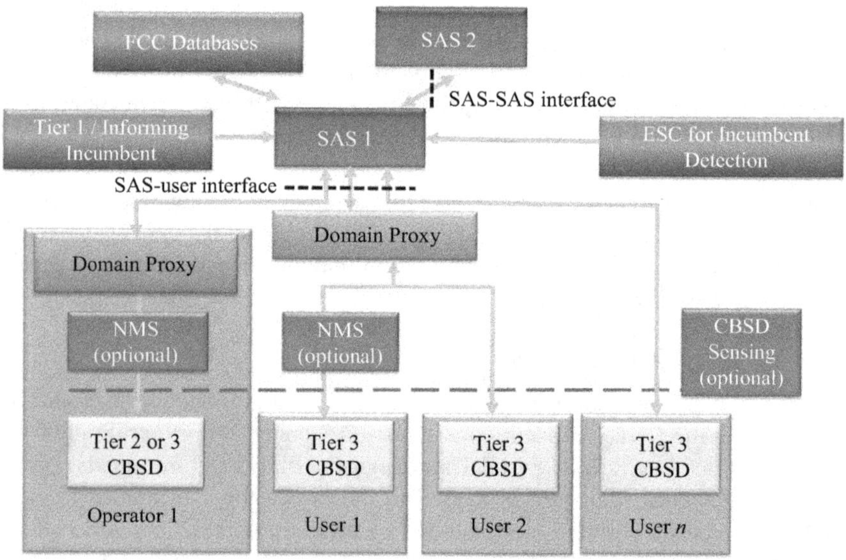

Figure 11.6 CBRS functional architecture

arena to a virtual BTS entity that covers the complete mall or sports arena. The latter option allows flexible self-control and interference optimizations in such a network. End User Devices (EUD), e.g., handsets are not considered as CBSDs. The SAS assigns spectrum and determines and enforces maximum power levels dynamically to certified CBSDs at a specific geographic location and time, controls the interference environment and enforces protection criteria and exclusion zones to protect higher priority users as well as takes care of registration, authentication and identification of user information and performs other functions as set forth in the FCC rules. As the IA users have primary spectrum rights at all times and in all areas over PA and GAA, all the CBSDs and EUDs must be capable of two-way communications across the entire 3.5 GHz band and discontinuing operation or changing frequencies at the direction of the SAS to protect IA.

It is mandatory for all the CBSDs to protect the IA users in the band. Based on nature and critical requirements of the federal incumbent the FCC adopted rules to require ESCs to detect incumbent radar activity in coastal areas and near inland military bases in and adjacent to the 3.5 GHz the band. When IA activity is detected, the ESC communicates that information to the SAS which if needed could order commercial tier users to vacate a spectrum resource within 60 s in frequency, location, or time which when in proximity to federal IA radar presents a risk of harmful interference. Federal IA protection will be introduced in two phases: First, a large area of the country outside the static exclusion zones will be available after the SAS is the FCC approved and commercially available. At the second phase, exclusion zones will be converted to protection zones through the ESC system enabling the rest of the country, including major coastal areas, to become available. An ESC consists of one or more commercially operated networks of sensing device-based or CBSD infrastructure-based sensors that would be used to detect signals from federal radar systems in the vicinity of the exclusion zones. Prospective ESC operators must have their systems approved through the same process as SASs and SAS administrators. The SAS would obtain the FCC information, e.g., about registered or licensed commercial users, exclusion zone areas requiring ESC from the FCC database. Functional architecture has an option for the informing incumbent in-case the federal IA wants to inform the SAS ahead of plans to use the spectrum in some area, e.g., related to planned training use of the spectrum.

11.4 Sharing model status in regulation and standardization

Next, the LSA and CBRS sharing concepts are discussed from the regulatory perspective including standardization, and the statuses of the two concepts in these fora are reviewed.

11.4.1 LSA

The LSA related regulatory and standardization efforts in Europe have so far focused on the 2.3–2.4 GHz band with MNOs being the LSA licensees and incumbents varying depending on the national situation. The use of LSA is a national matter that is decided by the NRA but a harmonized framework is seen to help to develop a harmonized approach and market for LSA in Europe. A detailed description of different activities and current status of LSA regulation and standardization are documented in [15]. An overview of that is presented in Figure 11.7. The EC has

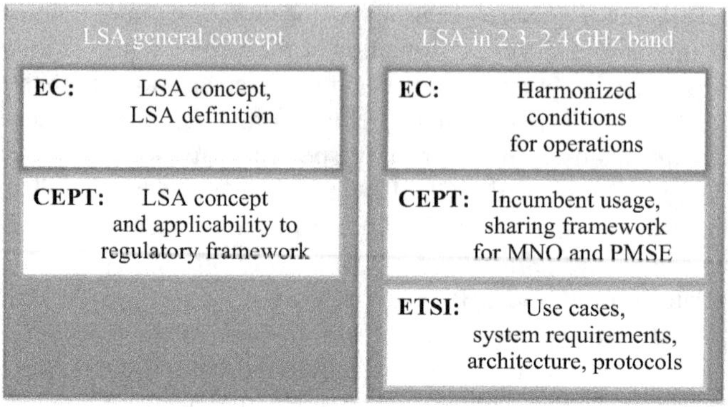

Figure 11.7 Overview of the activities in different regulatory and standardization bodies

worked on the general LSA concept first by introducing it and then by developing the official LSA definition [3]. Additionally, the EC has initiated studies on LSA in the 2.3–2.4 GHz band by issuing both a standardization Mandate to ETSI and a regulation Mandate to CEPT. In response to the regulation Mandate, CEPT has provided an overview of different incumbent services in different European countries and existing implementation examples [16]. CEPT has also developed the LSA framework with harmonized technical and regulatory conditions between mobile and Program Making and Special Events (PMSE), wireless cameras, which is the most common incumbent usage on the band [17]. The regulatory framework for LSA in Europe for the 2.3–2.4 GHz band is ready for national adoption but so far no deployments have taken place. However, trial activities are ongoing to ensure administrations on the feasibility of concept. The standardization status is summarized in [15]. ETSI has developed use cases, system requirements, architecture and functions, information flows, procedures, and interfaces for LSA operations for mobile broadband in the 2.3 GHz band, see [11,18,19]. Based on the studies performed in ETSI and CEPT, the EC will provide harmonized conditions for the operations on the 2.3–2.4 GHz. These will become binding to all EU Member States.

11.4.2 CBRS

The CBRS is a national concept proposed by the FCC in the US. The key policy messages of the PCAST report [4] were further strengthened in 2013 with Presidential Memorandum [20] stating that

> ... we must make available even more spectrum and create new avenues for wireless innovation. One means of doing so is by allowing and encouraging shared access to spectrum that is currently allocated exclusively for Federal use. Where technically and economically feasible,

> sharing can and should be used to enhance efficiency among all users and expedite commercial access to additional spectrum bands, subject to adequate interference protection for Federal users, ... we should also seek to eliminate restrictions on commercial carriers' ability to negotiate sharing arrangements with agencies. To further these efforts, while still safeguarding protected incumbent systems that are vital to Federal interests and economic growth, this memorandum directs agencies and offices to take a number of additional actions to accelerate shared access to spectrum.

Followed by intense discussion and consultation with the industry the FCC released Report and Order and Second Further Notice of Proposed Rulemaking to establish new rules for shared use of the 3,550–3,650 MHz band in April 2015 [5]. The FCC sees the opening of the 3.5 GHz Band as "a new chapter in the history of the administration of one of our nation's most precious resources – the electromagnetic radio spectrum." The framework defines a contiguous 150 MHz block at 3,550–3,700 MHz for mobile broadband (MBB) that the FCC calls Citizens Broadband Radio Service (CBRS). The 3,550–3,650 MHz spectrum is currently allocated for use by the US Department of Defense (DoD) radar systems and Fixed Satellite Services (FSS) while the 3,650–3,700 MHz spectrum incumbents are the FSS and the grandfathered commercial Wireless Broadband Services. FCC prefigures CBRS as an "innovation band" where they can assign spectrum to commercial MBB systems on a shared basis with incumbent radar and FSS systems and promote a diversity of heterogonous network technologies, particularly small cells. The technology neutrality of the CBRS will in particular play a role in the opportunistic General Authorized Access (GAA) tier opening up new opportunities.

The sharing framework in CBRS consists of three tiers: Incumbent Access (IA), Priority Access (PA) and General Authorized Access (GAA) as shown in Figure 11.3. It will be mandatory for all the CBRS users (PA and GAA users) to protect the IA users in the band. The PA users will obtain a PA license (PAL) from the FCC to operate up to 70 MHz of the 3,550–3,650 MHz spectrum segment and are protected from harmful interference from the GAA operations. PA users receive short-term priority authorization to operate within designated geographic areas with PALs such as 3 year 10 MHz unpaired channel in a single census track, awarded with competitive bidding. During the first application window only, an applicant may apply for up to two consecutive three-year terms for any given PAL. Licenses will be permitted to hold no more than four PALs in one census tract at one time. This will ensure availability of PAL spectrum to at least two licensed users in the geographical areas of highest demand. PALs are assigned specific frequencies within their service area, and their frequency assignment should not be dynamically controlled by the SAS. However, the center frequency may vary according to incumbent activity. At the end of its term, a PAL will automatically terminate and may not be renewed.

The third GAA tier will operate under a licensed-by-rule framework and will be allowed throughout the 150 MHz band without any interference protection from other CBRS users. This framework aims to facilitate the rapid deployment of compliant small cell devices while minimizing administrative costs and burdens on the public, licensees, and the FCC. GAA users may use only certified, FCC approved CBRS devices and must register with the SAS with information required by the rules, e.g., operator ID, device identification, and geolocation information.

CBRS Devices (CBSDs) which are fixed stations (access points) or networks of such stations will be assigned spectrum dynamically by the FCC selected SAS which could be multiple. User equipment, e.g., handsets, are not considered as CBSDs. SAS controls the interference environment and enforces exclusion zones to protect higher priority users as well as takes care of registration, authentication, and identification of user information. As the IA users have primary spectrum access rights at all times and in all areas over PA and GAA, all the CBRS users must be capable of operating across the entire 3.5 GHz band and discontinuing operation or changing frequencies at the direction of the SAS to protect IA. Automated channel assignment by a SAS will simply involve instructions to these users to use a specific channel, at a specific place and time, within 3,550–3,700 MHz. Based on nature and critical requirements of the federal incumbent the FCC adopted rules to require Environmental Sensing Capabilities (ESCs) to detect federal spectrum use in and adjacent to the 3.5 GHz the band.

The opportunistic GAA with no interference protection from other CBRS users is planned to provide a low-cost entry point into the CBRS band for a wide array of users and services first while PAL system operations have to wait auction process estimated to start after the US 600 MHz incentive auctions targeted for 2016. For the meanwhile, the FCC has encouraged multi-stakeholder groups to consider various issues raised by the rules. The WIreless iNNovation Forum (WInnForum) Spectrum Sharing Committee [21] with representatives from the mobile broadband, wireless broadband, Internet, and defense ecosystems has started initial standardization work on interfaces between an MBB system and a SAS work targeted to allow sharing of the CBRS within 2016. As a point of departure toward possible future alignment of the LSA and CBRS system evolutions WInnForum there is information exchange among the standards organizations (ETSI, 3GPP) on the 3.5 GHz band activities and network management system support for LSA. The US government has initially identified an additional 2 GHz of spectrum below 6 GHz owned by DoD and other users for future shared commercial use conditionally if the spectrum sharing at 3.5 GHz proves successful. This paves way to make licensed spectrum sharing a third mainstream way of licensing spectrum to commercial users complementing traditional exclusive licensing and unlicensed spectrum access. The FCC has vision to repeat Wi-Fi success through lowering the entry barrier QoS spectrum for new entrants and verticals, e.g., enterprise, utilities, healthcare, public safety, smart cities.

11.5 Business considerations

Along with the transformation toward sharing, the mobile communications industry is witnessing a variety of transitions that influence the way how business is done within the industry. Understanding the business environment around sharing for the

Table 11.1 Spectrum sharing related transformation trends within mobile broadband

Transformational trends within mobile broadband
• From exclusive spectrum and infrastructure to shared spectrum and infrastructure
• From exclusive licenses with countrywide coverage obligations to shorter term licenses in high demand areas
• From incumbent-operator sharing to also between-operators sharing
• From MNOs focusing on macro cells to a variety of service providers focusing on indoors and small cells
• From owning and controlling end-to-end network infrastructure to leveraging existing infrastructure and owning and controlling only where necessary
• From owning the value chain and partnering only where necessary to embracing open source and building ecosystems
• From installed base and legacy support to implementing cutting-edge technologies from inception to achieve market advantages
• From centralized and dedicated network infrastructure to orchestrated and virtualized network infrastructure
• From competing with Internet business to convergence with Internet business
• From monetizing connectivity and third party content to monetizing also digital ads, cloud, retail or content and give away broadband access for free or close to free

mobile broadband is important to develop viable spectrum sharing concepts. Table 11.1 outlines the key industry transitional trends within mobile broadband, relevant to LSA and CBRS.

The trends discussed above in the business context lead to an emerging and ecosystemic "sharing economy thinking" within mobile broadband, an economy that requires a layered and networked view to business models. To understand the transformation, a layered 4C business model consisting of connectivity, content, context, and commerce business models as presented in [22] depicts business opportunities and business models intertwined. There are different stakeholders and ways to make business in the different layers. Available mobile broadband *connectivity* services enable offering various *content* services, such as audio and video services. *Context* services become relevant in the abundance of content services, e.g., in the form of search engine services so that end users can find the content they are looking for, or in the form of providing context information such as spectrum availability information for MNOs in LSA or for PA users within CBRS. *Commerce* services, in turn, are based on platforms where the stakeholders involved with LSA or CBRS arrangements buy or sell in a continuous manner in real-time any of the lower layer services, or offer end users an opportunity to become prosumers.

At the general level, for incumbents sharing brings about an opportunity to enhance maintaining the rights to use spectrum in the longer run as they admit additional users in the band while continuing their own operations. It might also enable building additional sources of income, and enable utilizing MNOs infrastructure and thus reach savings in infrastructure investments. Regarding MNOs, sharing may help to avoid or decrease spectrum costs, open up new service revenue sources, increase infrastructure asset scalability, and help to utilize customer profiles. For other stakeholders in sharing, such as indoors communications service providers (CSPs) or integrators, cloud-based network-as-a-service concept may open up new business opportunities. Similarly, context information provisioning, e.g., regarding spectrum availability, may form a new class of services needed in the mobile broadband ecosystem.

11.5.1 LSA business aspects

There have been extensive research efforts on the business aspects of the LSA concept for sharing between mobile communications and incumbents, see e.g., [23–26]. The business ecosystem around LSA is mainly focused on the MNO, NRA, and incumbents. Two types of incumbents and MNOs are of relevance here, governmental incumbents (GIs) and private incumbents (PIs) as well as dominant and challenger MNOs. The prior business considerations for LSA are outlined in Figure 11.8.

Initial business benefits of the LSA for key stakeholders are summarized in [23]. Most notably, for NRAs LSA brings about an opportunity to promote innovation, guarantee fair access to spectrum, answer to the increased need for spectrum by MNOs and maintain reasonable price levels in the consumer markets. The incumbents' achievable benefits from LSA can be rooted to gaining a better position in possible spectrum reallocation negotiations and increasing turnover and profit potential, e.g., through the utilization of MNOs' infrastructure. Indeed, an incumbent can take either defensive or aggressive strategy regarding sharing [24], increase cost efficiency within existing businesses or to generate new revenue from new business opportunities. Similarly, they can take open or closed strategy regarding their system, focus on convergence of technologies, and businesses or on securing critical infrastructure.

For MNOs, rapid and cheap access to additional spectrum in a cost-efficient way and potentially without additional coverage obligations can bring benefits both

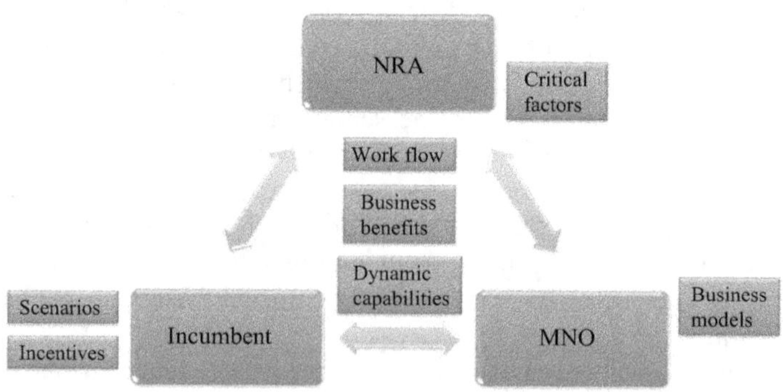

Figure 11.8 Business considerations for key stakeholders in LSA

from cost and revenue sides of the business model. MNO business models are presented in [25,26] for dominant and challenger MNOs. Dominant MNOs can better utilize the segmentation of their consumer and corporate customers when seeking growth or avoiding additional costs. On the operations side, dominant MNOs' customer data management and service design schemes as well as dynamic traffic management functionalities gain in importance in LSA compared to traditional business models. For challenger MNOs the opportunity is to win customers from dominant MNOs, especially with tailored or specific services to selected customers, through better customer experience, or in areas where the dominant MNOs have less focus.

To conclude on the business aspects of LSA, the main open items are the incentives for incumbents as the LSA concept is based on voluntariness. Furthermore, the business models for MNOs are open as the major MNOs prefer exclusive spectrum over shared spectrum and do not promote sharing. Therefore, the scalability of the LSA-based business models is more restricted if compared to CBRS [27], as it is mostly based on connection layer.

11.5.2 CBRS business aspects

The business ecosystem around CBRS is much more versatile compared to the basic LSA approach. The GAA layer opens the door to new players as analyzed in [28]. However, to date there is not much published material on business considerations regarding the CBRS. Figure 11.9 presents a simplified illustration of the business ecosystem around the CBRS concept and the key stakeholders. On the CBRS facilitator side, the regulation (FCC) and standardization organizations (WInnForum, etc.) have a key role in defining the rules and conditions and harmonized approaches to meet these requirements as well as the SAS and ESC capabilities for providing the spectrum availability information for the CBRS users. On the CBRS band users' side, the three layers of access rights to the 3.5 GHz band are held by different stakeholders. For example, co-channel incumbents include,

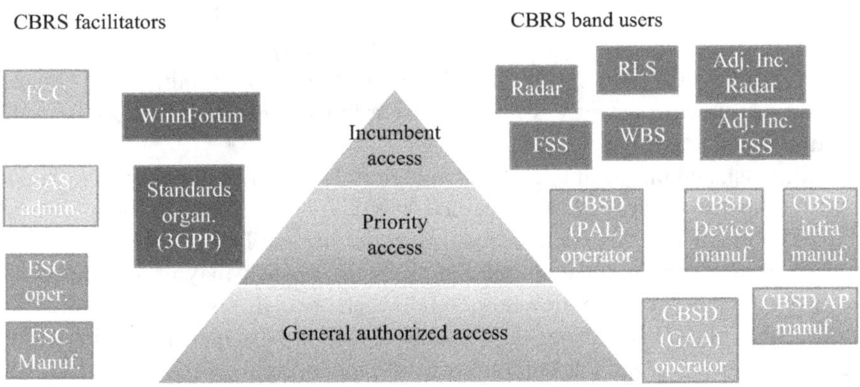

Figure 11.9 Key stakeholders in the CBRS business ecosystem

e.g., radar, FSS, radio location service (RLS), wireless broadband service, and radar and FSS as adjacent channel incumbents. As the entrants for providing CBRS, there are PA layer operators and GAA layer operators which could also be the same players, as well as the suppliers of the CBRS equipment.

In the systemic framework a change like the introduction of CBRS, all the stakeholders in the business ecosystem play a vital role in adopting of novel CBRS concept and spectrum sharing in general. In addition, when developing and analyzing the opportunity frame the three domains of regulation, business, and technology, affecting spectrum sharing concept should proceed in tandem. We have identified enabling, limiting, and challenging elements framing the business opportunities for the MNOs in the context of CBRS which are discussed next and listed in Table 11.2.

Business and technology elements can be identified as enablers for value co-creation. Fast growing demand and lack of exclusive spectrum combined with the drastic changes in the consumption habits will urge the adoption of novel more flexible and efficient spectrum management concepts. Framework radically unbundles investment in spectrum, network infrastructure and services which enable novel services and business models. Furthermore, different spectrum sharing schemes are high in regulators agenda with aims to lower the entry barrier to spectrum for new types of operators which could consider entering the wireless broadband business. Utilization of the LTE ecosystem scale and harmonization will reduce risk related technology maturity and provide tools to seamlessly integrate additional capacity to MNOs heterogeneous networks, e.g., through Carrier Aggregation (CA), LTE Unlicensed (LAA), and Self Organizing Network technologies. Big data analytics capabilities will play a major role in coping with the SAS dynamic and enabling low transaction costs.

Regarding limiting factors, sound, sustainable and harmonized regulatory environment can be the limiter that needs to be addressed before MNO can co-create and co-capture value from it with ecosystem partners. The limited spectrum availability in frequency, time, or location with potential restriction and uncertainties may negatively influence the MNOs outlook on shared use and the spectrum valuation. A specific technology item to be considered is the degree of business (MNOs) and mission (DoD) critical information needed to share and resulting need for the ESC system. In addition to MNO opportunities it is essential to consider reciprocal incentives for the current federal spectrum holders to further transition to CBRS.

Policy risk and uncertainty are the main elements of the competitive challenges in the competitive domain. Fragmented national and global market structure deprives economies of scale and scope, raising costs, and hampering innovation in the ecosystem. Furthermore, introduction of sharing models may impact the MNOs current exclusive spectrum licensing model and availability in the future. The complexity of the CBRS framework and the SAS, in particular, might impact the value of the spectrum and the required time of recovering the network investments. On the competence domain MNOs need to pay attention to dynamic capabilities needed to deploy, manage, and optimize multi-layered heterogeneous networks

Table 11.2 Analysis of enablers, limiters, and challengers for MNOs in CBRS business

	Business opportunity framing elements
Enablers	• Regulators considering shared spectrum framework in the EU and the US • Shared spectrum allocation improves overall spectrum use efficiency • Lack of exclusive spectrum triggers new spectrum access approaches • Consumers MBB consumption habits are changing toward asymmetric multi-device usage • Unbundles investment in spectrum, network infrastructure, and services • Additional lower cost capacity to cope with asymmetric traffic and improve performance • May lower entry barriers for challenger MNOs and new type of operators • Better QoS spectrum may increase dense urban area business • Additional GAA capacity for offloading • Harmonized LTE technology base leverage heterogeneous networks asset optimization and offers scale • Big data and analytics capabilities with Internet domain
Limiters	• Need for global and national regulation outside of the US may slow down entry – Harmonization is a precondition to scale and enable potential benefit fully. • Standardization of SAS functionalities for IMT technologies needed • Limited spectrum availability and predictability limit MNO business opportunities • Real incentives for the federal incumbents unclear or missing • Federal incumbent special requirements in particular related to security and need for sensing • Degree of information sharing of business critical (MNOs) and secret information (Federal incumbent) and needed ESC system
Challenges	• Impact on exclusive spectrum licensing model and availability in the future • Uncertainty and risks related to regulation in timing, term, licenses, and flexibility creates exposure and risk for an MNO to proceed with the investment. • Attractive and dynamic spectrum market with potentially lower transaction costs. • May increase and change competition. New operator types, and from other business domains. • Increased technical and operational complexity (SAS) with related capital and operational costs • New competencies and capabilities needed for network management and optimization • Timely availability of terminals and potential impact on cost and complexity

under sharing conditions. Traditional MNOs support for the 3.5 GHz spectrum in their networks is paramount to encourage chip and device manufacturers to support the whole 3.5 GHz band introduction with competitive terminals. Attractive and dynamic spectrum market with potentially lower transaction costs may increase and change competition, e.g., through introducing new operator types local and from other business domains.

In summary, in order to realize the business potential of the CBRS, MNOs have occasion to simultaneously co-create and co-capture value with ecosystem players in a competitive business environment where co-operation (spectrum) and competition (customers and services) exist parallel to each other. MNOs are in unique position to leverage additional multi-tiered capacity CBRS concept offers. Faster access to QoS licensed small cell optimized spectrum without mandatory coverage obligations will help them to timely cope with booming asymmetric data needs. Additional scalable and flexible spectrum resource leveraged with LTE technology enablers will enable MNOs to better retain and grow existing customer base with changing demand and consumer habits, offer differentiating services and explore new vertical segments.

Regarding the business models within the CBRS, some considerations concerning the MNOs and CSPs business models can be presented. For MNO, the business opportunity in CBRS resembles that of LSA: avoid costs or get extra capacity. MNOs' customers, however, can now be better profiled rather than just segmented, as is the case of LSA. We expect that the role of customer experience management is even more pronounced in CBRS, and also pricing and charging schemes of the MNOs could be expected to be built to utilize enhanced customer profiling. An interesting case is the new emerging indoors "micro" operators or CSPs that can in specific "closed" locations such as shopping malls or campuses provide superior local services. Thus, context-specific customer segments could be served by CSPs with tailored, bundled service packages such as combined connectivity and specific content. In closed contexts connectivity can be offered for free to consumers, as the corporate customers such as advertisers or other third parties could cover the costs of the service, and corporate customers in turn could be interested to buy "wholesale" connectivity from the local CSP. Also, it could be anticipated that MNOs and CSPs would not necessarily always compete against each other, but depend on each other. For example, MNOs could sometimes provide their exclusive spectrum-based connectivity services or billing services for the CSPs. When discussing business model scalability within CBRS article [23] concludes that CBRS concept facilitates business model scalability not only at connectivity layer business models, but it also provides enhanced opportunities for content- and context-based business models to emerge successfully.

11.6 Results

11.6.1 LSA reference implementation

The world's first reference implementation of the LSA concept in 2.3–2.4 GHz band was developed in Finland in 2013 as depicted in Figure 11.10 for sharing between LTE and PMSE service and originally introduced in [12] and reported in [16]. It is a combination of elements that are either commercially available or specifically designed for the LSA concept. Commercial equipment in the LSA trial environment

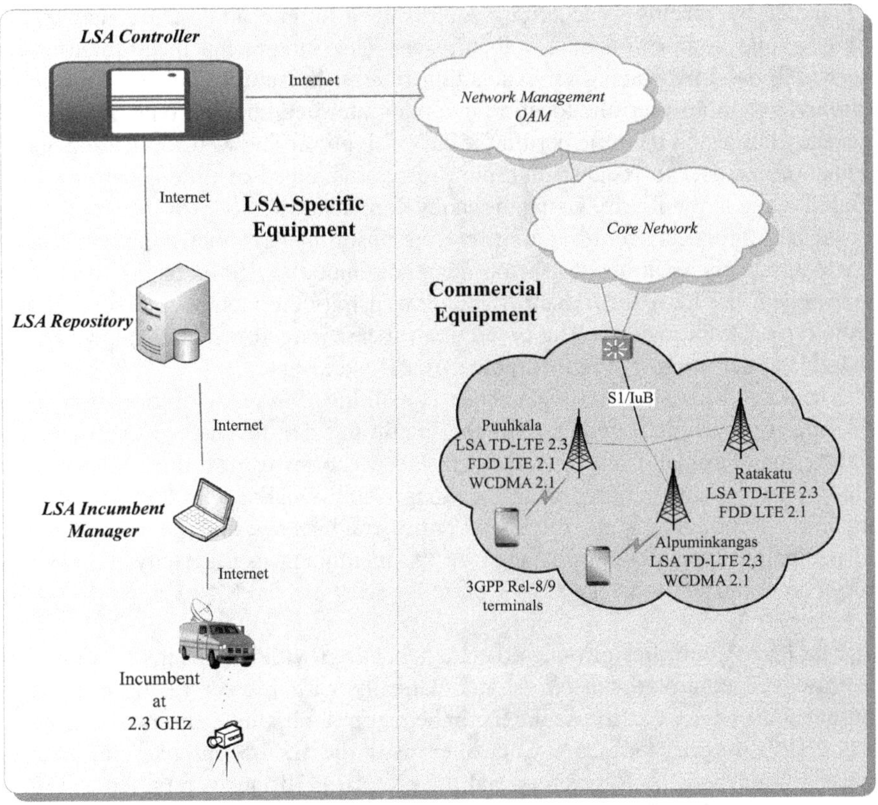

Figure 11.10 Original Finnish LSA trial setup

includes a cellular LTE network that consists of 3GPP compliant radio accesses and a real core network. Since 2013, the LSA trial environment has been expanded to contain three Time-Division LTE (TD-LTE) evolved NodeBs (eNBs) with five sectors and several small cells located at Centria in Ylivieska, Finland. These eNBs operate in the 2.36–2.40 GHz LSA frequency band. The trial environment also contains two Frequency Division Duplexing LTE (FDD-LTE) eNBs with four sectors that provide coverage to the same area using the 2.1 GHz band. The eNBs provide support for handover between TD and FDD, and load balancing. All eNBs are connected to LTE core network and are managed by a commercially available Operations, Administration and Management (OAM) system. Off-the-shelf LTE multi-band user equipment (UE) that support dual-mode LTE and provide TD-FDD seamless handover are used. The incumbent spectrum users are PMSE video links, whose protection distances are reported in [29].

In addition to the commercial equipment, the Finnish LSA trial environment includes LSA-specific components developed for the trials including LSA Incumbent Manager, LSA Repository, and LSA Controller. The LSA Incumbent Manager is a specific tool developed to help the PMSE service provider to reserve frequencies by placing evacuation requests in a simple and secure manner via mobile phone application or a web browser. The incumbent spectrum user can reserve the band by placing an evacuation request by setting the following information via web browser: the location in a map interface, the link type, the transmit frequency range and the time period for the reservation. The web application allows the incumbent user to request multiple protections ahead of time according to its planned usage in the future. Using the mobile application, the incumbent may make an evacuation request according to its current position. The mobile application uses GPS to locate the incumbent user on demand. Once the details of the reservation are obtained, the LSA Incumbent Manager transmits the incumbent user information to the LSA Repository. The protection process may also include algorithms to mask the actual usage of the incumbent user if necessary.

The LSA Repository is a database containing up-to-date information about LSA spectrum bands together with the conditions for the use of each band. It collects, maintains and manages data on the spectrum use on the LSA band for different times and geographical locations. Additionally, the LSA Repository contains information, e.g., on the LSA Controllers, licenses, and the incumbent link and equipment types. Upon a change on the incumbent user activity, LSA Repository communicates it to the LSA Controller allowing the MNO network to react the changes.

The LSA Controller provides the LSA licensee with the means to access the spectrum and control the network automatically with respect to the changes in incumbent user activity. In case the incumbent appears in the operational area of the LSA LTE network, the LSA Controller uses the incumbent user information received from the LSA Repository and the protection distances reported in [29] to calculate the affected eNBs/sectors on the LSA band. Accordingly, the LSA Controller sends deactivation/activation commands to the OAM. When the incumbent user has removed its evacuation request from the incumbent manager or the predetermined evacuation period has ended, the LSA Controller will send an activation request to the OAM.

The Finnish LSA trials have iteratively evolved to showcase more complicated LSA setups since its introduction in 2013. Several public trials in 2014–2015 presented enhanced features including tracking of mobile incumbent, enhanced incumbent protection algorithms, and LSA Controller integrated as self-organizing networking (SON) feature in the network management system. The latest LSA trial features were reported in [14] including enhanced LSA Controller power control concept algorithm and incumbent protection so that the aggregate interference from the LTE network to the incumbent is taken into account.

The functions of the different LSA components, described above, can be presented as the different phases of the LSA band evacuation process as presented in Figure 11.11. First, the incumbent spectrum user makes an evacuation request to the LSA Incumbent Manager which in turn submits this information to the LSA Repository. The LSA Repository stores the information into the database and forwards it to the corresponding LSA Controller. Based on the incumbent user information

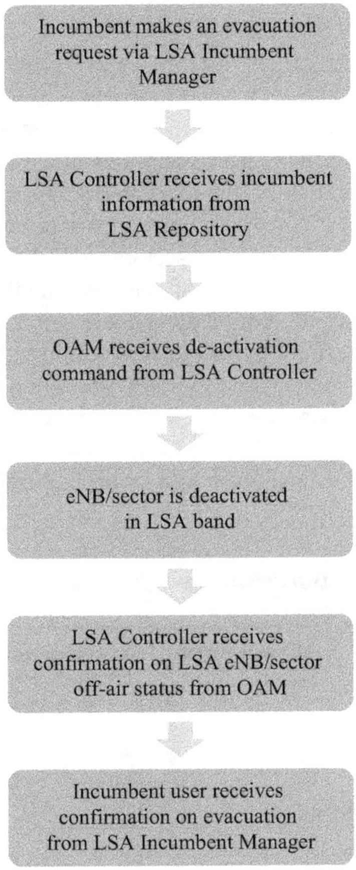

Figure 11.11 Process flow for LSA band evacuation

received from the LSA Repository, the LSA Controller calculates which eNBs or sectors on the LSA band are affected. The LSA Controller submits de-activation command to the OAM accordingly. The OAM then executes de-activation radio plan for the affected eNBs/sectors on the LSA band by either locking the affected LSA eNBs/sectors thereby turning off their air interfaces or using graceful shutdown where the power of the LSA eNBs/sectors is decreased gradually before locking turning it off completely. In both cases, UEs are handed over to other networks to continue their operations unaffected. The OAM then finishes the radio plan execution and begins the LSA eNB/sector status check whose result is sent to the LSA Controller. As soon as all needed LSA eNBs/sectors have reached off-air status, the LSA Controller completes evacuation and submits the evacuation completed information to the LSA Repository. Finally, the incumbent spectrum user receives a confirmation on the evacuation from the LSA Incumbent Manager.

11.6.2 LSA performance evaluation

The LSA reference implementation, described above, has been tested, and first performance measurements of the LSA band evacuation process have been reported in [16] to evaluate the involved time scales for the different phases of the process flow using the original LSA setup. The phases and the corresponding time stamps for the phases are presented in Table 11.3. The measurements consider the locking of eNB/sectors in two separate cases: normal evacuation where one eNB/sector is locked or emergency evacuation where all 3 eNBs/5 sectors are locked. The results in Table 11.3 indicate that in the case of the evacuation of one sector it takes a bit over half a minute from submitting the evacuation request until the LSA band has been cleared, and approximately the same time until the evacuation confirmation seen by the incumbent in the LSA Incumbent Manager. When the evacuation of all the sectors with a single command in the LSA trial environment is considered (i.e., emergency evacuation), the clearance of the band takes almost the same time, however, it takes over 10 s more for the confirmation to arrive to the incumbent. In more detailed analysis of the time scales,

Table 11.3 Table 11.3 Measurement times for LSA band evacuation process

	Measurement point	Evacuation using eNB/sector locking (1 sector)		Emergency evacuation using single radio plan (3 eNBs/5 sectors)	
		Time (s)	Std. Dev. (s)	Time (s)	St. Dev. (s)
1. The incumbent makes an evacuation request via LSA Incumbent Manager	The LSA Incumbent Manager	0		0	
2. LSA Controller receives incumbent information from LSA Repository	The LSA Controller	1.135	0.532	1.044	0.267
3. OAM receives the de-activation command from LSA Controller	OAM	9.807	2.664	9.029	1.881
4. eNB/sector on the LSA band is deactivated	LSA band	31.778	2.572	33.800	1.833
5. The LSA Controller receives confirmation on the LSA eNB off-air status from OAM	The LSA Controller	59.360	3.457	73.059	1.493
6. Incumbent user receives a confirmation on the evacuation to the LSA Incumbent Manager	The LSA Incumbent Manager	61.000	3.464	76.400	4.128

the measurement results can be divided into the delays from the LSA research platform and the delays coming from the commercial LTE OAM. The original delay for one eNB/sector of 61 s consists of 20.6 s from research and 40.4 s from commercial platforms. The original delay for emergency evacuation of 76 s consists of 33 s from research and 43 s from commercial platforms. Delays in the commercial equipment side arise from the OAM starting the radio plan provisioning until the time that the LTE transmit power disappeared from the channel, after which the radio plan provisioning still continued for 9–10 seconds. The delay from the commercial equipment can be minimized by using pre-validated radio plans to shorten the execution time of the provision operation. It can also be noted that provisioning a multi-site de-activation radio plan takes only about 3 more seconds to complete than a single site command, which is promising for the management of larger LSA networks.

The measurements reported above were from the original LSA setups from 2013 to 2014. New results were reported in [14] for the new LSA trial setup with enhanced LSA Controller with improved incumbent protection features. In the new LSA setup, the LSA Controller was implemented as SON solution integrated into a commercial network management system including enhanced power control algorithms for optimizing the protection zones for incumbent protection while maximizing the LSA band availability for the MNO. The performance measurements of the enhanced LSA setup reported in [14] showed that for evacuating the first cell took 21 s from making the evacuation request until the band was cleared. Additional 13 s were needed until the confirmation of the evacuation was visible to the incumbent. Average evacuation times for protection zone optimization were reported to be 24 s and graceful power reconfiguration of 58 s which are considered to be adequate particularly for the PMSE use case in the 2.3–2.4 GHz band.

The Finnish LSA trial environment is expandable to demonstrate the three-tier sharing model as well. The first public CBRS trials were presented at IEEE GLOBECOM 2015 conference showcasing the different levels of access rights and how the LTE network can react to incumbent activity.

11.7 Future outlook and conclusions

This chapter has summarized the status of the European LSA concept in the 2.3–2.4 GHz band and the US three-tier spectrum sharing concept for providing CBRS in the 3.5 GHz band using the SAS from regulation, business, and technology perspectives. The main development of the LSA concept so far has focused on the use of the 2.3–2.4 GHz band by mobile communication systems on a shared basis providing predictable QoS conditions with incumbents that vary depending on the national setup. The feasibility of LSA in the 2.3–2.4 GHz band for sharing has been demonstrated successfully in Finland and the regulatory framework is ready for adoption. For national deployment, a more detailed analysis of the specific incumbents in the countries is needed including the derivation of sufficient and practical protection criteria. The next application area of LSA to be studied in Europe after the 2.3–2.4 GHz band is the 3.6–3.8 GHz band where the studies have been started.

The incumbents in that band are different and a database approach alone could become insufficient when protection of non-informing or non-registered usage needs to be considered.

The US three-tier CBRS model is a more complex sharing model that goes beyond LSA by introducing the third tier of opportunistic access providing a rich ecosystem and more dynamic operations. A common functional architecture for the LSA and the CBRS seems to be possible when respective design is accepted and first initial steps have been taken through liaisons between the standards organizations (ETSI, 3GPP, and WInnForum). The CBRS supports dynamic features, which are not required in the first phase of the LSA though dynamic sharing for LSA is already discussed as possible LSA extension and re-use of SAS definitions seems to be possible for LSA. LSA and CBRS concept have comparable functionality related to IA tier informing incumbents and the SAS can be seen as evolution of the LSA Repository. As part of the LSA evolution future LSA Repository would require comparable functionality to SAS, extended toward a Flexible Spectrum Manager. Furthermore Domain Proxy and LSA Controller have comparable functionality and may be located in SAS/LSA Repository domain or in operator domain as support of external LSA Controller is already identified as feature in LSA concept in [2]. Mobile network operator networks for PA and GAA spectrum have benefits when Domain Proxy is assigned to MNO domain, e.g., through own interference management between several BS/AP/CBSD of the MNO, while operator-less GAA spectrum usage have benefits when no Domain Proxy is used.

References

[1] Ahokangas, P., Matinmikko, M., Yrjölä, S., Okkonen H., and Casey, T. "'Simple rules' for mobile network operators' strategic choices in future cognitive spectrum sharing networks". *IEEE Wireless Communications*. 2013;**20**(2): 20–26.

[2] ECC. *Licensed shared access*. ECC Report 205. Feb. 2014.

[3] RSPG. *RSPG opinion on licensed shared access*. RSPG13-538. Nov. 2013.

[4] The White House. *Realizing the full potential of government-held spectrum to spur economic growth*. President's Council of Advisors on Science and Technology (PCAST) Report, Jul. 2012.

[5] FCC. *Report and order and second further notice of proposed rulemaking (in the 35 GHz band)*. Apr. 2015. Available from URL [Accessed October 2015] http://transition.fcc.gov/Daily_Releases/Daily_Business/2015/db0421/FCC-15-47A1.pdf

[6] ECC. *Light licensing, license-exempt and commons*. ECC Report 132. Jun. 2009.

[7] Mustonen, M., Matinmikko, M., Palola, M., Yrjölä, S., and Horneman, K. "An evolution towards cognitive cellular systems: licensed shared access (LSA) for network optimisation". *IEEE Communications Magazine*. 2015;**53**(5): 68–74.

[8] Mustonen, M., Matinmikko, M., Roberson, D., and Yrjölä, S. "Evaluation of recent spectrum sharing models from the regulatory point of view". *Proceedings*

of the First International Conference on 5G for Ubiquitous Connectivity; Levi, Finland, Nov. 2014, pp. 11–16.

[9] Matinmikko, M., Okkonen, H., Palola, M., Yrjölä, S., Ahokangas, P., and Mustonen, M. "Spectrum sharing using licensed shared access (LSA): the concept and its work flow for LTE-advanced networks". *IEEE Wireless Communications Magazine*. 2014;**21**(2):72–79.

[10] Mustonen, M., Chen, T., Saarnisaari, H., Matinmikko, M., Yrjölä, S., and Palola, M. "Cellular architecture enhancement for supporting European licensed shared access (LSA) concept". *IEEE Wireless Communications*. 2014;**21**(3):37–43.

[11] ETSI. *System Architecture and High Level Procedures for operation of Licensed Shared Access (LSA) in the 2300 MHz–2400 MHz band*. TS 103 235. 2015.

[12] Matinmikko, M., Palola, M., Saarnisaari, H., *et al.* "Cognitive radio trial environment: first live authorized shared access-based spectrum-sharing demonstration". *IEEE Vehicular Technology Magazine*. 2013;**8**(3):30–37.

[13] Matinmikko, M., Palola, M., Mustonen, M., *et al.* "Field trial of licensed shared access (LSA) with enhanced LTE resource optimization and incumbent protection". *Proceedings of the 2015 IEEE International Symposium on Dynamic Spectrum Access Networks (DySPAN)*; Stockholm, Sweden, Sep.–Oct. 2015.

[14] Yrjölä, S., Hartikainen, V., Tudose, L., Ojaniemi, J., Kivinen, A., and Kippola T. "Field trial of licensed shared access (LSA) with enhanced LSA controller power control concept algorithms". *Proceedings of Wireless Innovation Forum Conference on Wireless Communications Technologies and Software Defined Radio*; Reston, VA, Mar. 2016.

[15] Mustonen, M., Matinmikko, M., Palola, M., Rautio, T., and Yrjölä, S. "Analysis of requirements from standardization for Licensed Shared Access (LSA) system implementation". *Proceedings of the 2015 IEEE International Symposium on Dynamic Spectrum Access Networks (DySPAN)*; Stockholm, Sweden, Sep.–Oct. 2015.

[16] CEPT. *Technological and regulatory options facilitating sharing between wireless broadband applications (WBB) and the relevant incumbent services/applications in the 2.3 GHz band*. CEPT Report 56. Mar. 2015.

[17] CEPT. *Technical sharing solutions for the shared use of the 2300–2400 MHz band for WBB and PMSE*. CEPT Report 58. Jul. 2015.

[18] ETSI. *Mobile broadband services in the 2300–2400 MHz frequency band under Licensed Shared Access regime*. ETSI TR 103.113 v 1.1.1, Jul. 2013.

[19] ETSI. *System requirements for operation of Mobile Broadband Systems in the 2300 MHz–2400 MHz band under LSA*. ETSI TS 103 154, Sep. 2014.

[20] The White House. *Presidential memorandum: expanding America's leadership in wireless innovation*. Jun. 2013.

[21] WInnForum Spectrum Sharing Committee. Available from URL [Accessed October 2015] http://groups.winnforum.org/page/spectrum-sharing-committee

[22] Yrjölä, S., Matinmikko, M., Ahokangas, P., and Mustonen, M. "Licensed shared access to spectrum" in Matyjas, J. D., Kumar, S., and Hu, F. (eds.) *Spectrum Sharing in Wireless Networks: Fairness, Efficiency, and Security*. Boca Raton, FL: CRC Press; 2016. pp. 139–163.

[23] Matinmikko, M., Okkonen, H., Yrjölä, S., *et al.* "Business benefits of licensed shared access (LSA) for key stakeholders" in Holland, O., Bogucka H., Medeisis, A. (eds.). *Opportunistic Spectrum Sharing and White Space Access: The Practical Reality*. Hoboken, NJ: John Wiley & Sons; 2015. pp. 407–424.

[24] Ahokangas, P., Matinmikko, M., Yrjölä, S., Mustonen, M., Luttinen, E., and Kivimäki, A. "Business scenarios for incumbents in licensed shared access (LSA)". *Proceedings of the Ninth International Conference on Cognitive Radio Oriented Wireless Networks and Communications (CROWNCOM)*; Oulu, Finland, Jun. 2014, pp. 407–412.

[25] Ahokangas, P., Matinmikko, M., Yrjölä, S., *et al.* "Business models for mobile network operators in licensed shared access (LSA)". *Proceedings of the 2014 IEEE International Symposium on Dynamic Spectrum Access Networks (DySPAN)*; McLean, VA, Apr. 2014, pp. 263–270.

[26] Ahokangas, P., Matinmikko, M., Atkova, I., Minervini, L. F., Yrjölä, S., and Mustonen, M. "Coopetitive business models in future mobile broadband". *Proceedings of the Sixth Workshop on Coopetition Strategy – Coopetition Strategy and Practice*; Umeå, Sweden, May 2014.

[27] Yrjölä, S., Ahokangas, P., and Matinmikko, M. "Evaluation of recent spectrum sharing concepts from business model scalability point of view". *Proceedings of the 2015 IEEE International Symposium on Dynamic Spectrum Access Networks (DySPAN)*; Stockholm, Sweden, Sep.–Oct. 2015.

[28] Yrjölä, S., Matinmikko, M., Mustonen, M., and Ahokangas P. "Analysis of dynamic capabilities in the citizens broadband radio service". *Proceedings of Wireless Innovation Forum Conference on Wireless Communications Technologies and Software Defined Radio*; Reston, VA, Mar. 2016.

[29] CEPT. *Broadband wireless systems usage in 2300–2400 MHz*. ECC Report 172. Mar. 2012.

Chapter 12
Epilogue: wireless beyond 5G

Angeliki Alexiou[1]

12.1 A vision for wireless beyond 5G

Assuming success, 5G will not just introduce a new way of thinking in wireless networks design and optimization, but a new era, where "Wireless Connectivity as a Utility" will completely transform the way people, things, and (any) application devices communicate, store, and process information. As speed (data rate), data volume, distance, agility (latency), security, and reliability become "non-issues," "infinite" capacity, and unlimited connectivity initiate an ecosystem paradigm shift, according to which network infrastructure as a whole is always available, scalable, and functioning "in the background," whereas wireless access is ubiquitous and technology agnostic (independent). This new utility-like Quality of Experience clearly shifts the center of gravity from the network (infrastructure) side toward the user end (network edge) and from radio and networking technologies to applications.

The proliferation of a plethora of new applications, involving a diverse usage landscape of vertical industries, breaks socio-technological barriers like never before: wireless connectivity, by becoming available always, everywhere, and for all, can be taken for granted and instead of just a technological benefit is now considered a social and human right. Moreover, Wireless Connectivity as a Utility will catalyze the market place, by helping realize Industry 4.0 vision: the digitization of the manufacturing sector. Driven by disruptions, such as big data processing, low computational power communications, big data analytics based intelligence, and advanced human–machine interaction, wireless connectivity is anticipated to play a cornerstone role in helping creating growth in every sector of the economy, society, and culture.

But will the success of 5G signify the end of wireless networking research as we know it? Is the impressive evolution journey, which started more than 30 years ago with the introduction of mobile communications and wireless local

[1]Department of Digital Systems, University of Piraeus, Piraeus 18534, Greece

access about to get into a "maturing"/saturation phase? Or is it exactly the opposite? The optimistic scenario would predict that 5G will open up novel, green field research directions that only under the prism of the new wireless thinking can unleash their true potential.

In this context, networks beyond 5G would be expected to provide unprecedented performance excellence, not only by targeting data rates in the Terabit-per-second (Tbps) regime but also by inherently supporting a large dynamic range of novel usage scenarios and applications that combine these extreme data rates with agility, reliability, zero response time, and artificial intelligence. Virtual presence, 3D printing, Cyber Physical Systems for Intelligent Transport, and Industry 4.0 are only a few examples of several highly challenging anticipated use cases. Trying to articulate a beyond 5G vision in today's terms, it would definitely include three main attributes:

- Tbps wireless links, always and everywhere available;
- Real convergence between optical (reliability) and wireless (flexibility);
- Artificial Intelligence applications for Cyber Physical Systems (to succeed IoT).

Although 5G seems more than willing to embrace several game changing design principles, such as virtualization and softwarization, in order to enhance scalability, flexibility, and efficient resources utilization, it can be easily understood that fundamental performance limitations related to available bandwidth, transmission, and processing delay and cost and energy consumption still define the envelope of 5G capabilities. To break these technological barriers in networks beyond 5G, one needs to bring little explored resources and technologies (e.g. *THz communications*) to validation and exploitation by directing research towards de-risking technological concepts, components, architectures, and systems concepts (e.g. *Big Data based processing/optimization, extreme resources sharing and commoditization*). Innovative joint investigation, assessment, and design of theoretical models, aligned, and supported by experimental parameter extraction and validation are required for this reason.

12.2 Expectations and challenges for wireless beyond 5G

Given the 5G wireless evolution trends and looking ahead into the future in 2030 and beyond several core questions arise, opening up new directions, rising expectations, and motivating exciting research and innovation. The following questions were asked to wireless research leaders, world renowned for the innovations on understanding the fundamentals and bringing novel concepts and technologies to realization.

1. Does basic/fundamental science have a role to play in the future of wireless? Which disciplines are potential candidates?
2. What unexplored technologies could play a key role in wireless beyond 5G?

3. How paradigm shifts in the wireless industry value chain brought about by 5G are expected to catalyze future wireless? Could one predict the emergence of new players and new niche markets in the beyond 5G era that could dramatically transform ICT industry landscape?

Their thought-provoking statements are given below in place of closing remarks for this book.

As emphasized in the 5GMF White Paper on "5G Mobile Communications Systems for 2020 and beyond", 5G will realize two key concepts. One of the two key concepts of 5G consists in "Satisfaction of End-to-End quality" required in all usage scenarios. It makes users feel satisfied with the quality, whatever applications are used anytime and anywhere. The 2nd key concept is the ultimate "Extreme Flexibility", in order to satisfy End-to-End quality of services in wider range of use cases in flexible, secure and efficient manner.

As described clearly in Recommendation ITU-R M.2083 ("IMT Vision—Framework and overall objectives of the future development of IMT for 2020 and beyond"), a wide range of emerging services including eMBB (enhanced mobile broadband), mMTC (massive machine type communications) and URLLC (ultra-reliable and low latency communications) would be supported in 5G era. These paradigm shifts in the wireless industry value chain brought about by 5G would metamorphose current mobile markets.

Dr. Kohei Satoh
Executive Manager on Standardization, Association of Radio Industries and Businesses (ARIB),
Secretary General of the 5G Mobile Communications Promotion Forum (5GMF) and Chairman of APT Wireless Group (AWG)

***

Both material science and quantum science may be expected to play an increasing role. Batteries still require further research in conjunction with wireless RF/laser power transfer and energy harvesting, where the communications capabilities depend both on the fullness of the information-buffer and energy-buffer as well as on the channel quality. The most promising results might emerge in visible-light communications, where solar-charged calculators paved the way. Material science is also expected to facilitate the commercialization of foldable display and keyboards.

Quantum leaps are also required in the more applied science of chips, which are about to reach atomic scales, where the weird phenomena of quantum physics apply. Moore's law has been fueled by massive investments in the semi-conductor industry, but apart from quantum computing numerous other options will have to be explored in the upper echelons of material science beyond silicon and graphene. Security is of paramount importance and the only innately secure solutions rely on quantum key distribution, which is poised to enter commercial reality, as the first instantiation of quantum communications.

Prof. Lajos Hanzo, FREng, FIEEE, FIEE, DSc,
RS Wolfson Fellow
Chair of Telecommunications
School of ECS, University of Southampton

***

Many of the 5G enablers will further evolve and mature throughout the 2020s, especially the millimeter wave and massive MIMO technologies. Eventually, interest will likely extend to the above 100 GHz bands as well. Understanding, characterizing and exploiting these bands necessitates new research in wave sciences and in engineering. Moreover, there are a number of exciting new concepts and technologies under study. For instance, in the physical layer, Faster-than-Nyquist (FTN) signaling is an important technology to watch for in 6G. Developing an even more dynamic, flexible, and agile radio access network (RAN) will continue to be a major goal. The HetNets paradigm has been a key enabler in both 4G and 5G; the RAN architecture will likely become even more heterogeneous, perhaps with the LAP/MAP/HAP (low/medium/high-altitude platform) and drone-BSs concepts. In a complementary way, we will likely see a more progressive and liberal use of spectrum towards dynamic access, facilitated with real-time auction-based pricing technologies. This may result in disruptive business models, such as crowdfunded wireless access through micro-operators. It is not difficult to imagine that the big data and machine learning will play a big role in the 6G discussions. For instance, application delivery based on individual users' preferences, satisfaction and context, possibly at varying prices, can result in win–win situations for both operators and users.

5G marks the start of a new era in wireless. The road towards 5G has been very exciting. The road towards 6G will be even more exciting…

Prof. Halim Yanikomeroglu
Department of Systems and Computer Engineering
Carleton University, Canada

Index

www.ingramcontent.com/pod-product-compliance
Lightning Source LLC
LaVergne TN
LVHW011904200926
848484LV00002B/531

9781785610615